Railway Engineering Technology

This book presents advances in railway engineering, including railway wheels, wheel failure mechanisms, innovations in railway track designs, and axle failures. It covers rail–wheel traction characteristics under various environmental conditions during full-scale test conditions, including braking systems. It provides an understanding of the applications of digital twin technology in railway systems, predictive maintenance, performance optimization, and overall system efficiency. It also emphasizes sustainable practices, energy-efficient technologies, and environmental impact assessments.

Features:

- Comprehensively reviews various aspects of railway engineering technology.
- Discusses major failure mechanisms and various types of braking systems.
- Covers the modernization of railways and high-speed trains.
- Explores emerging topics such as digital twin technology, sustainable practices, and autonomous technologies.
- Includes case studies of railway component failures and accidents.

This book is aimed at researchers and professionals in railway and mechanical engineering as well as transportation systems.

Railway Engineering Technology

Developments and Innovations

Edited by
V. M. Ravindra Kumar, K. P. Vineesh,
and Sunil Kumar Sharma

CRC Press
Taylor & Francis Group
Boca Raton London New York

CRC Press is an imprint of the
Taylor & Francis Group, an **informa** business

First edition published 2026
by CRC Press
2385 NW Executive Center Drive, Suite 320, Boca Raton FL 33431

and by CRC Press
4 Park Square, Milton Park, Abingdon, Oxon, OX14 4RN

CRC Press is an imprint of Taylor & Francis Group, LLC

ISBN: 9781032836508 (hbk)
ISBN: 9781041114369 (pbk)
ISBN: 9781003659983 (ebk)

DOI: 10.1201/9781003659983

Typeset in Times
by codeMantra

Contents

Preface

Wheel, rail, and axle play an important and primary role in railways. They work together to ensure smooth motion and stability to maintain train alignment on the tracks. This book provides fundamental knowledge of railway wheels, axles, and track manufacturing process; heat treatment; the effect of heat treatment on their performance; developments in their materials; key mechanical properties; track dynamics; dynamic analysis of the train-track compound system; railway track design; meta-model-based multi-objective optimization for a rail vehicle's dynamic response; common failures; and failure modes. More importantly, it also provides the details of railway transition zones.

A braking system ensures controlled acceleration, maintains a prescribed speed, and stops the train whenever required. The railway braking system is more complex than that of road vehicles because it needs to control multiple linked vehicles. This book provides knowledge of tread, disc, and regenerative braking systems. It details railway friction materials used for braking, their characteristics, and their relative advantages and disadvantages. It gives insights into the importance and benefits of regenerative braking systems, technological developments, and innovations in these systems. Estimating wheel running temperatures in tread braking is crucial to mitigate failures and improve wheel life. This book discusses all existing analytical models and presents a detailed procedure for accurately estimating wheel running temperatures. It also provides details of on-field studies to measure live wheel running temperatures. Most importantly, it presents various case studies of wheel and rail damage or failures reported across the globe. These case studies will help readers understand and avoid such failures in the future.

Digital twin technology provides a virtual model to simulate and predict behaviour in real time. It is evolving and gaining importance in railways to improve safety, enable predictive maintenance, and reduce train delays. This book covers key aspects of digital twins in railways, real-time monitoring, and operational safety. Cybersecurity in railways is an innovation that can mitigate safety risks for railway passengers and staff. This book provides knowledge of cyber threats in railways, case studies of past cybersecurity attacks, and various innovative cybersecurity measures implemented in the railway sector. Overall, this book provides great knowledge of railway systems, developments, and innovations.

About the Editors

V. M. Ravindra Kumar is a faculty member at the School of Mechanical Engineering at VIT-AP University. He received his Ph.D. from the Indian Institute of Technology, Kharagpur. He worked on a major railway engineering problem, "Locomotive Wheel Failure from Excessive Wheel Gauge Change," and successfully identified the root causes of the problem. His research areas include railway engineering, numerical modelling of manufacturing processes, and 3D and 4D printing. He has published research articles in various national and international journals and contributed to book and book chapters. He has the experience of working in sponsored research projects.

K. P. Vineesh is a faculty member in the Department of Mechanical Engineering at the National Institute of Technology Calicut. He has 8 years of experience in teaching, research, and academic administration. He obtained his Ph.D. from the Indian Institute of Technology Kharagpur. His major research areas are railway engineering, machine design, metal additive manufacturing, and vehicle dynamics. He has several peer-reviewed international journals in his core research areas and holds two patents.

Sunil Kumar Sharma is a Professor and Chair of Rail Vehicle Technology at the Institute of Railway Vehicles and Railway Technology, Technische Universität Dresden, Germany, and an Assistant Professor at Gati Shakti Vishwavidyalaya, Vadodara, a Central University established by the Ministry of Railways, Government of India. He received his Ph.D. from the Indian Institute of Technology, Roorkee. He has worked at the Non-destructive Evaluation and Structural Health Monitoring Laboratory at C.N. University, South Korea. His research interests include vehicle dynamics, contact mechanics, mechatronics, and real-time software-enabled control systems for high-speed rail vehicles. He has published several research articles in national and international journals, contributed to book chapters, and edited a book. He is among the top 2% of scientists in a global list compiled by Stanford University, USA.

Contributors

Samson Abiodun Aasa
Department of Automotive Engineering
University of Ibadan
Ibadan, Nigeria

Karel van Dalen
Faculty of Civil Engineering and Geosciences, Department of Engineering Structures
Delft University of Technology
Delft, The Netherlands

Azad Duppala
Department of Mechanical Engineering
Aditya Institute of Technology and Management
Tekkali, Andhra Pradesh, India

Saikat Dutta
School of Mechanical Engineering
University of Leeds
Leeds, United Kingdom

Holger Fricke
"Friedrich List" Faculty of Transportation and Traffic Science
Institute of Railway Vehicles and Railway Technology, Technische Universität Dresden
Dresden, Germany

Abhishek Gannarapu
Analog Devices Inc
Vancouver, WA, United States

Sai Sarath Gonuguntla
Department of Electronics Engineering
Technische Universität Chemnitz
Chemnitz, Germany

Badana Govindarajulu
Department of Civil Engineering
Aditya Institute of Technology and Management
Tekkali, Andhra Pradesh, India

Boris I
Department of Smart Manufacturing
Changwon National University
Gyeongsangnam-do, South Korea

Abdulsamad Ibrahim
University of Science and Technology of China
China

Avni Jain
Faculty of Civil Engineering and Geosciences, Department of Engineering Structures
Delft University of Technology
Delft, The Netherlands

Shilpi Lavania
Department of Electronics & Communication Engineering
Institute of Engineering and Technology, Dr. Bhimrao Ambedkar University
Agra, Uttar Pradesh, India

Jaesun Lee
School of Mechanical Engineering
Changwon National University
Gyeongsangnam-do, South Korea

Andrei Metrikine
Faculty of Civil Engineering and Geosciences, Department of Engineering Structures
Delft University of Technology
Delft, The Netherlands

Prajapati Naik
School of Engineering (Mechanical Engineering)
NIST University
Berhampur, Odisha, India

Srihari Palli
Department of Mechanical Engineering
Aditya Institute of Technology and Management
Tekkali, Andhra Pradesh, India

Manish Pandey
Indian Railway institute of Transport Management
Ministry of Railways
Lucknow, India

Smitirupa Pradhan
School of Mechanical Engineering,
KIIT Deemed to be University
Bhubaneswar, Odisha, India

Seepana Praveenkumar
Ural Federal University
Yekaterinburg, Russian Federation

Sivasankara Raju Rallabandi
Department of Mechanical Engineering
Aditya Institute of Technology and Management
Tekkali, Andhra Pradesh, India

Pavankumar Rejeti
Department of Mechanical Engineering
Aditya Institute of Technology and Management
Tekkali, Andhra Pradesh, India

Sunil Kumar Sharma
School of Technology
Gati Shakti Vishwavidyalaya
Vadodara, Gujarat, India

Yogesh Kumar Singla
Department of mechanical Engineering
Navajo Technical University
Crownpoint, New Mexico

Bendi Venkataramana
Department of Information Technology
Aditya Institute of Technology and Management
Tekkali, Andhra Pradesh, India

Ajay Kumar Verma
Department of Computer Science Engineering
Hindustan College of Science and Technology
Mathura, Uttar Pradesh, India

K. P. Vineesh
Department of Mechanical Engineering
National Institute of Technology Calicut
Calicut, Kerala, India

Vineesh Vishnu
Department of Mechanical Engineering
Providence College of Engineering
Chengannur, Kerala, India

1 Railway Wheels Failure Mechanism and Locally Resonant Bandgap for Vibration Isolation

Boris I and Jaesun Lee

1.1 INTRODUCTION

Railway wheels are fundamental to the rail transportation system and are tasked with supporting the entire weight of a train, maintaining balance, and ensuring smooth, safe movement. However, these wheels operate in a high-stress environment with continuous exposure to extreme mechanical loads, friction, and fluctuating temperatures. Over time, these conditions contribute to a range of degradation processes, making railway wheels vulnerable to various forms of failure.

Understanding these failure mechanisms is essential, as they directly impact safety, maintenance costs, and operational efficiency. Railway wheel failures typically manifest as wear, cracks, rolling contact fatigue (RCF), and thermal-induced stresses. Each failure mechanism has unique causes and consequences, influenced by variables such as speed, wheel and track material, environmental exposure, and load conditions. By analysing these factors and their effects on wheel longevity, engineers can develop strategies to predict, prevent, and mitigate failures, enhancing the durability and reliability of railway systems.

This chapter examines the primary failure mechanisms in railway wheels, exploring wear processes, crack formation, fatigue, and thermal degradation. The goal is to shed light on how these mechanisms occur, their progression, and the preventative measures that can reduce failure risks. Additionally, vibration—a critical contributor to wear and cracking patterns—is discussed in the context of its impact on wheel integrity. As we know, railway wheels are among the most critical components in the rail transportation system, bearing the full weight of the train while providing smooth motion. Over time, these components endure significant stresses due to dynamic loading, high friction, and adverse environmental conditions. As a result, railway wheels are susceptible to various forms of degradation and failure, making their study essential for railway operations' safety, reliability, and cost-effectiveness.

Over the years, railway wheels have not just changed; they have evolved. From simple cast-iron designs to sophisticated, high-performance materials, they have been engineered to withstand immense forces [1]. However, the failure mechanisms remain a persistent concern. Failures can occur due to material fatigue, wear, thermal

DOI: 10.1201/9781003659983-1

stresses, or RCF, among other factors. Such failures lead to costly repairs and downtime and pose significant safety risks, including derailments.

Railroad wheels must endure throughout their lifespan. Thus, any decrease in the longevity of railroad wheels should be taken seriously, and the root causes must be determined. It is common for railroad wheels to become worn out from continuous contact with the track and subsequently be deemed unfit for use until the end of their lifespan. In addition, various factors such as operational conditions (axle load, braking loads, type of braking, operating speed, and frequency of braking), braking materials, maintenance, and environmental conditions also significantly contribute to causing failures.

1.1.1 Wheel Failures

One obvious mechanism for wheelset wheel failure is the constant interaction of the wheel with the rail. Continuous traction/friction between rail and wheel leads to the change of the geometry of both rail and wheel. Ramalho [2], in his study, experimentally showed that the amount of wear in the wheel is significantly higher than in rail, except in pure rolling cases where the amount of the removed material was similar in rail and wheel. Moreover, the rise of the contact pressure and lower tangential speed increases the wear loss. Pombo et al. [3] indicated that the inclination angle of the wheel tread region and average vehicle speed significantly influenced wear. Another study, provided by [4], showed that wear volume reduced with an increase in speed and increased with an increase in tonnage. Jin et al. [5] demonstrated how irregularities in rail joints and impact loading can practically influence wheel wear. The study's findings revealed a notable impact of these two factors on wear. When severe polygonal wear occurs at higher operational speeds, the fluctuation of normal forces resulting from the polygonal worn wheels is more intense on ballasted tracks than on slab tracks.

Railway wheels often encounter vibrations that can cause failures despite their uniformity. High-speed train wheels can develop polygonal wear, resulting in periodic changes in the wheel radius along the circumference. At first, there are minimal variations in the wheel radii, but over time, the contact with the rails changes from a smooth circle to a polygonal shape. According to [6], specific frequencies of frictional self-excited vibration can lead to 18th-order polygonal wear. The damping of the fastener significantly influences the frictional self-excited vibration in the wheelset-track system of high-speed trains. Increasing the damping of the fastener can effectively decrease wheel polygonal wear. However, the vertical stiffness of the fastener has little impact on the frictional self-excited vibration. Altering the fastener's vertical stiffness can make it challenging to suppress wheel polygonal wear. Kang et al. [7] pointed out that tightly curved saturated creep forces on two wheels can lead to unstable vibrations that cause ninth-order wheel polygonal wear.

1.1.2 Cracks

Throughout their lifespans, structural components endure different types of loads, such as random, static, and constant loads [8–13], which may sometimes exceed or fall short of expected levels. Various research studies have investigated how structural elements react to different loading conditions, leading to the development of cracks [14–18]. Since a number of contributing forces exist, cracks can appear on railway wheels, which are also larger in size under normal circumstances. Such cracks might be thermal cracks present on the tread surface, on the wheel rim, or at the rim-disc interface. The formation and growth of cracks need to be the most important and the most severe only in the case of railway wheels that are used together with tread brakes. For this reason, compressive residual stresses are imparted in the wheel tread zone during the manufacturing process intentionally to avoid crack creation. However, stresses that are retained after initial tread braking in use will change the orientated compressive stresses into tensile hoop stresses that are useful to aid crack spreading. The amount of these service-induced residual stresses that cause the tensile hoop stresses to develop is influenced by the number of times the tread in the aircraft is applied and therefore strands during use.

Caprioli et al. [19] has also shown how 'pure rolling contact' on the tread surface with the presence of 'circumferential tensile residual stress', resulted in no thermal cracking even at relatively larger wheel/rail normal contact force. Also, the tangential force that was loaded on the tread surface with circumferential tensile residual stress caused at ambient temperature also did not create a thermal crack even though there was a larger wheel/rail normal contact force. The origin of thermal cracks possessed a distinct loading pattern that was alternately applied tensile stresses resulting from repeated application of the tread braking and then a rolling contact tangential force which seemed to repeat around one hundred times. There was no substantial effect of the normal wheel/rail contact force that links with Hertz contact pressure on thermal cracks.

Thus, the thermal stresses brought about when a tread is being braked and rolling contact are together more responsible than any other factor for the formation of these cracks [18, 19]. Furthermore, previous studies have revealed that extreme drag braking, brake block failures, unbalanced braking, and poor maintenance can cause excessive thermal stresses in the tread portion [20]. Known tensile stresses together with cyclic mechanical load on railway wheel rims further extend the existing cracks to a greater depth and may propagate across the rim area and the disc area.

The repetitive mechanic force applied onto railway wheels, though highly recommended, is often associated with some amounts of plastic deformation [21–28]. When the accumulated plastic strain exceeds the available deformation capacity of the material, ratcheting and crack formation stage can occur. These cracks tend to form in characteristic regions, such as the wheel tread [24]. Fatigue cracks are then initiated and propagate perpendicular to the direction of the plastic deformation, usually becoming deeper as they propagate towards the centre of the wheel. Furthermore, they might also traverse around the circumferential direction, causing material disintegration on a bigger scale. It has been shown that the microstructure of the material plays a significant role in the propagation of fatigue cracks [23]. Additionally, pipes are regarded as the home of cracks, and some may contain cracks lying beneath the surface of the wheel which is an example of the most dangerous type, as they do not make its presence known. Finally, in rare instances, it's stated that crack development caused by corrosion can lead to ultimate fatigue failure [29–34].

1.1.3 Vibration

The wheels and tracks generate vertical vibrations in railway moving units. Before reaching the car body, these vibrations are absorbed by the car's suspension. Vertical body vibrations are mainly triggered by vertical and lateral track geometries as well as by wheel road flanges sound after excessive wear. Over time, some wheel profiles get damaged due to repeated stress, which causes vibrations. Vibration of this nature is noticeable in the predominant eigenmodes of rigid system masses, including vertical and pitch modes [35–51].

For reliable estimation of dynamic characteristics, as well as for regular condition assessment, the analysis of vibration in mechanical plants and engineering constructions is essential. In this case, consideration of the impact of vibration on the stability of operational elements is important. Airplanes, vessels, high-speed rail, and other large machinery are often made up of hundreds of components that include vibration sources such as engines, motors, and other systems. These elements or systems are interconnected directly or via support elements with the frame, body, or cart of the structure, which transmits vibration to the main system, causing the equipment's vibration and noise.

The support structure transmits certain vibration modes, which considerably influence the dynamic behaviour of the equipment. For example, ships have many pipeline systems, and the vibration in these systems comes from electromechanical equipment, fluids in the pipes, and other stimuli. Studies have shown that a very high percentage of pipecrack vibration occurs in the vicinity of 3 kHz. This study found that high-frequency (>500) vibration is primarily caused by the fluid in the pipeline and is then transmitted through the support structure to the hull. Vibration not only interferes with the stable functioning of instruments or other machinery and the health and mental state of the personnel but also generates radiation noise in the hull that could affect the stealth of the vessel and greatly affect the life span and the combat performance of submarines. In this context, it becomes necessary to restrain support structure vibration transmission for ships as well as in mechanical devices [52–60].

Vibration reduction can be achieved using either passive or active control techniques. Passive control methods are dominant because active vibration control systems have several disadvantages such as high cost, complexity of system, and low stability. Three basic types of systems use passive control techniques: vibration isolation, damping, and absorption. These control systems are easy to operate as they do not require energy input from an external source. Vibration isolation involves installing isolators between the range of the source of vibration and the object being controlled in order to restrict the level of vibration transmitted. It consists of metal spring elements, rubber abutments, air springs, and floating rafts. Metal spring vibration isolators have low natural frequencies, which is a great advantage as it increases the isolation of low-frequency vibrations. Moreover, this type of vibration isolator can adapt well to its surroundings. However, the downside to this type of isolation material is that there is no damping in metal springs which induces sway vibrations and affects the overall performance of vibration isolation. There is internal resistance in rubber abutments have internal resistance, which gives them strong damping structures, allowing them to absorb high-frequency vibrations while also maintaining low natural frequencies. However, they are vulnerable to low temperatures and high temperatures and deteriorate quickly under such conditions. Their exposure to grime and oil also reduces their load-bearing capacity. Likewise, these air springs have adjustable stiffness and low self-vibration frequencies along with great performance in isolation. However, they demonstrate very low adaptability to the environment and low load-bearing capacity while they do require a lot of airtightness as well as high maintenance costs and large space. Floating rafts have high mechanical resistance and are extremely effective in vibration isolation, but they require a large area.

Typically, passive methods are used for vibration isolation and have proven to be efficient in practical applications. However, these methods have limitations, particularly at lower frequencies. To address this issue, resonators can be employed. By engineering the resonator to have a natural frequency at lower frequencies, it becomes possible to generate a bandgap, which involves Bragg scattering [1–12] and local resonance [13–23] to create a bandgap, which can be observed in the dispersion curve and transmittance spectrum, respectively. Of course, the local resonance metamaterials are preferable for low-frequency range over Bragg scattering as the size of the single cell is much smaller than the wavelength, and there is no need to utilize the periodicity. For example, in [16], the authors proposed a lightweight and compact resonator. The test findings indicated that two resonators reduced the vibration amplitude by a factor of 1,400. Compared with the classical mass law approach, the combined mass of the two resonators is approximately 1.78 g, which is less than 0.6% of the total mass of the host structure. Another example is the resonator proposed by [52]. Numerical analysis indicated that the proposed metamaterial creates two bandgaps related to the two-layered structure. The authors found that the tensile stress of the magnitude of attached mass significantly affects the bandgap and makes it broader, which makes this design flexible and adjusts the bandgap location. In addition to the tensile stress, [18] found that the occurrence and width of the two are closely related to the mass ratio between the masses of the resonator and the host plate. In [19], the authors proposed the locally resonant metastructure with a weight less than 0.3% of the host plate and showed that only one cell attenuates the vibration amplitude. The authors in [20] provide another solution for broadening the bandgap. Since the bandgaps are related to the size of the resonator, several resonators, each with its bandgaps, can be attached to the plate and cover the relatively large frequency range. The study showed four resonators and a total bandgap of 40–167 Hz.

Highlighted examples show perspectives for locally resonant absorbers offering highly flexible resonator designs. The materials that can be used vary from silicone, whose durability varies widely, and steel to eco-friendly materials like cork [23]. In this case, studying the lightweight low-frequency vibration isolator for thin copper tubes is interesting since the thin-walled copper tubes are widely used in heat exchangers, and failure could occur under vibration.

1.2 DETAILED FAILURE MECHANISMS

1.2.1 Wear Mechanisms

Wear is the progressive loss of material on the wheel surface due to constant contact and friction with the rail. This phenomenon is primarily influenced by the load on the wheels, track conditions, and operational speed [19–25]. The main types of wear include (see Table 1.1):

TABLE 1.1
Types of Wear Mechanisms and Contributing Factors

Wear Type	Description	Contributing Factors
Adhesive Wear	Material transfer due to sliding contact	Curved tracks, wheel slippage
Abrasive Wear	Gradual wear from hard particles on the rail	Debris, embedded particles
Surface Fatigue Wear	Crack formation under repeated rolling loads	High axle loads, operational speed

- Adhesive Wear: Occurs when the wheel slides along the rail rather than rolling, causing material to transfer from one surface to the other. This type is prevalent in curved tracks where slippage is more common.
- Abrasive Wear: Results from hard particles on the rail or embedded into the wheel surface, which act as abrasives and gradually wear down the wheel surface.
- Surface Fatigue Wear: Involves surface cracks that form under repeated rolling loads. As these cracks propagate, they may lead to material loss or spalling, affecting wheel geometry and safety [26–34].

Key factors that accelerate wear include high axle loads, frequent braking, and the use of certain materials with low resistance to abrasion. Understanding these mechanisms aids in selecting appropriate wheel materials and tracking lubrication practices to minimize wear.

1.2.2 Rolling Contact Fatigue

RCF is a critical failure mechanism resulting from the cyclic stress and strain imposed on the wheel surface under repeated rolling contact. This process leads to subsurface cracking and eventual failure. Factors affecting RCF include:

- Stress Concentration: High loads and uneven track surfaces can cause localized stress, initiating subsurface cracks.
- Contact Pressure: The wheel/rail interface undergoes high pressure, especially in high-speed and heavy-haul applications. When this pressure exceeds the material's yield strength, it initiates fatigue damage.
- Load and Speed Variations: Accelerated wear and increased RCF occur at higher operational speeds and axle loads, making high-speed and heavy-rail applications particularly susceptible.

RCF can lead to spalling and surface cracks that compromise wheel integrity. Detecting these cracks early, through technologies like ultrasound or eddy current testing, allows maintenance teams to address issues before they worsen [35–37].

TABLE 1.2
Causes and Effects of Thermal Cracking in Railway Wheels

Cause	Effect	Mitigation Measures
Thermal gradients	Residual tensile stress	Optimized braking system, better wheel material
Repeated braking cycles	Material fatigue and crack propagation	Use of disc brakes, maintenance of braking systems

1.2.3 Thermal Cracking

Thermal cracking arises due to the significant temperature changes experienced by the wheel tread during braking. These cracks primarily form because of various factors as shown in Table 1.2.

- Thermal Gradients: Frequent or severe braking causes temperature spikes on the wheel tread, inducing thermal expansion. As the wheel cools, contraction follows, creating residual tensile stresses [38–41].
- Material Fatigue: Repeated cycles of heating and cooling gradually weaken the wheel material. Tensile stresses from braking can exacerbate these fatigue issues, especially if compressive residual stresses (introduced during wheel manufacturing) are overcome.
- Influence of Braking System: Tread braking systems, particularly those using cast-iron blocks, increase the likelihood of thermal cracks compared to disc brakes, which distribute heat more evenly.

In severe cases, thermal cracks can propagate into the rim and disc areas, leading to wheel disintegration. Material selection, brake pad material, and brake maintenance are crucial to preventing thermal cracking [42–49].

1.2.4 Polygonal Wear and Out-of-Roundness

Polygonal wear describes an uneven wheel wear pattern where the wheel cross-section resembles a polygon rather than a perfect circle. This wear type typically results from:

- Vibrational Resonance: Self-excited vibrations in high-speed trains can cause polygonal wear by inducing periodic contact forces between the wheel and rail.
- Instability in Wheel–Rail Interaction: On curved tracks, the alignment between wheel and rail can lead to uneven wear patterns. Increased creep forces in the wheel–rail interface also contribute to this effect.
- Operational Speed and Load: Higher speeds and loads amplify the intensity of polygonal wear by increasing the rate of material loss along certain wheel sections.

Polygonal wear disrupts wheel geometry, leading to excessive vibrations, noise, and a bumpy ride, potentially damaging both the wheel and rail. Strategies like proper alignment, vibration damping, and periodic re-profiling of wheels help mitigate this issue.

1.2.5 Crack Propagation and Fatigue Failure

Cracks in railway wheels develop from surface imperfections and grow under repeated stress cycles. The main causes and progression include:

- Initiation Phase: Cracks start as small surface defects or inclusions in the wheel material, exacerbated by wear, thermal stresses, or external damage.
- Propagation Phase: Once initiated, cracks propagate radially and circumferentially, especially when tensile stresses or thermal cycles act on the wheel.
- Fatigue Failure: If cracks reach critical size, they can cause catastrophic failure. Crack growth direction is often aligned with stress patterns, typically radiating outward or along the wheel's circumference [50–52].

Fatigue life is highly dependent on material choice, loading conditions, and operational maintenance. Implementing regular inspections and using advanced materials with high fatigue resistance can extend wheel life significantly.

1.2.6 Corrosion and Environmental Factors

Although less common, corrosion can still affect railway wheels, especially in harsh environmental conditions:

- Moisture and Salt Exposure: Wheels exposed to moisture, particularly in coastal or humid areas, may experience surface corrosion. Corrosive environments can weaken the wheel material, leading to quicker initiation of cracks [53–58].
- Corrosion Fatigue: The combination of corrosion and cyclic loading accelerates crack growth, particularly on the wheel rim and tread.

Using corrosion-resistant materials, protective coatings, and regular cleaning in vulnerable areas are effective countermeasures (Table 1.3).

TABLE 1.3
Environmental Factors Contributing to Corrosion and Countermeasures

Environmental Factor	Impact on Wheels	Countermeasures
Moisture	Surface corrosion	Protective coatings, regular cleaning
Salty air (coastal areas)	Accelerated corrosion fatigue	Use of corrosion-resistant materials
Temperature fluctuations	Material stress and crack growth	Regular inspections, maintenance

1.3 VIBRATIONS AS A CONTRIBUTING FACTOR

1.3.1 Self-Excited Vibration

Self-excited vibrations occur when the system's own dynamics cause vibrations, often at specific resonance frequencies. In railway wheels, these vibrations play a significant role in the development of polygonal wear and cracks:

- Resonance and Wear Patterns: When a wheel rotates at certain speeds, it can reach a resonance frequency where vibrations amplify. This resonance causes cyclic loading on specific sections of the wheel, leading to polygonal wear—an uneven wear pattern resembling a polygonal shape around the wheel circumference [59–64].
- Impact on Crack Formation: Resonance-induced vibrations also increase the frequency and intensity of loading at localized spots, accelerating crack initiation and propagation. The cyclic stress can cause micro-cracks to grow into larger fractures, especially in high-speed and heavy-haul applications.
- Operational Influence: Factors like speed, load, and track irregularities influence the onset of self-excited vibrations. For example, higher speeds and heavy loads increase the likelihood of encountering resonant frequencies, while track imperfections amplify the effects.

These self-excited vibrations require careful monitoring, as they are a primary driver of both wear and crack growth. Detecting and addressing resonance frequencies can reduce maintenance needs and improve wheel longevity [65–68].

1.3.2 Damping and Control

To counteract the detrimental effects of vibrations, strategies that enhance damping or control resonance are employed in railway wheel and track designs. Key strategies include:

- Use of Damping Materials: By incorporating damping materials in wheel assemblies, vibrations can be absorbed and dissipated. For example, rubber pads or resilient materials placed between the wheel and axle assembly can absorb a portion of the vibrational energy, reducing the impact on the wheel's structural integrity.
- Track Design and Maintenance: Tracks with better stability and reduced irregularities can help minimize vibrational forces on wheels. Maintaining smooth track surfaces and regularly inspecting and replacing worn rails can reduce resonance and vibrations.
- Active and Passive Damping Systems: Passive systems like rubber isolators or air springs help absorb vibrations without external energy. Active systems, although more complex, use sensors and actuators to counteract vibrations in real time, significantly improving vibration control.

- Re-Profiling and Balancing Wheels: Regular re-profiling and balancing of wheels prevent uneven mass distribution, reducing the likelihood of resonant vibrations. This practice is essential for high-speed trains, where even minor imbalances can lead to amplified vibrations.

1.4 DISPERSION CURVE AND EFFECTIVE MASS DENSITY

Figure 1.1 shows the resonator proposed in this paper. It consists of the frame, membrane, and mass block, with a bandgap from 100 to 150 Hz.

The resonator's material is silicone, which gives it flexibility in the bandgap frequencies. In other words, by changing the durability of the silicone, the bandgap can be shifted to lower or higher frequencies since durability affects Young's modulus. The frame is also made of silicone and attached to the copper tube. The material parameters of the silicone and copper are presented in Table 1.4.

The dispersion curve calculation is based on the finite element method (FEM) with the Floquet boundary condition implemented in COMSOL; the proposed resonator is treated as an infinite chain of resonators with a lattice length of 50 mm. The mathematical representation of the Floquet boundary conditions is described by the following equation:

$$u_1 = u_2 e^{jk\Delta r}, \tag{1.1}$$

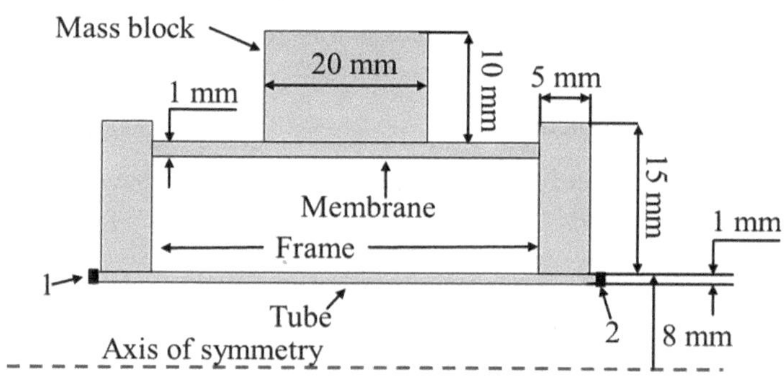

FIGURE 1.1 Proposed resonator attached to the tube in 2D axisymmetric representation. The Floquet boundary condition applied to the boundaries.

TABLE 1.4
Material Properties

	Young's Modulus	Density, kg/m³	Poisson Ratio	Isotropic Loss Factor
Silicone	Shore A 70 (~5.5 MPa) [24]	1,100	0.44	0.09
Copper	110 GPa	8,960	0.35	

where u_1, u_2 - displacement at red boundaries, k - wavenumber, Δr - distance between boundaries. Solving the eigenfrequency problem for the wavenumber k in the range of $[0, \pi/L]$, where $L = 50$ mm is a lattice length, give the set of frequencies for the given wavenumber and circumferential wavenumber $m = 0, 1, 2$, which define the wave propagation in the out-of-plane direction.

1.4.1 Mesh Convergence

The mesh convergence is an essential step in each FEM simulation since it affects the resulting error. For that, a parametric study was performed for quadrilateral third-order mesh sizes varying from 0.15 to 2 mm. Figure 1.2 depicted the relative error of the current study, which was calculated as the relative difference between the eigenfrequency value calculated with a specific mesh size and the eigenfrequency value calculated with a 0.15 mm mesh size (Equation 1.2):

$$\text{Error} = \frac{f_i - f_{0.15}}{f_{0.15}} * 100\%, \tag{1.2}$$

where the f_i – eigenfrequency value at specific mesh size, $f_{0.15}$ – eigenfrequency value at mesh size equal 0.15 mm.

The 0.5% value is chosen as an optimum relative error. Based on the result, all four eigenvalues start to converge at a 1 mm mesh size, so this value was chosen as a mesh size for further simulations.

1.4.2 Dispersion Curves

After establishing the mesh size, the dispersion curves were calculated. The Dell Precision 7960 workstation was used for this, and the calculation took 7 minutes and 15 seconds (4942 DoF). In Figure 1.3a, the calculated dispersion curves are presented.

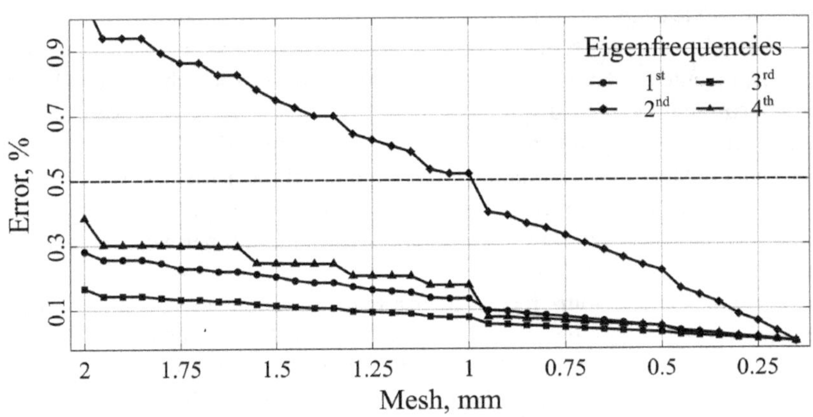

FIGURE 1.2 Mesh convergence of the 1st, 2nd, 3rd, and 4th modes.

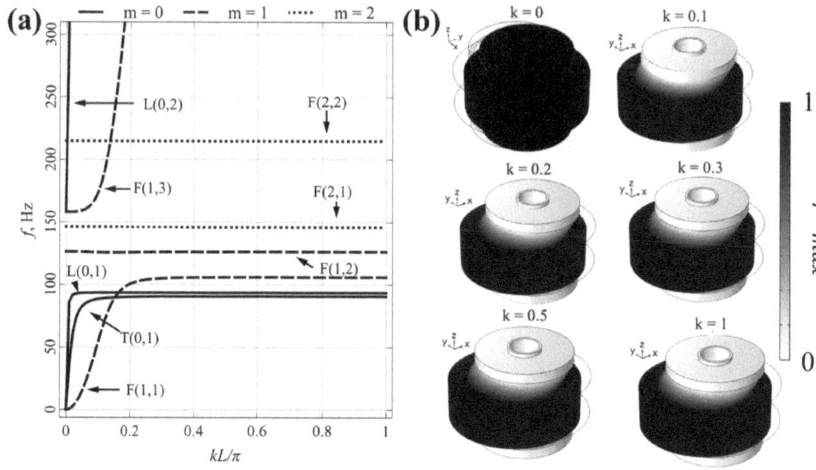

FIGURE 1.3 (a) Dispersion curve of the proposed resonator and (b) the mode shape of the F(1,1) for different values of wavenumber k. The first value in brackets is related to the circumferential wavenumber n, second to the number of the mode (ex., T(0,1) – $n=0$, first torsional mode). The characters L, T, and F correspond to the longitudinal, torsional, and flexural waves respectively.

The results indicate that the array of resonators significantly impacts the elastic wave propagation through copper tubes. When looking at the frequency f of the first torsional mode T(0,1), it was expected to increase linearly with the wavenumber k, suggesting a constant group velocity in the case of the bare tube. However, after introducing an array of resonators, the behaviour of the torsional mode changed dramatically; the group velocity was no longer constant (slope on the dispersion curve indicates the variation of the group velocity), and it only existed within a specific frequency range. This observation holds actual for other types of waves as well.

Additionally, another conclusion can be made by looking at the mode shape. As an example, the mode shapes of the first flexural wave F(1,1) are depicted in Figure 1.3b. As the wavenumber k and frequency f increase, the dispersion curves become flatter, and the displacement u starts to localize in the mass block, while the tube experiences approximately zero displacements. This behaviour is observed across all waves, particularly the high-order modes like F(2,1) and F(2,2), which mainly consist of displacements of the resonator. This influence of the array of resonators on the elastic waves leads to the conclusion that at circumferential wave number n greater than 2, only the modes of the resonators will be present; in other words, these modes will be absorbed by the array.

Such complex behaviour forms a bandgap between waves. Based on the Figure 1.3a, the bandgaps between L(0,1) and L(0,2), F(1,1) and F(1,2), F(1,2) and F(1,3) exist in 100–150 Hz range. The bandgap between T(0,1) and T(0,2), which is not presented in Figure 1.3a since it appears at frequencies higher than 300 Hz, is relatively large. The F(2,1) and F(2,2) can be neglected for the above reasons. By taking into account that the dispersion curve of F(1,2) is flat, the total bandgap exists in the 106–156 Hz range.

1.4.3 Effective Mass Density

The description of how a subwavelength resonator affects a vibrational host body can be obtained by finding its effective mass. The effective mass can be determined using the hybridization technique [25] to describe the behaviour of the resonator. Newton's second law is used to define the effective mass, and it is presented in the following manner:

$$m_{ij}^{\text{eff}} = \rho_{ij}^{\text{eff}} V = \frac{\langle F_i \rangle}{\langle a_j \rangle}, \tag{1.3}$$

$$\langle F_i \rangle = \frac{1}{S} \int_S \sigma_{ik} n_k \, ds, \tag{1.4}$$

$$\langle a_j \rangle = \frac{1}{V} \int_S a_j \, ds, \tag{1.5}$$

where ρ_{eff} – effective mass density, V – unit cell volume, F – force, a – acceleration, S – boundary area, σ – local stress, n – boundary unit normal vector.

Figure 1.4 illustrates the effective mass normalized by the bulk mass. It's worth noting that the effective mass has negative values at bandgap frequencies and significantly differs from the bulk mass, but it is equal to the bulk mass in frequencies out of bandgap.

The resonance behaviour of the curve, highlighted in the zoomed-in Figure 1.4b, is a significant observation. The effective mass starts to grow exponentially closer to the lower frequency of the bandgap, reaches the bandgap, dramatically falls to negative values, increases almost vertically, and then drops to negative values closer to the end of the bandgap. This complex behaviour has important implications for the bandgaps between $F(1,1)$ and $F(1,2)$ and $F(1,2)$–$F(1,3)$.

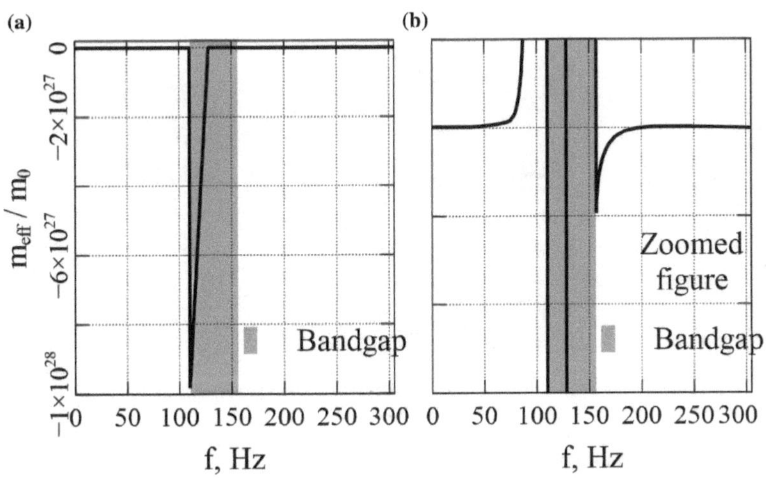

FIGURE 1.4 (a) The normalized effective mass and (b) zoomed-in.

1.5 EXPERIMENT

The resonator depicted in Figure 1.5 was created using silicone, with the material's characteristics detailed in Table 1.1. It was produced using a silicone 3D printer from Sandraw (Taichung, Taiwan) with the same dimensions as the previous section, except for a slightly reduced diameter (approximately 15.6 mm) to make a tight connection between the tube and resonator and avoid using glue. According to the manufacturer, the silicone 3D printing process has a resolution of 0.4 mm, expected to result in the bandgap occurring at the frequencies we anticipate. The resonator has a mass of 56 g, representing 6% of the mass of the copper tube.

In the previous section, the proposed resonator was assumed to be an infinite chain of resonators. Since it is impossible to create an infinite number of resonators and copper tubes, the real-life metamaterials based on Bragg scattering limit the number of cells [1–12], and the experiments show good agreement with simulation with ten cells in a row. In the case of local resonance, it can also be done by limiting the number of resonators to 10. However, it increases the total mass, contradicting the original idea of creating a lightweight vibration suppressor. Considering local resonance, only one resonator is considered enough to create the bandgap.

1.5.1 Experimental Setup

Figure 1.6 illustrates the experimental setup. The copper tube, measuring 1,430 mm with a 1 mm wall thickness and a diameter of 16mm, was lying on the thin aluminium frame with silicon pads and clamped at both ends. The resonator was attached to the tube at three different points to avoid the nodes of the modeshape.

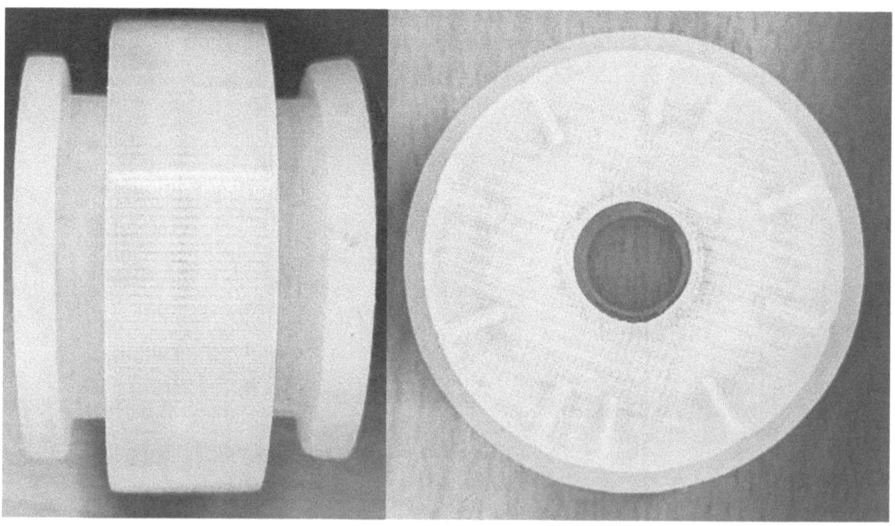

FIGURE 1.5 Proposed resonator.

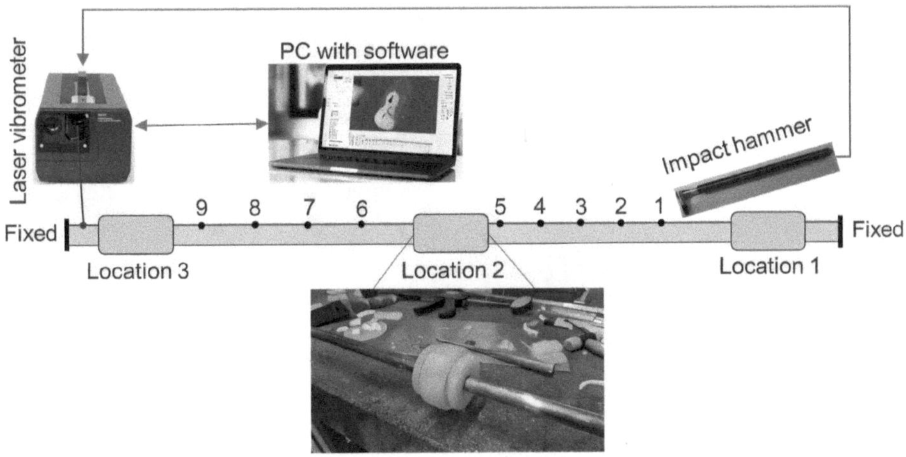

FIGURE 1.6 The experiment setup.

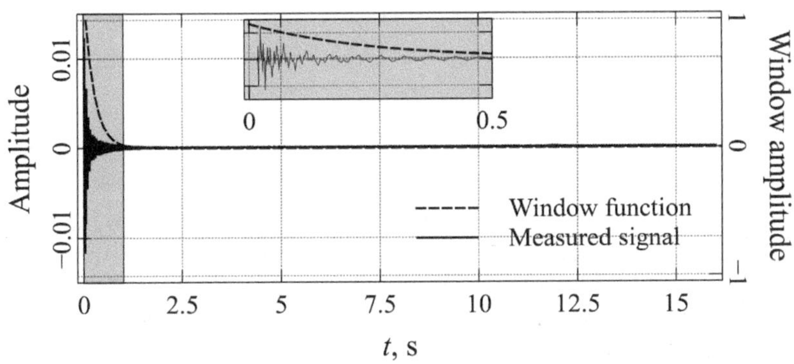

FIGURE 1.7 The example of time track collected by laser vibrometer and exponential window function.

An impact hammer, PCB 084A17, was utilized to stimulate the tube's vibration at nine points (rover hammer technique), with a 100 mm interval between points. This impact hammer was connected to the Optomet (Darmstadt, Germany) laser Doppler vibrometer and initiated the measurement. The entire process, including the start-stop measurement and post-processing of the data, was controlled by a PC with the Optoscan software.

1.5.2 Data Processing

After triggering the measurement by hammer impact, the software started to write the measured data and applied the window function to make the frequency spectrum less noisy (Figure 1.7).

The spectrum is represented as a frequency response function (FRF). The FRF is a ratio between the linear system *y*(*t*) response on the excitation *x*(*t*):

$$\mathrm{FRF} = \frac{Y(f)}{X(f)} = \frac{\langle G_{YX}(f)\rangle}{\langle G_{XX}(f)\rangle} = \frac{\langle G_{YY}(f)\rangle}{\langle G_{XY}(f)\rangle}, \tag{1.6}$$

$$\langle G_{YX}(f)\rangle = \frac{\sum_k X_k(f)\cdot Y_k^*(f)}{N_{av}}, \tag{1.7}$$

$$\langle G_{XX}(f)\rangle = \frac{\sum_k X_k(f)\cdot X_k^*(f)}{N_{av}}, \tag{1.8}$$

$$x(t)\overset{\mathrm{FFT}}{\rightarrow}X(f),\ y(t)\overset{\mathrm{FFT}}{\rightarrow}Y(f), \tag{1.9}$$

where *X*(*f*), *Y*(*f*) – fast Fourier transform of the *x*(*t*) and *y*(*t*) respectively, $\langle G_{XX}(f)\rangle$, $\langle G_{YX}(f)\rangle$ – average auto power spectrum and cross power spectrum, respectively, $N_{av} = 4$ – number of repetitions of excitation.

1.6 RESULTS

The experimental results are presented in Figure 1.8. Firstly, the bare tube was measured to identify the resonant peaks. The measurements at point 1 show four resonant peaks. At point 5,6,9, the second peak disappears, possibly because the measurement points overlapped with the nodes of the second mode. Previously, the bandgap was predicted at frequencies 106–156 Hz, which matches the third resonant peak. Attaching the resonator to the tube, the third peak was eliminated at all measurement points, and the first and second peaks attenuated after introducing the resonator. Unfortunately, the fourth peak is difficult to identify, and after attaching the resonator, the amplitude increased.

Figure 1.9 depicts the results at point 3 in the log scale. Such representation helps to calculate the attenuation level. The first peak relates to the silicone pad, which produces a bouncing mode. The rest of the peak of the bare tube is associated with the first three flexural natural frequencies; after attaching the resonator bouncing mode, the amplitude drop is equal to −8 dB for the first flexural — −14 dB. The most dramatic drop has a second flexural mode, equal to −30 dB. For the third flexural mode, it is difficult to calculate the amplitude drop, but comparing peaks, the drop equals −2 dB. It must be noted that the resonator creates additional peaks, which are related to the first and third flexural modes (*F*(1,1) and *F*(1,2) in Figure 1.3a), so the amplification at this frequency is approximately 6 dB.

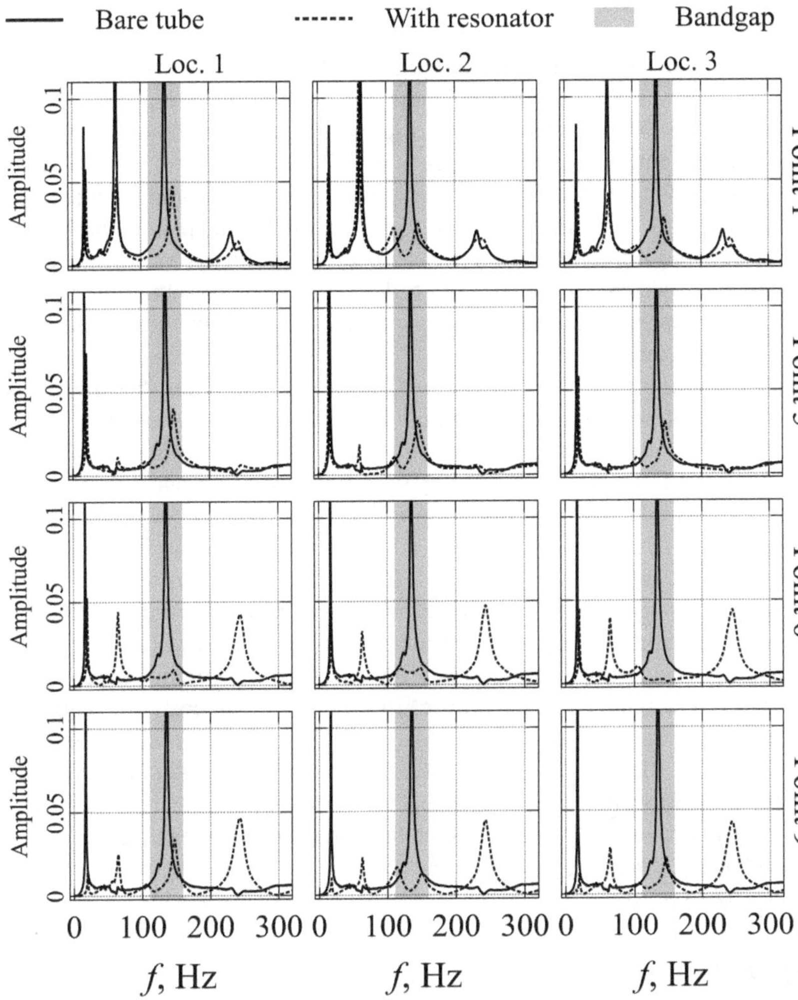

FIGURE 1.8 The measured data in four different points and three different locations of the resonator.

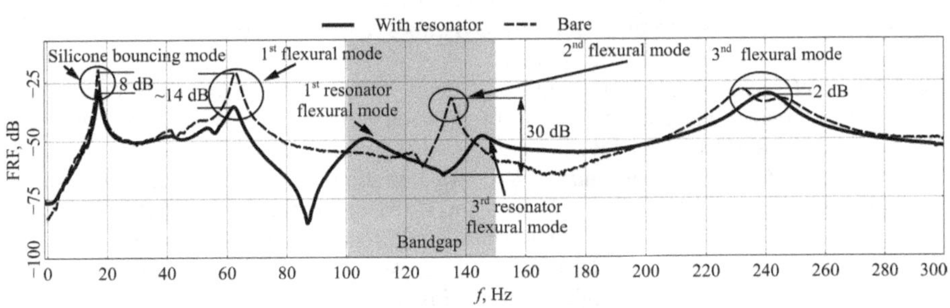

FIGURE 1.9 The FRF for bare tube and tube with resonator measured at point 3.

1.7 SUMMARY

The proposed resonators demonstrate a high level of agreement with the predicted results, effectively eliminating the resonant peak while weighing only 6% of the copper tube's mass. This aligns with the original objective of creating a lightweight vibration absorber for thin-walled copper tubes. The experimental data confirm that the observed bandgap is due to local resonance rather than Bragg scattering. The resonator's simple design, allowing adjustment of the bandgap frequency by using silicone of varying durability properties, provides significant versatility and utility. This adaptability enables resonators tailored for specific vibration suppression ranges, opening up opportunities for customized solutions in industries requiring precise vibration control.

By understanding these failure mechanisms—wear, fatigue, thermal cracking, and the effects of self-excited vibrations—engineers and railway operators can implement preventative strategies to extend wheel life, reduce maintenance costs, and enhance operational safety. The study of resonators has shown a promising avenue for developing effective vibration suppression solutions that could benefit railway applications, structural engineering, and other domains where weight and material properties are crucial factors. The potential for tailored resonators to address specific frequency ranges opens doors for more efficient and reliable systems in machinery design, structural applications, and construction. Through strategies like vibration damping, material selection, optimized design, and proactive maintenance practices, the rail industry can mitigate wheel failure risks, reduce repair costs, and enhance rail safety. Ultimately, a comprehensive approach to railway wheel maintenance, focusing on material resilience, design improvements, and vibration control, is essential for advancing railway system durability and reliability. Continued research and technological advancements will enhance our understanding and management of wheel failure mechanisms, leading to safer, more reliable, and cost-effective transportation and engineering systems.

REFERENCES

1. Bracciali, A. "Railway Wheelsets: History, Research, and Developments." *International Journal of Railway Technology* 5, no. 1 (2016): 23–52.
2. Ramalho, A. "Wear Modelling in Rail–Wheel Contact." *Wear* 330 (2015): 524–532.
3. Pombo, J., A. Ambrosio, P. Ambrósio, and J. Pereira. "A Study on Wear Evaluation of Railway Wheels Based on Multibody Dynamics and Wear Computation." *Multibody System Dynamics* 24 (2010): 347–366.
4. Zhong, W., Z. Sun, W. Zhang, and X. Cheng. "Experimental Investigation between Rolling Contact Fatigue and Wear of High-Speed and Heavy-Haul Railway and Selection of Rail Material." *Wear* 271, no. 9–10 (2011): 2485–2493.
5. Jin, X., X. Zhang, M. Lei, and Z. Wu. "An Investigation into the Mechanism of the Polygonal Wear of Metro Train Wheels and Its Effect on the Dynamic Behaviour of a Wheel/Rail System." *Vehicle System Dynamics* 50, no. 12 (2012): 1817–1834.
6. Zhao, X. N., H. S. Yang, W. Q. Hao, and H. L. Wu. "Study on the Mechanism for the Wheel Polygonal Wear of High-Speed Trains in Terms of the Frictional Self-Excited Vibration Theory." *Wear* 426 (2019): 1820–1827.

7. Kang, X., Y. Zhu, W. Liu, and J. Yin. "Study on Wheel Polygonal Wear of Metro Trains Caused by Frictional Self-Excited Oscillation." *Tribology Transactions* 64, no. 6 (2021): 1108–1117.
8. Peixoto, D. F. C., and P. M. S. T. de Castro. "Fatigue Crack Growth of a Railway Wheel." *Engineering Failure Analysis* 82 (2017): 420–434.
9. Nejad, R. M., and F. Berto. "Fatigue Fracture and Fatigue Life Assessment of Railway Wheel Using Non-Linear Model for Fatigue Crack Growth." *International Journal of Fatigue* 153 (2021): 106516.
10. Nejad, R. M., and F. Berto. "Fatigue Crack Growth of a Railway Wheel Steel and Fatigue Life Prediction under Spectrum Loading Conditions." *International Journal of Fatigue* 157 (2022): 106722.
11. Peixoto, D. F. C., and P. M. S. T. de Castro. "Mixed Mode Fatigue Crack Propagation in a Railway Wheel Steel." *Procedia Structural Integrity* 1 (2016): 150–157.
12. Guagliano, M., and L. Vergani. "Experimental and Numerical Analysis of Sub-Surface Cracks in Railway Wheels." *Engineering Fracture Mechanics* 72, no. 2 (2005): 255–269.
13. Zeng, D., W. Zhang, X. Wan, and L. Cui. "Investigation of the Crack Initiation of Subsurface Rolling Contact Fatigue in Railway Wheels." *International Journal of Fatigue* 130 (2020): 105281.
14. Hu, Y., Q. Liu, Y. Lin, and X. Chen. "Microstructure Evolution of Railway Pearlitic Wheel Steels under Rolling-Sliding Contact Loading." *Tribology International* 154 (2021): 106685.
15. Shariati, M., M. Mirzaei, and R. M. Nejad. "An Applied Method for Fatigue Life Assessment of Engineering Components Using Rigid-Insert Crack Closure Model." *Engineering Fracture Mechanics* 204 (2018): 421–433.
16. Rezende, A. B., J. C. Pereira, A. P. Silva, M. F. Amaya, and L. C. Casteletti. "Wear Behavior of Bainitic and Pearlitic Microstructures from Microalloyed Railway Wheel Steel." *Wear* 456 (2020): 203377.
17. Beretta, S. "More Than 25 Years of Extreme Value Statistics for Defects: Fundamentals, Historical Developments, Recent Applications." *International Journal of Fatigue* 151 (2021): 106407.
18. Handa, K., and F. Morimoto. "Influence of Wheel/Rail Tangential Traction Force on Thermal Cracking of Railway Wheels." *Wear* 289 (2012): 112–118.
19. Caprioli, S., T. Vernersson, and A. Ekberg. "Thermal Cracking of a Railway Wheel Tread Due to Tread Braking: Critical Crack Sizes and Influence of Repeated Thermal Cycles." *Proceedings of the Institution of Mechanical Engineers, Part F: Journal of Rail and Rapid Transit* 227, no. 1 (2013): 10–18.
20. Meizoso, A. M., J. M. Meñaloa, and J. G. Giménez. "Life Prediction of Thermally Cracked Railway Wheels: Crack Growth under Mixed Mode Stress Fields." In *Tenth International Wheelset Congress: Sharing the Latest Wheelset Technology in Order to Reduce Costs and Improve Railway Productivity; Preprints of Papers*. Barton, ACT: Institution of Engineers, Australia, 1992.
21. Cvetkovski, K., and J. Ahlström. "Characterisation of Plastic Deformation and Thermal Softening of the Surface Layer of Railway Passenger Wheel Treads." *Wear* 300, no. 1–2 (2013): 200–204.
22. Ławrynowicz, Z. "Plastic Deformation and Softening of the Surface Layer of Railway Wheel." *Advances in Materials Science* 15, no. 4 (2015): 5–13.
23. Walther, F., and D. Eifler. "Local Cyclic Deformation Behavior and Microstructure of Railway Wheel Materials." *Materials Science and Engineering:* A 387 (2004): 481–485.

24. Johnson, K. L. "The Strength of Surfaces in Rolling Contact." *Proceedings of the Institution of Mechanical Engineers, Part C: Mechanical Engineering Science* 203, no. 3 (1989): 151–163.
25. Wagner, V., F. Walther, and D. Eifler. "Cyclic Deformation Behaviour of Railway Wheel Steels in the Very High Cycle Fatigue (VHCF) Regime." *International Journal of Fatigue* 33, no. 1 (2011): 69–74.
26. Zambrano, O. A., G. Tressia, and R. M. Souza. "Failure Analysis of a Crossing Rail Made of Hadfield Steel After Severe Plastic Deformation Induced by Wheel-Rail Interaction." *Engineering Failure Analysis* 115 (2020): 104621.
27. Lansler, E., S. Olofsson, A. Ekberg, and J.-O. Josefson. "Influence of Plastic Deformations on Growth of Subsurface Rolling Contact Fatigue Cracks in Railway Wheels." *Proceedings of the Institution of Mechanical Engineers, Part F: Journal of Rail and Rapid Transit* 220, no. 4 (2006): 461–473.
28. Zhang, Q., Y. Chen, W. Guo, and H. Zhao. "Tension-Shear Multiaxial Fatigue Damage Behavior of High-Speed Railway Wheel Rim Steel." *International Journal of Fatigue* 133 (2020): 105416.
29. Ren, X., G. Ding, Z. Liu, and Q. Liu. "Corrosion Induced Fatigue Failure of Railway Wheels." *Engineering Failure Analysis* 55 (2015): 300–316.
30. Andreiko, I. M., V. V. Kulyk, and O. P. Ostash. "Resistance of Steels of Railroad Wheels to Corrosion-Fatigue Fracture." *Materials Science* 47 (2012): 608–612.
31. Gómez-Guarneros, M. A., J. C. Peña, T. Saucedo-Orozco, and R. Almanza-Robles. "Corrosive Effects of Artificial Rainwater on Rolling Contact Fatigue on Wheel and Rail Materials." *Wear* 530 (2023): 205046.
32. Hu, Y., L. Zhang, M. Kang, Y. Xu, and D. Tang. "Corrosion Fatigue Lifetime Assessment of High-Speed Railway Axle EA4T Steel with Artificial Scratch." *Engineering Fracture Mechanics* 245 (2021): 107588.
33. Jin, X., C. Huang, W. Sun, and X. Yang. "Mechanisms and Countermeasures of Out-of-Roundness Wear on Railway Vehicle Wheels." *Journal of Southwest Jiaotong University* 53, no. 1 (2018): 1–14.
34. Lin, P. T., H. C. Gea, and Y. Jaluria. "A Modified Reliability Index Approach for Reliability-Based Design Optimization." *Journal of Mechanical Design* 133, no. 4 (2011): 044501.
35. Newton, I. "*The Principia: Mathematical Principles of Natural Philosophy.*" Translated by I. Bernard Cohen and Anne Whitman. Berkeley, CA: University of California Press, 1999.
36. Lin, P. T., and B. R. Lin. "Fuzzy Automatic Contrast Enhancement Based on Fuzzy C-Means Clustering in CIELAB Color Space." In *The 12th IEEE/ASME International Conference on Mechatronic and Embedded Systems and Applications (MESA 2016)*, Auckland, New Zealand, 2016.
37. Cheng, Z., and Z. Shi. "Vibration Attenuation Properties of Periodic Rubber Concrete Panels." *Construction and Building Materials* 50 (2014): 257–265.
38. Zhou, J., G. Huang, X. Liu, and G. Zhao. "A Nonlinear Resonator with Inertial Amplification for Very Low-Frequency Flexural Wave Attenuations in Beams." *Nonlinear Dynamics* 96 (2019): 647–665.
39. Wei, W., D. Chronopoulos, and H. Meng. "Broadband Vibration Attenuation Achieved by 2D Elasto-Acoustic Metamaterial Plates with Rainbow Stepped Resonators." *Materials* 14, no. 17 (2021): 4759.
40. Yang, Z., Z. He, X. Zheng, and H. Wang. "Membrane-Type Acoustic Metamaterial with Negative Dynamic Mass." *Physical Review Letters* 101, no. 20 (2008): 204301.
41. Naify, C. J., M. W. Frecker, and T. L. Van Buren. "Transmission Loss of Membrane-Type Acoustic Metamaterials with Coaxial Ring Masses." *Journal of Applied Physics* 110, no. 12 (2011): 124903.

42. Li, J., X. Fan, and F. Li. "Numerical and Experimental Study of a Sandwich-Like Metamaterial Plate for Vibration Suppression." *Composite Structures* 238 (2020): 111969.
43. Liu, Y., D. Chronopoulos, and J. Yang. "Vibration Power Flow and Wave Transmittance Analysis of Inerter-Based Dual-Resonator Acoustic Metamaterial." In *International Conference on Wave Mechanics and Vibrations*. Cham: Springer International Publishing, 2022.
44. Banerjee, A. "Influence of the Torsional Vibration of the Periodically Attached Perpendicular Beam Resonator on the Flexural Band of a Euler-Bernoulli Beam." *Physics Letters A* 384, no. 29 (2020): 126757.
45. Jensen, J. S. "Phononic Band Gaps and Vibrations in One-and Two-Dimensional Mass–Spring Structures." *Journal of Sound and Vibration* 266, no. 5 (2003): 1053–1078.
46. Shao, H.-B., J.-Y. Zhan, and G.-Q. Li. "Simulation and Experimental Investigation of Low-Frequency Vibration Reduction of Honeycomb Phononic Crystals." *Chinese Physics B* 27, no. 12 (2018): 126301.
47. Meng, H., X.-Q. Li, and X. Li. "3D Rainbow Phononic Crystals for Extended Vibration Attenuation Bands." *Scientific Reports* 10, no. 1 (2020): 18989.
48. Yang, H.-Y., C. Xiang, and G.-J. Yu. "Low Frequency Elastic Waves and Vibration Control Mechanism of Innovative Phononic Crystal Thin Plates." *Physica B: Condensed Matter* 667 (2023): 415189.
49. Peng, H., P. F. Pai, and H. Deng. "Acoustic Multi-Stopband Metamaterial Plates Design for Broadband Elastic Wave Absorption and Vibration Suppression." *International Journal of Mechanical Sciences* 103 (2015): 104–114.
50. Huang, T.-Y., C. Shen, and Y. Jing. "Membrane-and Plate-Type Acoustic Metamaterials." *The Journal of the Acoustical Society of America* 139, no. 6 (2016): 3240–3250.
51. Sun, L., J. Li, Y. Ren, and Y. Bai. "Membrane-Type Resonator as an Effective Miniaturized Tuned Vibration Mass Damper." *AIP Advances* 6, no. 8 (2016): 085212. https://doi.org/10.1063/1.4961469
52. Gao, C., D. Halim, and X. Yi. "Study of Bandgap Property of a Bilayer Membrane-Type Metamaterial Applied on a Thin Plate." *International Journal of Mechanical Sciences* 184 (2020): 105708.
53. Sharma, S. K., R. C. Sharma, J. Lee, and H.-L. Jang. "Numerical and Experimental Analysis of DVA on the Flexible-Rigid Rail Vehicle Carbody Resonant Vibration." *Sensors* 22, no. 5 (2022): 1922.
54. Vishwakarma, P. N., P. Mishra, and S. K. Sharma. "Characterization of a Magnetorheological Fluid Damper: A Review." *Materials Today: Proceedings* 56, no. 5 (2022): 2988–2994.
55. Sharma, R. C., S. Sharma, N. Sharma, and S. K. Sharma. "Linear and Nonlinear Analysis of Ride and Stability of a Three-Wheeled Vehicle Subjected to Random and Bump Inputs Using Bond Graph and Simulink Methodology." *SAE International Journal of Commercial Vehicles* 15, no. 1 (2022): 65–79.
56. Bhardawaj, S., R. C. Sharma, and S. K. Sharma. "Development and Advancement in the Wheel-Rail Rolling Contact Mechanics." *IOP Conference Series: Materials Science and Engineering* 691, no. 1 (2019): 012034.
57. Sharma, S. K., and J. Lee. "Crashworthiness Analysis for Structural Stability and Dynamics." *International Journal of Structural Stability and Dynamics* 21, no. 04 (2021): 2150039.
58. Sharma, R. C., M. Avesh, S. K. Sharma, S. Sharma, S. Palli, and N. Sharma. "Linear and Non-Linear Stability Analysis of a Constrained Railway Wheel Axle." *International Journal of Vehicle Structures & Systems* 12, no. 2 (2020): 128.
59. Sharma, R. C., S. K. Sharma, N. Sharma, and S. Sharma. "Analysis of Ride and Stability of an ICF Railway Coach." *International Journal of Vehicle Noise and Vibration* 16, no. 3–4 (2021): 127–144.

60. Goswami, B., A. Rathi, S. Sayeed, P. Das, R. C. Sharma, and S. K. Sharma. "Optimization Design for Aerodynamic Elements of Indian Locomotive of Passenger Train." In *Advances in Engineering Design*, edited by A. Prasad, S. Gupta, and R. Tyagi, vol. 61. Lecture Notes in Mechanical Engineering. Singapore: Springer, 2019.
61. Dong, W., J. Liu, Z. He, and J. Li. "Characteristics of Band Gaps of a Metamaterial Plate with Membrane-Type Resonators Based on the Energy Approach." *Thin-Walled Structures* 191 (2023): 110930.
62. Ma, F., Y. Cai, and J. H. Wu. "Ultralight Plate-Type Vibration Damper with Designable Working Bandwidth and Strong Multi-Peak Suppression Performance." *Journal of Physics D: Applied Physics* 54, no. 5 (2020): 055303.
63. Wang, X., Z. Shi, and K. Zhan. "A Broadband Metamaterial Damper Design Based on Synergetic Coupling among Multi-Cells." *Applied Acoustics* 206 (2023): 109303.
64. Yu, R., Q. Li, K. Zhang, and Z. Ren. "An Integrated Load-Bearing and Vibration-Isolation Supporter with Decorated Metamaterial Absorbers." *International Journal of Mechanical Sciences* 253 (2023): 108406.
65. Zhou, W., L. Wang, and Z. Ma. "Actively Controllable Flexural Wave Band Gaps in Beam-Type Acoustic Metamaterials with Shunted Piezoelectric Patches." *European Journal of Mechanics-A/Solids* 77 (2019): 103807.
66. Ciaburro, G., and G. Iannace. "Membrane-Type Acoustic Metamaterial Using Cork Sheets and Attached Masses Based on Reused Materials." *Applied Acoustics* 189 (2022): 108605.
67. Gent, A. N. "On the Relation between Indentation Hardness and Young's Modulus." *Rubber Chemistry and Technology* 31, no. 4 (1958): 896–906.
68. Liu, X. N., G. K. Hu, C. T. Sun, and G. L. Huang. "Wave Propagation Characterization and Design of Two-Dimensional Elastic Chiral Metacomposite." *Journal of Sound and Vibration* 330, no. 11 (2011): 2536–2553.

2 Railway Axles

Sunil Kumar Sharma, Holger Fricke, Samson Abiodun Aasa, and Srihari Palli

2.1 INTRODUCTION

Railway axles are an integral part of train systems and operations, though in a more conceived way, they are often ignored compared to other prominent components such as wheels, tracks or bogies. In essence, axles constitute an essential segment of the wheelset – two wheels that are mounted on their ends. These axles are expected to rotate together with the wheels and carry the entire weight of the train thus transferring loads of the great weight of the train body to the railway track. Safety, strength and operational efficiency of a vehicle axle are vital since failure of an axle for some reasons may invite serious disasters for instance derailment and large destruction. Because of this specific reason, railways axles design, railways axles production and railways axles maintenance are critical activities in engineering of railways so as to maximally promote safety and ensure reliable, efficient operation. Railway axles, wheels, and rail help run railway trains smoothly. Proper functioning is essential for a long life. Earlier studies have analysed the performance of railway wheels along with heat treatment and respective failures [1–5].

The role of railway axles is much more than what meets the eye. It is worth noting that just like the growth of the railways, the development of axles has always been influenced by progress in industrial development and materials science. In the days of horse-drawn carts and the earliest steam engines, railroad axles, though needed, were very basic and made of either wood or cheap iron. However, as trains got stronger and faster, the demand for better, tougher axles became apparent. Wrought iron and later, steel axles became common and this was due to their better strength and durability as compared to wood and iron. In modern railway systems, the axle is one critical piece of equipment, the development of which is executed in compliance with the world's highest standards of safety and performance [1–5]. They have to withstand high lots, including moving trains-dynamic load, moving forces while accelerating and braking, and loads caused by operational conditions. Axles experience various kinds of loads including flexural, shear, torsional, thermal and many more and since these loads are there all the time, fatigue sets in. Therefore, material selection and the machinery used are crucial to ensure that these components can withstand harsh environments and perform reliably throughout their lifespan. Railway axle manufacturing entails a series of procedures that are well planned and implemented. Usually, a steel forging of the relative shape of the railway axle is done that is further machined out by machining methods e.g., turning, drilling, and grinding to certain shapes and flatness required. They also incorporate processes of heat treatment, where quenching and tempering are performed to thus fit some toughness

DOI: 10.1201/9781003659983-2

of the axles and also make them wear-resistant and fatigue-resistant. Each stage is then followed up with a process of verifying the integrity of the axles using both conventional and advanced non-destructive techniques such as leak testing using ultrasonics and magnetic particle tests amongst others.

Apart from that, the construction of railway axles has gone through a number of changes concerning axle weight, strength, and other functional characteristics [6–10]. Steel in the present day has remained the main material due to its cost and the good balance of strength and toughness of the material. The developments in recent times have also opened the range of high tensile alloys and sometimes even composites to weight reduction efforts of the axle with some performance characteristics equivalence or even improved in most cases.

Railway axles perform more functions than merely connecting two wheels; they also play an integral role in keeping the train balanced and its movement smooth. In this regard, axles should always provide accurate locating of wheels relative to each other so that the train can run within the boundaries of the required direction. A slight inconsistency or any deformations will cause adverse effects both to the wheels and to the tracks, will make riding uncomfortable, and in the most critical scenarios lead to derailing. Therefore, inspections of the conditions of axles and their maintenance should be very frequent during the operation of these trains. As the quest for next-generation rail-friendly construction goes on, more exotic designs are introduced alongside more conventional types of axles such as solid or tubular ones. For example, new, lighter axles are being developed that help reduce the unsprung mass of the trains which, in turn, improves energy efficiency and reduces the wear between wheel and rail. This is especially applicable as the changes in the railway industry are synonymous with increasing pressure to go green and cut down on costs.

2.2 MANUFACTURING OF RAILWAY AXLES

2.2.1 Overview of Manufacturing Processes

The construction of railway axles usually consists of several major technological operations, or processes of forging, machining, and heat treatment and inspection/testing [11, 12]. The mechanical properties of steel grades commonly used in railway axles vary significantly, as shown in Figure 2.1. These properties are crucial for determining the performance and durability of the axles under operational stresses

It is these very processes that serve the purpose of forming the axle and increasing its mechanical and physical properties in preserving itself while under the working conditions of the transport operation.

a. Forging Process: Forms the basic shape of the axle.
b. Axle Turning Machining: Used in the final stage to accurately define details about the axle's shape and size.
c. Reheating Process: Changes metal features for better strength/durability enhancement.
d. Quality Assurance and Control: Confirms that the axle complies with the quality and assurance requirements.

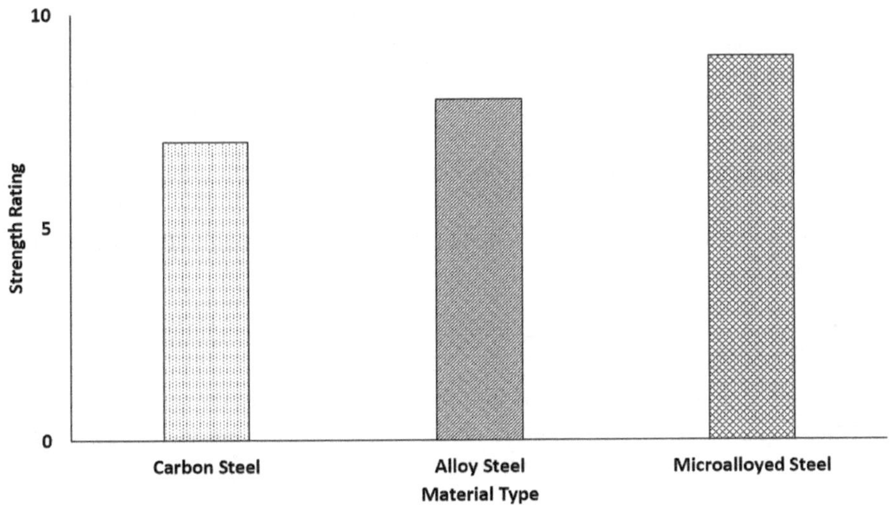

FIGURE 2.1 Strength ratings of different steel grades used in railway axles.

2.2.2 Forging Process

The forging process comes first among the processes of making an axle. In this process, a blank that has been heated to a reasonable temperature is forced into an approximately ax capture using mechanical or hydraulic presses that have a lot of power [13–17]. In such a manner, the microstructure of the material is created in a better way in order to improve the effective properties of the metal.

- Steps in Forging
  a. Billet Preparation: The material that is used is steel and most often, steel material is available in lumpy rounded forms which is what is referred to as a steel billet. This steel billet is cut to the required length and it is then heated in a furnace to the extent that it can reach temperatures of up to and including around 1,200°C.
  b. Pre-Forming: After heating up the steel which is the aim of pre-forming, the steel blank will be inserted into its respective die to begin the process of pre-forming.
  c. Forging: The pre-formed billet or pipe is placed in a die, and the assisted high-pressure forces (that can be to the tune of several thousand tonnes), are applied, which bends or folds the billet into the axle shape.
  d. Trimming: After the process of forging has been completed, trimming will occur so that any material that is referred to as flash that exists beyond the periphery of the forged part is entirely removed.
- Equipment Used:
  a. Hydraulic Presses: With extra strength used to compress and enable assist the other axial-related features

   b. Furnaces: The main aim for weathering or heating in the furnace is to achieve temperatures that would be ideal for forging the bezels.
   c. Trimming Machines: These machines are configured to operate on overs or excess material removed during forging at the end of the trial.
- Quality Control Measures
   a. Visual Inspection: To observe if there are any cracks or excessive flash on the surface.
   b. Dimensional Checks: To check if the axle is within the specified tolerances.

2.2.3 Machining Process

In order to obtain the long axial working lengths, forging is used in conjunction with some machining operations in order to obtain smooth finishes and accurate measurements. Among the core cutting operations, one can mention shaping, cutting, and roughing of surfaces.

- Turning, Drilling, and Grinding
   a. Turning: Involves the process of modifying the axle's form and diameter.
   b. Drilling: Issues where holes are needed for bearings or other sensors.
   c. Grinding: It is concerned with smoothing and working to a fine size of the surfaces being worked upon.
- Tolerances and Surface Finish
   a. Tolerances: Because axles usually have a high number of revolutions, they are mild tolerances (±0.01 mm, for instance) to ensure a proper fit.
   b. Surface Finish: The surface finish of certain parts of an axle is obtained through the processes of grinding and polishing to minimize wear and friction when the part is in operation.

The importance of these processes is related to the finishing of the axle's shape which requires processes such as turning, drilling, and grinding, and facilitates precise dimensional accuracy that are both important for axle's suitability for use as presented in Table 2.1.

TABLE 2.1
Overview of Machining Processes in Railway Axle Manufacturing, Including Turning, Drilling, and Grinding

Machining Process	Purpose	Key Equipment	Tolerances
Turning	Shaping the axle diameter	Lathes	±0.01 mm
Drilling	Creating holes	Drill presses	±0.05 mm
Grinding	Smoothing surfaces	Grinding machines	Surface finish 1–2 μm

2.2.4 Heat Treatment

The purpose of heat treatment is to change the axle attitude hardness, toughness, and fatigue of the material. This process is carried out by heating up the axle at a certain temperature for a certain time period and then allowing it to cool under certain restrictions.

- Types of Heat Treatments:
  1. Quenching: Rapidly cooling the axle in water or oil to increase hardness.
  2. Tempering: Reheating the quenched axle to a lower temperature to reduce brittleness while maintaining strength.
  3. Normalizing: Heating the axle and allowing it to cool in the air to refine the grain structure and improve toughness.
- Effects on Mechanical Properties:
  1. Quenching: Increases hardness but can make the material brittle.
  2. Tempering: Balances hardness and toughness, reducing brittleness.
  3. Normalizing: Improves uniformity of mechanical properties.

Heat treatment significantly alters the mechanical properties of axles, as shown in Table 2.2. Processes like quenching and tempering are particularly effective in enhancing the hardness and toughness of the axle material.

2.2.5 Inspection and Testing

As with any machined part, the inspection, and testing must be done to ascertain that the axle design, and general specifications are all adhered to urban decay [18–20]. That means performing not only measurements of the parts but also a range of non-destructive techniques.

- Non-destructive Testing Methods:
  1. Ultrasonic Testing: Uses high-frequency sound waves to detect internal defects such as cracks or voids.
  2. Magnetic Particle Inspection: Detects surface and near-surface defects using magnetic fields and iron particles.
  3. Dye Penetrant Testing: Identifies surface cracks by applying a dye to the surface and observing where it seeps into defects.

TABLE 2.2
Effects of Different Heat Treatment Processes on the Mechanical Properties of Railway Axles

Heat Treatment Type	Purpose	Effects on Properties
Quenching	Increase hardness	Hardens but increases brittleness
Tempering	Reduce brittleness	Balances hardness and toughness
Normalizing	Improve toughness	Refines grain structure, improves ductility

TABLE 2.3
Overview of Inspection Methods Used in Railway Axle Manufacturing and Their Purposes

Inspection Method	Purpose	Typical Equipment Used
Ultrasonic testing	Detect internal defects	Ultrasonic flaw detectors
Magnetic particle inspection	Identify surface defects	Magnetic yokes, iron particles
Dye penetrant testing	Find surface cracks	Dyes, developer sprays
Coordinate measuring machine	Measure dimensions	CMM devices
Surface roughness testing	Assess surface smoothness	Roughness testers

- Dimensional and Surface Inspections:
  1. Coordinate Measuring Machines (CMM): Used for precise measurement of axle dimensions.
  2. Surface Roughness Testers: Measure the smoothness of the axle's surface.

Ensuring the quality and integrity of railway axles requires a range of inspection and testing methods, as outlined in Table 2.3. These methods are critical for detecting defects that could compromise axle performance and safety.

Every manufacturing process is strategically planned to assure that the railroad axles have metallurgical, mechanical and structural properties. Thus, it will be possible to manufacture axles that are reliable and suitable for the operational environment of railways, by managing the interplays of design, manufacture and inspection processes.

2.3 MATERIAL AND TYPES OF RAILWAY AXLES

The type of material and the nature of the axle are still the most important aspects which influence performance, safety, and service of railway axles. Depending on the perpendicular criteria of mechanical properties, weight, cost, and applications different materials and designs of the axles are chosen. In this section, the commonly employed materials for railway axles and their types, and their effectiveness in a number of railways applications have been discussed.

2.3.1 Common Materials Used

The material of construction of railway axles should be such that it is capable of being depressed and formed into a railway axle and has very good mechanical properties like tensile resistance, toughness, fatigue, and machinability. The material most often used for railway axles is steel because it has a good ratio of weight and strength as well as cost-effectiveness [21–27]. Nevertheless, with the improvement of material science in recent years, designers have been looking for and using new alloys and special materials for improved efficiency of the axle.

- Steel Grades and Compositions: The first are carbon steels and the second are alloy steels of the steel grade most offered for this particular application, these materials have been limited as to their use as they have been designed to withstand tough mechanical stress. A qualitative comparison of yield strength, tensile strength, toughness, and hardness is provided in Figure 2.2 for several steel grades. As a result, micro-alloyed steels are superior for high-performance applications. Steel grade selection for railway axles is important because the strength, machinability and fatigue resistance are different for various axles. Table 2.4 is a list of the more common steel grades outlining their composition and other properties as well.

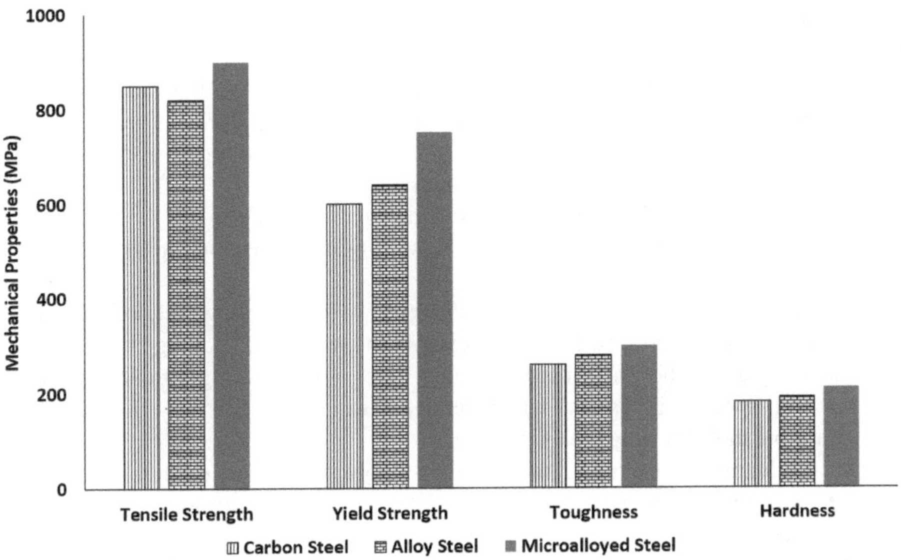

FIGURE 2.2 Comparison of mechanical properties of carbon steel, alloy steel, and micro-alloyed steel used in railway axles.

TABLE 2.4
Comparison of Steel Grades Used in Railway Axles, Their Compositions, and Key Properties

Steel Grade	Composition	Key Properties
Carbon steel	Fe, C (up to 0.7% carbon)	High strength, good machinability
Alloy steel	Fe, C, Cr, Ni, Mo, V	Enhanced toughness, fatigue resistance
Micro-alloyed steel	Fe, C, small amounts of Nb, Ti, V	Improved strength-to-weight ratio, weldability

- Material Selection Criteria:
  1. Mechanical Properties: Use of materials with high tensile strength, yield strength and toughness.
  2. Fatigue Resistance: Resistance to cyclic stress caused by loading without failure.
  3. Cost and Availability: Materials that are cheap and also available are used.
  4. Manufacturability: Materials which are simple to forge and can be machined and heat treated.

2.3.2 Types of Railway Axles

Depending on the purpose and characteristics of each railway vehicle, there are different types of railway axles in existence. Among them, solid axles, hollow axles, lightweight axles and so forth are the main types.

- Solid Axles: In many railways normal running gear with solid axles is appropriate. Constructed of one piece of solid steel, these axles also offer great resistance factors. Such axles are normally installed in freight cars and seriously overloaded railway usages.
  - Advantages: High strength, simple design, cost-effective.
  - Disadvantages: Heavier, higher unsprung mass, which can affect ride quality.
- Hollow Axles: Hollow axles are designed with a hollow section in the structure and therefore have a lesser weight while still having the necessary rigidity. These are used in many cases with passenger coaches and in high-speed rail systems where reducing weight is necessary in order to improve the performance of the system.
  - Advantages: Lighter, less unsprung weight, better ride quality.
  - Disadvantages: Difficult to manufacture, less strength as compared to solid axles.
- Lightweight Axles: Modified materials, aluminium alloys of high strength or other composites that suitably fit in the strength, weight and performance parameters. Such axles are utilized in the new and high-speed trains where reduction in weight is crucial.
  - Advantages: Disproportionate reduction in weight, energy efficient, less track damage.
  - Disadvantages: Costly, require specialized processes of manufacturing.

2.3.3 Comparison of Different Types

Almost all different types of axles are optimized for some given area of application, but at the same time, they do have some disadvantages. The Table 2.5 summarizes the most crucial properties of solid and hollow as well as lightweight axles. Based on the final application requirements, which include strength, weight and cost, five different types of railway axles are used. In Table 2.5, some common types of axles

TABLE 2.5
Comparison of Different Types of Railway Axles, Their Applications, Advantages, and Disadvantages

Axle Type	Applications	Advantages	Disadvantages
Solid axles	Freight cars, heavy-duty rail	High strength, cost-effective	Heavy, high unsprung mass
Hollow axles	Passenger trains, high-speed rail	Reduced weight, improved ride quality	Complex manufacturing, reduced strength
Lightweight axles	High-speed trains, modern rail systems	Significant weight reduction, enhanced performance	High cost, advanced materials required

and their typical applications, as well as the advantages and disadvantages of these types are presented.

All the applicants must be considered when determining the type of axle to use in enthusiast railways:

- Solid: Optimum use in applications for cranes and other load-lifting structures where high weight is not an issue.
- Hollow: Best performance will be achieved on passenger trains where reducing weight and increasing ride comfort are key objectives.
- Lightweight Axles: These are ideal for modern rail which emphasizes speed.
  - Pros and Cons:
- Solid Axles: These are simple and robust; however, their weight is more and its adaptability to modern day requirements is less.
- Hollow Axles: New engineering complexities have higher manufacturing complexity and they have dynamic advantages due to weight reduction.
- Lightweight Axles: Their advantages include paradigms of modern engineering performance; however, their costs and associated materials are also advanced.

2.4 MECHANICAL PROPERTIES OF RAILWAY AXLES

The mechanical properties of railway axles are crucial for their performance and longevity. In addition, they are designed to tolerate heavy weights, dynamic loads as well as environmental conditions that rail axles are normally subjected to. Knowing these properties assists in picking out the various kinds of materials, their forms to manufacture and the processes to undertake.

2.4.1 Key Mechanical Properties

Mechanically, tensile strength, yield strength, toughness, and hardness are the most important attributes of railway axles. These parameters define the load-bearing capacity, resistance to any permanent shape change, and cyclic load endurance characteristics of the axle.

- Tensile Strength: Tensile strength is defined as the greatest longitudinal stress in the axle substance which occurs when the axle is in the process of being drawn apart up to the point of breaking [21]. Axles have to be able to have high tensile strength in order to withstand forces that are experienced when in normal operation such as acceleration, braking and being loaded.
- Yield Strength: It is the maximum amount of stress applied after which a structural material is said to be irreversibly deformed or in the case of the axle irreversibly deforms its volume in context when using railway axles of great yield strength to ensure that there is no permanent deformation of the axle under loads during usage.
- Toughness: Toughness is the property of a material to withstand impact without fracturing while is deformed through plastic means. It speaks of the impact resistance of the axle which aids in moments of shock or sudden loads which may arise from forces such as collision or track irregularities.
- Hardness: this defines a material's ability to withstand penetration, surface pressure and impact use against its focal surface, Hardness will relieve the axle materials of rapid surface erosion retaining its effectiveness especially when placed under extreme environmental conditions.

It is important to grasp the mechanical features of railroad axles for proper perceptions of their efficiency and life span abuse during a course of different operational situations. The key mechanical properties such as tensile strength, yield strength, toughness and hardness are encapsulated in the Axles application together with their definitions and relevance in Table 2.6.

2.4.2 Factors Affecting Mechanical Properties

Factors like type of materials used for the manufacturing of railway axle and their geometrical configuration, production technique, heat treatment, etc. affect the

TABLE 2.6
Key Mechanical Properties of Railway Axles, Their Definitions, and Importance

Mechanical Property	Definition	Importance for Axles
Tensile strength	Maximum stress before breaking	Resists operational forces like acceleration and braking
Yield strength	Stress at which permanent deformation occurs	Prevents permanent axle deformation under load
Toughness	Energy absorption before fracturing	Withstands impacts and sudden forces
Hardness	Resistance to surface indentation and wear	Extends service life by reducing wear

mechanical ability of the components like axles. These factors need to be appreciated so as to design adequately the optimum performance of an axle.

- Material Composition: The additives that are used in producing axle materials alter the physical/mechanical properties of the axle. Carbon, chromium, nickel and molybdenum are the elements commonly incorporated into steels in order to increase their strength and toughness and resistance to corrosion. For example, carbon addition usually enhances tensile strength and hardness of a steel but decreases toughness.
- Manufacturing Process: Both the methods of forging and the methods of machining contribute greatly to the internal features of the axle and hence its strength and resistance to fatigue are of concern. These processes should also be controlled in order to meet the required standards of the properties. For example, it is evident that the forging technique of producing the components, the degree of grain structure of the components is reinforced resulting in increased strength and ductility while the controlling of the parameters of the machining approach ensures that there are no deviations from the required dimensions and surface quality [21–27].
- Heat Treatment Effects: Processes of heat treatment including quenching, tempering and normalizing, certain thermal inelastic deformation is essential owing to the other mechanical properties that pertain to the axles. Hardening and tempering ensure an increase and strength is improved which causes restrictions. The process of tempering helps recover the materials that have become brittle due to hardening. Normalizing makes uniform both strength and ductility while making the grain structure finer.

Several factors influence the mechanical properties of railway axles, affecting their performance and durability. Table 2.7 highlights key factors such as material composition, manufacturing processes, and heat treatment, illustrating how each factor impacts properties like strength, toughness, and hardness.

TABLE 2.7
Factors Influencing the Mechanical Properties of Railway Axles, Their Impacts, and Examples

Factor	Impact on Properties	Examples
Material composition	Alters strength, toughness, hardness	Higher carbon content increases strength
Manufacturing process	Affects internal structure and surface finish	Forging improves grain structure
Heat treatment	Modifies hardness, toughness, ductility	Quenching increases hardness

2.4.3 Standards and Specifications

Railway axles must meet specific standards and specifications to ensure safety, reliability, and performance. These standards describe attributes such as mechanical strength, size and the tests appropriate for these components used in various railroad functions.

- International Standards: The International standards code various aspects regarding railway axles, which include characteristics, testing, materials and fabrication. Herein are some of the common standards observed:
  - ISO 1005-1. It has standardized practice on the application of railway axles. Mechanical properties, dimensions, and testing methods are the topics summarized with regard to railroad axles.
  - EN 13261: presents further instructions on the acceptable quality and procedures of railway axle materials and properties as well as manufacturing techniques practiced in the European railways.
  - AAR M-101: details standards applicable to the materials of axle components, their mechanical properties and methods for most North American railroads.

  Railway axles must adhere to specific standards that ensure their safety, reliability, and performance. Table 2.8 lists the key standards applicable in different regions, detailing their focus on material properties, mechanical performance, and testing methods.
- Industry-Specific Requirements: Various applications within the rail industry impose various axle performance conditions. For example:
  - High-Speed Trains: Require Routine replacements with heavy fatigue resistance, lightweight axles to enhance efficiency on high speed.
  - Freight Trains: Require excessive durable axles that withstand excessive operation loads and rough climates.
  - Urban Transit Systems: An emphasis on the axles, which are reliable enough to be operated for a long period with the least possibility of breakdown is essential.

TABLE 2.8
Key Standards for Railway Axles Across Different Regions, Their Focus Areas, and Applications

Standard	Region	Focus
EN 13261	Europe	Material and mechanical properties, manufacturing
AAR M-101	North America	Material, properties, inspection methods
ISO 1005-1	International	Mechanical properties, dimensions, testing methods

2.5 FAILURES OF RAILWAY AXLES

The breakage of railway axles may lead to the derailment of the train, expensive damage, and may be associated with the loss of human life. Therefore, knowledge of the failure modes of the components, their reasons and the FMEA methods is essential for avoidance of such cases. The subsequent section discusses the most common failure of railway axles and the methods employed to analyse and prevent these failures.

2.5.1 Common Failure Modes

There are many types of loading and environmental interactions that could lead to various types of failures of the railway axles. The more common ones include fatigue failures, corrosion and impact failures. Specifically, regarding the distribution of failure modes, as seen in Figure 2.3, there are more fatigue failures than other types, which points towards the need for improvements in fatigue resistance of materials and designs.

- Fatigue Failures: Fatigue is classified as a cumulative stress – life and that means that cyclic stresses that are not greater than the ultimate tensile strength of a material will damage the material over an extended period. In general, in railway axles fatigue failures are due to acute internal corners, surface indefects or notches acting as stress risers. Many deficiencies often

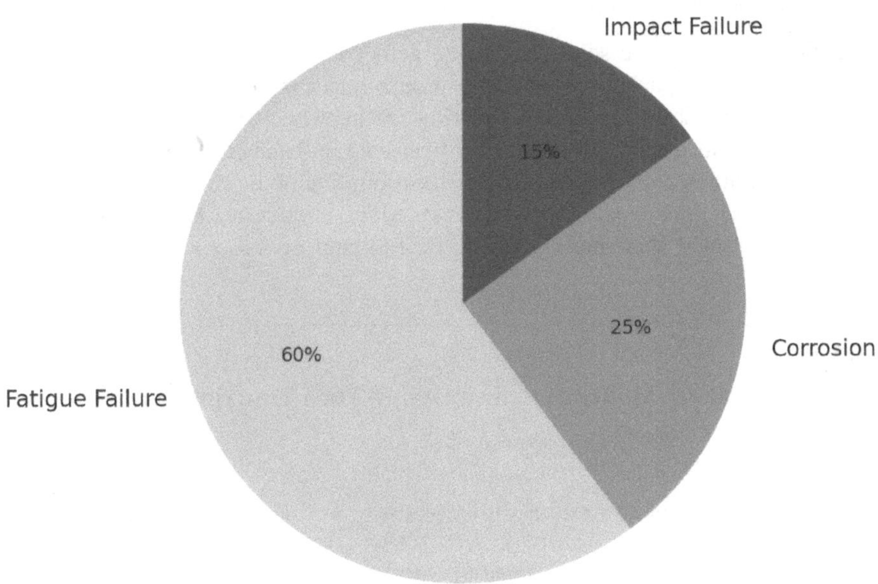

FIGURE 2.3 Distribution of common failure modes in railway axles, including fatigue, corrosion, and impact failures.

fail under cyclic loading due to crack initiation and long micro-cracks that grow and eventually lead to failure.
- Corrosion: Corrosion results from the interaction between the axle material and the elements of the environment, such as water and salts. The axle material can corrode, impairing the strength of the axle, such that makes it more prone to failure through other modes, for example, fatigue. Pitting corrosion, where many tiny holes or pits appear on the material surface, is highly critical because it has the potential to be a stress riser and thus cause more rapid normal growth of fatigue cracks.
- Impact Failures: Impact failures are those that occur when the axle encounters a single, high-intensity load which has the capacity to be delivered at blast-like collisions or harsh impacts caused by irregularities experienced on the track. If the material is unable to dissipate the energy well, the composite can crack or even break, since such powerful forces can be present about the axle. The wrinkling and hanging fatigue is more present than fractures caused by impact but this type of fracture is more dramatic and shocking.

However, fatigue, corrosion and impact failures are the most visible parameters of failure and performance of railway axles. Railroad Axles Table 2.9 presents the general sketch of the factors responsible for these types of failures, starting with corrosion, and impact failures, along with their descriptions and primary causes.

2.5.2 Causes of Failures

In practice, cost-effective solutions can be achieved by optimizing tool usage and organizing common parameters and peer groups into analogous vertices.

To reduce the skidding effect occurring in the wheel-slide clutch, and subsequently improve its design, it is necessary to effect rapid and effective changes, avoid stress concentration, and properly arrange the mounting of the clutch parts. Such concerns are remedied using a number of advanced techniques available on the market. It can be illustrated that manufacturing defects and operational stresses are two

TABLE 2.9
Common Failure Modes of Railway Axles, Their Descriptions, and Primary Causes

Failure Mode	Description	Primary Causes
Fatigue Failure	Progressive crack growth under cyclic loading	Stress concentrations, cyclic stresses
Corrosion	Material degradation due to environmental exposure	Moisture, salts, poor protective coatings
Impact Failure	Cracking or fracturing due to sudden forces	Collisions, track irregularities, operational impacts

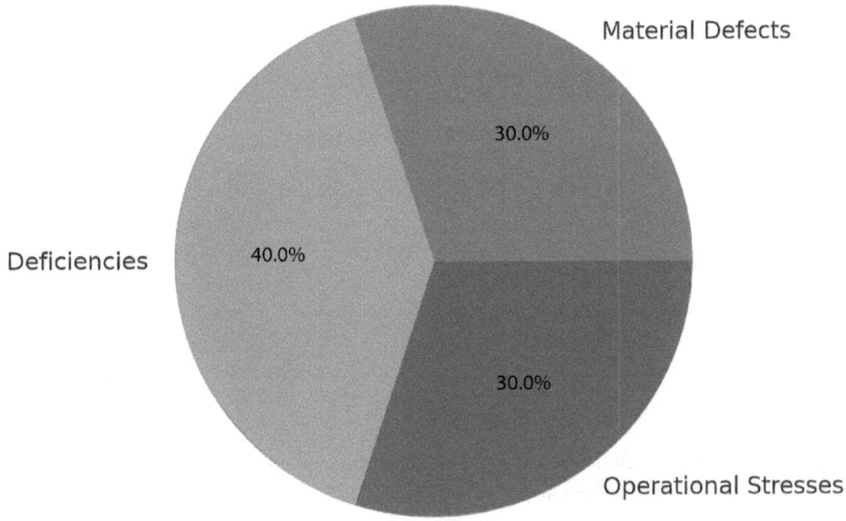

FIGURE 2.4 Distribution of failure causes in railway axles, including material defects, manufacturing deficiencies, and operational stresses.

major causes of axle failures. Thus, the importance of quality control measures and maintenance becomes evident, as both these measures will assist in ensuring that the axles have a longer operational service life (Figure 2.4).

- Material Defects: The presence of inclusions, voids, and uneven grain structure in the axle materials can lead to falls in its strength as well as its fatigue life leading to a scenario of potential failure. Weaknesses such as these are often expressed in the form of cracks which may occur as a result of manufacturing defects or occupational errors among inadequately trained personnel involved in the processes. The processes of training a work-force as well as the processes of manufacture are interlinked because ill-trained work-force in manufacturing will result in: poor procedures of handling materials, poor methods of production, and insufficient attempts to inspect key flaws that may result in the axle being unsafe for use during application.
- Manufacturing Deficiencies: The presence of manufacturing deficiencies like a not sufficiently forged part because the temperature was not high enough or due to machining mistakes as well as some failures in heat treatment processes that leave the axle brittle. An illustrative example of this case is the unacceptable mechanical properties of the axle which was caused by poor thermal treatment, or a soft temper and poor hardness causing the axle to fail when subjected to normal operation [21–27].

It is also useful to know how components fail in order to prevent the same or similar failures in future scenarios. The most frequent causes of axle failures include material

TABLE 2.10
Causes of Railway Axle Failures, Their Descriptions, and Impacts on Axles

Cause	Description	Impact on Axles
Material defects	Inclusions, voids, uneven grain structure	Reduced strength, crack initiation points
Manufacturing deficiencies	Improper forging, machining errors, heat treatment issues	Weaknesses, reduced fatigue resistance
Operational stresses	Overloading, excessive speeds, poor maintenance	Increased risk of fatigue and impact failures

defects, manufacturing deficiencies and operational stresses and their effects on axle integrity and performance are shown in Table 2.10.

2.5.3 Failure Analysis Techniques

It is imperative to conduct failure analysis with the objective of gaining insight into the problem that leads to axle failure and to determine the measures that can be taken in order to avert recurrence. These include fractography, metallography, or any non-destructive testing techniques.

- Fractography: Fractography is the science and art where a fractured surface is thoroughly studied in order to ascertain the failure type and cause. If engineers were to look at crack initiation points or crack propagation features and even fracture features, they would be able to tell whether the component failed by fatigue, impact failure or any other means.
    - Metallography (Metallurgical Analysis): Metallurgical investigation includes investigation of the microstructure, analysis of chemical composition and the study of the properties of the axle material. The techniques include scanning electron microscopy (SEM), energy-dispersive X-ray spectroscopy (EDS) and microhardness tests for defect detection, grain structure assessments and heat treatment assessments.
    - Non-Destructive Testing: In this case, non-destructive testing (NDT) methods ultrasound testings, magnetic particle inspection, dye penetrant testing is employed which does not put the axle at risk of surface or subsurface defects. These methods are very important to periodic checks and repairs that help prevent the development of problems that would have debilitating effects.

In order to achieve a clear picture of why railway axles fail, some specific analysis techniques need to be utilized. Table 2.11 provides details regarding some of the core techniques for failure analysis including but not limited to fractography,

TABLE 2.11
Failure Analysis Techniques for Railway Axles, Their Purposes, and Common Methods Used

Analysis Technique	Purpose	Common Methods
Fractography	Identify mode and cause of failure	Visual examination, SEM
Metallography	Assess material properties and defects	SEM, EDS, microhardness testing
Non-destructive testing	Detect surface and subsurface defects	Ultrasonic testing, magnetic particle inspection, dye penetrant testing

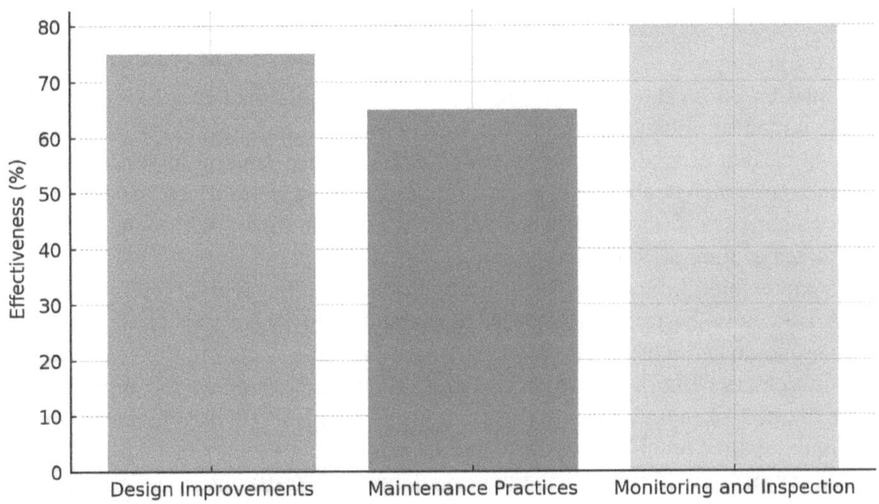

FIGURE 2.5 Effectiveness of various preventive measures for railway axle failures, including design improvements, maintenance practices, and monitoring.

metallurgical analysis and NDT, their uses and some methods that are often used for these purposes.

2.5.4 Preventive Measures

Axle failure is mainly due to the absence of appropriate maintenance. Other preventive measures, as illustrated in Figure 2.5, include design changes, up-to-date maintenance practices, and regular inspection that vary in success. Practicing early detection and performing more advanced maintenance activities are the best methods as they considerably lower the chances of failures occurring

TABLE 2.12
Preventive Measures for Railway Axle Failures, Their Descriptions, and Associated Benefits

Preventive Measure	Description	Benefits
Design improvements	Optimize geometry, use high-strength materials	Reduced stress concentrations, improved fatigue resistance
Maintenance practices	Regular inspections, cleaning, lubrication	Early detection of wear and damage, extended axle life
Monitoring and inspection	Use of sensors and condition monitoring systems	Real-time data, proactive maintenance actions

- Design Improvement: The existing configuration of the axle will be strengthened based on the results of analysis of the fatigue load conditions, which is aimed at enhancing the radial internal stress concentration at the bearing located around the shaft gear and also resisting bending moments. Such measures include the use of stronger materials than 'normal steel' and having non-corrosive coatings that prevent stress concentrations by the use of optimized geometric configurations.
- Maintenance Strategies: Each product demands that regular maintenance practices be carried out in order to prevent failures from occurring. Such maintenance includes all day-to-day activities like checking the equipment for defects Clinging of dirt and greasing the machine parts. Observing the evolution of maintenance policies, there is room for improvement on how maintenance practices are ensuring the reliability of the axle.
- Monitoring and Inspection: Traditional emission testing and visual inspections for cracks in axles, while useful in low-tension environments, have become less effective over time, offering limited new insights. However, leveraging already recorded data (big data) through digitalization 5.0 allows for more proactive monitoring and early detection of issues like crack formation and fatigue patterns. This approach enhances the ability to address potential problems before they escalate into major failures.

Implementing effective preventive measures is crucial for reducing the risk of axle failures and enhancing the overall safety and reliability of railway operations. Table 2.12 outlines key preventive measures, such as design improvements, maintenance practices, and continuous monitoring, along with their descriptions and benefits.

2.6 INNOVATIVE MATERIALS FOR RAILWAY AXLES

As the constraints on railway systems grow, the demands for axles have been evolving towards not only high strength and durability but also low weight and high efficiency. There have been efforts to find new materials due to the quest for improving the overall performance of trains while reducing the energy that is used while cutting down on the maintenance expenses. This chapter seeks to look out for the ongoing new material evolution to the railway axles and innovations and case history who have applied them to the present rail systems.

2.6.1 New Material Developments

Great strides in the development of various materials in recent years have led to the inclusion of metal matrix composites and alloys that have remarkable strength and outperform the steel methodology of axles. The purpose of these advanced materials is to improve mechanical properties such as strength, toughness as well as fatigue resistance while keeping the weight of the axles as low as possible.

Alloys with High Strength: Micro-alloyed steels and advanced high-strength steels (AHSS) are also finding applications in the manufacturing of railway axles. These alloys are specially developed to have a fine-grained structure and are believed to improve the strength, weight, and resistivity of the material by using synthesizing other compound minerals like niobium, titanium, and vanadium.

- Micro-alloyed Steels: Microalloying entails the addition of small amounts of alloying elements in steels with the aim of strengthening the matrix with little or no increase in the weight of the steels. They have been appropriately produced in a combination of strength, toughness, and weldability which is necessary for applications on high-performance railway axles.
- Advanced High Strength Steels (AHSS). These ensure that the tensile strength and the yield strength achieved are such that axles that are both light in weight and yet durable can be utilized in the designs. They are good in high-speed and heavy load applications where the performance requirements are demanding.
  - Composite Materials: In place of traditional materials, weight-saving composites such as carbon fibre-reinforced polymers (CFRP) are being evaluated for application in railway axles. Although composites are still in the experimental phase as regards to the application in the common railways, they have the potential to alter axle design for the better as regard to weight and fatigue resistance constraints.
- Carbon Fibre Composite Reinforced Polymer Materials (CFRP): These materials are particularly recognized due to their qualities of lightness and resistance to fatigue. The use of CFRP axles would probably facilitate reductions of up to 50% in axle masses, which would mean considerable energy efficiency and lesser track damage.

Advancements in material science have introduced innovative materials that may significantly enhance the performance of railway axles. Table 2.13 summarizes these materials, including their compositions and key benefits, highlighting the advantages they bring to modern rail systems.

TABLE 2.13
Innovative Materials for Railway Axles, Their Compositions, and Key Benefits

Material	Composition	Key Benefits
Micro-alloyed steels	Fe, C, V, Nb, Ti	Enhanced strength, toughness, good weldability
Advanced high-strength steels (AHSS)	Fe, C, Cr, Ni, Mo	High tensile and yield strengths, reduced weight
Carbon fibre-reinforced polymers (CFRP)	Carbon fibres, resin matrix	Exceptional strength-to-weight ratio, fatigue resistance

2.6.2 Benefits of Innovative Materials

The components of innovative design materials in the construction of the railway axle eliminates most of the drawbacks posed by the steel axles and helps in addressing the requirements of the current dynamics of railway systems.

- Weight Reduction: Lightweight axles lead to a reduction of unsprung mass which in turn enhances train dynamics, reduces the wear on the track and makes energy efficiency better. This also leads to decreased weight of the axles, so better operational development of trains is achievable through improved acceleration and braking that encourages the notion of cost saving.
- Enhanced Durability: The incorporation of high strength alloys and composites into the axle provides a very critical advantage durability due to its resistance to fatigue that comes with repetitive cycles of loading to the axle. Enhanced durability means a longer service life, less frequency of the replacements and lower operating costs.
- Improved Performance: New materials allow designing axles with the best characteristics to suit certain needs like high-speed railway or heavy freight. For instance, impact and abrasion resistance of the axle may be enhanced by the use of AHSS and the girdle may make use of CFRP which has been quite efficient for high-speed applications due to its stiffness and low mass.

The use of CFRP composite materials in railway axles is also advantageous in terms of weight savings, better performance, and increased durability, all resulting in an improvement in train dynamics and a decrease in cost of operation. These advantages have been outlined in the Table 2.14. However, in order to achieve these benefits, considerable design changes might be required to improve the CFRP axle shape so that the operational efficiency and power of the axle are ensured.

TABLE 2.14
Benefits of Using Innovative Materials in Railway Axles, Their Descriptions, and Impacts on Railway Operations

Benefit	Description	Impact on Railway Operations
Weight reduction	Lighter axles reduce unsprung mass	Improved train dynamics, energy efficiency
Enhanced durability	Increased resistance to fatigue and wear	Longer service life, reduced maintenance costs
Improved performance	Optimized strength, stiffness, and fatigue resistance	Better acceleration, braking, and safety

2.6.3 Case Studies and Applications

New advanced materials are at the time of writing already being implemented into contemporary rail systems, where some of the case studies prove successful.

- Implementation in Modern Rail Systems: In Europe and Japan, high-strength alloy axles have been implemented in high-speed trains and there have been notable improvements in performance and safety. For instance, the AHSS axles have been incorporated on existing fleets of high-speed rail in Europe and axle weight has significantly been reduced, fatigue life improved and thus the overall reliability of the external system has been enhanced. However, as the experience of very long-term use is important and is demonstrated by the inefficient dismantling of almost 1,900 axles in the fleet of the ICE-T in Germany. This brings one to the stresses of the factors as time, the type and the intensity of the testing are vitally important to anticipate unforeseen problems concerning reliability and safety of the axles and other details for the lifetime of the structure.
- Performance Comparisons: Extensive performance studies relative to the traditional steel axles or micro-alloyed steel or CFRP axles demonstrate adequate improvement in performance parameters. As a case in point, pre-production prototypes of CFRP axles fitted on trains had shown as much as a 30% decrease in energy consumption owing to their lower weight, as well as better performance with regard to corrosion and fatigue life degradation.

2.7 SUMMARY

Railway axles are vital parts that assure the safety and efficiency and reliability of the rail systems since they are the ones that connect the wheels and their essential task is to keep the whole structure steady when trains are moving. The design and materials of the selection of the axles are important deciding factors for the development of railway systems because they have to comply to certain strength and endurance factors. Part of the manufacturing process for railway axles includes forging, turning, heat treatment and quality control checks of the cutoffs so that the safety of the parts is ensured. Steel spans a wide range of benefits including strength, toughness and importantly, cost which means this carries over as the default material for axles. At the same time, however, it is important to know the most common failure modes of wheel-axle system components, including hinges, contacts and fingers, in order to help form lessons learned effective countermations. The morphology of axles is unlikely to remain the same over the years as key trends towards their modification have already developed in the current context. The manipulation of material composition, particularly alloys, will aid the shift as TMT's economy driven x and y provides the desired modification.

In snippet, the future is harsh on the railway industry. If you build railway axles, you might focus on radical development of current materials and manufacturing processes, while enhancing performance. In time, furthermore, the gap in perspective designs will be satisfied as there will be greater demands on lower grade steel as well as development on eco-friendly materials. Cost will still remain a concern, yet flexibility on design will offer solutions. The means of incorporation will also have advanced enabling the integration of various forms of smart technology like sensors, cameras and the analytics that come along with them into axles, wheelsets and bogies.

Despite these advancements, the railway axle industry faces challenges, including the high cost of new materials like composites, which may require novel manufacturing techniques. Ensuring that new materials and designs meet rigorous safety standards also poses a challenge, often slowing down innovation. Furthermore, the adoption and integration of new technologies into existing systems can be a slow and complex process for the rail industry. Nevertheless, the future of railway axles holds significant promise, with advancements in materials and design poised to enhance safety, efficiency, and sustainability. By embracing these innovations, the rail industry can effectively address the demands of modern rail transport.

REFERENCES

1. Vakkalagadda, M. R. K., K. P. Vineesh, A. Mishra, and V. Racherla. 2015. "Locomotive Wheel Failure from Gauge Widening/Condemning: Finite Element Modeling and Identification of Underlying Mechanism." *Engineering Failure Analysis* 57: 143–155.
2. Vineesh, K. P., M. R. K. Vakkalagadda, M. Dev, B. K. Rao, and V. Racherla. 2017. "Gauge Widening of Passenger Coach Wheel Sets in Indian Railways: Observed Statistics and Failure Analysis." *Engineering Failure Analysis* 71: 105–119.
3. Vineesh, K., M. Vakkalagadda, M. Dev, B. Rao, and V. Racherla. 2020. "Effect of Periodic Wheel Tread Reprofiling on Wheel Gauge Evolution in the Wheelsets of Tread-Braked Coaches: Finite Element Modeling and Field Observations." *Proceedings of the Institution of Mechanical Engineers, Part F: Journal of Rail and Rapid Transit* 234 (6): 678–686.

4. Desam, V., M. R. K. Vakkalagadda, V. Racherla, and K. P. Vineesh. 2023. "Wheel Gauge Evolution of Railway Wheels: Effect of Wheel Heat Treatment." *AIP Conference Proceedings* 2745 (1): 030001.
5. Abenesh, K. S., M. R. K. Vakkalagadda, V. Racherla, and K. P. Vineesh. 2022. "Finite Element Modeling of Railway Wheel Gauge Change During Braking: Effect of Lateral Shift of Brake Block." *Materials Today: Proceedings* 56 (Part 3): 1447–1451.
6. Sharma, S. K., and A. Kumar. 2016. "The Impact of a Rigid-Flexible System on the Ride Quality of Passenger Bogies Using a Flexible Carbody." In *Proceedings of the Third International Conference on Railway Technology: Research, Development and Maintenance*, p. 87. Civil-Comp Press, Stirlingshire, UK.
7. Sharma, S. K., and A. Kumar. 2017. "Ride Performance of a High Speed Rail Vehicle Using Controlled Semi Active Suspension System." *Smart Materials and Structures* 26 (5): 055026.
8. Sharma, S. K., and A. Kumar. 2018. "Impact of Longitudinal Train Dynamics on Train Operations: A Simulation-Based Study." *Journal of Vibration Engineering & Technologies* 6: 197–203.
9. Sharma, S. K., and A. Kumar. 2018. "Disturbance Rejection and Force-Tracking Controller of Nonlinear Lateral Vibrations in Passenger Rail Vehicle Using Magnetorheological Fluid Damper." *Journal of Intelligent Material Systems and Structures* 29 (2): 279–297.
10. Sharma, S. K., and A. Kumar. 2018. "Ride Comfort of a Higher Speed Rail Vehicle Using a Magnetorheological Suspension System." *Proceedings of the Institution of Mechanical Engineers, Part K: Journal of Multi-Body Dynamics* 232 (1): 32–48.
11. Zhai, Z., S. Zhu, X. Yuan, Z. He, and C. Cai. 2023. "Nonlinear Effects of a New Mesh-Type Rail Pad on the Coupled Vehicle-Slab Track Dynamics System under Extremely Cold Environment." *Proceedings of the Institution of Mechanical Engineers, Part F: Journal of Rail and Rapid Transit* 237 (3): 285–296.
12. Garg, V. K., and R. V. Dukkipati. 1984, eds. "Tangent Track Response of Railway Vehicles." In *Dynamics of Railway Vehicle Systems*, pp. 177–224. Academic Press, Toronto, ON.
13. Goodall, R. M., and W. Kortüm. 2002. "Mechatronic Developments for Railway Vehicles of the Future." *Control Engineering Practice* 10 (8): 887–898.
14. Craciun, C., C. Cruceanu, and M. Spiroiu. 2015. "Braking Process Influence on Vertical Load of Railway Vehicles." *Acta Technica Corviniensis-Bulletin of Engineering* 8 (2): 17–22.
15. Iwnicki, S., M. Spiryagin, C. Cole, and T. McSweeney, eds. 2019. *Handbook of Railway Vehicle Dynamics*. CRC Press, Boca Raton, FL.
16. Stribersky, A., A. Kienberger, G. Wagner, and H. Müller. 1998. "Design and Evaluation of a Semi-Active Damping System for Rail Vehicles." *Vehicle System Dynamics* 29 (S1): 669–681.
17. Gao, G., W. Guan, J. Li, H. Dong, X. Zou, and W. Chen. 2017. "Experimental Investigation of an Active–Passive Integration Energy Absorber for Railway Vehicles." *Thin-Walled Structures* 117 (August): 89–97.
18. Cruceanu, C., C. I. Crăciun, and I. C. Cruceanu. 2015. "Effects of Mechanical Wheel Slide Protection Devices Action on Railway Vehicles Braking Process." *Applied Mechanics and Materials* 809: 1085–1090.
19. Sun, W., J. Zhou, D. Gong, and T. You. 2016. "Analysis of Modal Frequency Optimization of Railway Vehicle Car Body." *Advances in Mechanical Engineering* 8 (4): 1–12.
20. Guo, J., H. Shi, R. Luo, and P. Wu. 2019. "Parametric Analysis of the Car Body Suspended Equipment for Railway Vehicles Vibration Reduction." *IEEE Access* 7: 88116–88125.

21. Zong, L.-H., X.-L. Gong, S.-H. Xuan, and C.-Y. Guo. 2013. "Semi-Active H∞ Control of High-Speed Railway Vehicle Suspension with Magnetorheological Dampers." *Vehicle System Dynamics* 51 (5): 600–626.
22. Majka, M., and M. Hartnett. 2008. "Effects of Speed, Load and Damping on the Dynamic Response of Railway Bridges and Vehicles." *Computers & Structures* 86 (6): 556–572.
23. EN 16286-1:2013. 2013. *Railway Applications - Gangway Systems between Vehicles - Part 1:* Main Applications.
24. Dumitriu, M. 2016. "Influence of Suspended Equipment on the Carbody Vertical Vibration Behaviour of High-Speed Railway Vehicles." *Archive of Mechanical Engineering* 63 (1): 145–162.
25. Sun, Y., D. Gong, and J. Zhou. 2016. "Study on Vibration Reduction Design of Suspended Equipment of High Speed Railway Vehicles." *Journal of Physics: Conference Series* 744 (1): 012212.
26. Goodall, R. 1997. "Active Railway Suspensions: Implementation Status and Technological Trends." *Vehicle System Dynamics* 28 (2–3): 87–117.
27. Suda, Y., and Y. Michitsuji. 2023. "Improved Curving Performance Using Unconventional Wheelset Guidance Design and Wheel-Rail Interface – Present and Future Solutions." *Vehicle System Dynamics* 61 (7): 1879–1913.

3 Rail–Wheel Interaction

Yogesh Kumar Singla and Samson Abiodun Aasa

3.1 INTRODUCTION

The interfacing of the rail and the wheel forms an integral part of railway engineering, determining the productive, safe and efficient operation of railways. This is because it encompasses forces and dynamics occurring at the interface of the rail and wheel, which affects many factors within the system, including but not limited to, the traction and stability of a train, and the wear and tear of both the track and rolling stock surfaces currently in use. Such an 'interfacial' approach directly correlates with the improvement of the projecting life of both rails and wheels, decreasing the cost of the maintenance services and increasing the safety as well as comfort of the operation of the railways [1].

3.1.1 Significance in Railway Systems

Interaction between the wheel and rail is of great importance to the functionality and efficiency of the railway systems. It influences the ride quality, the safety aspects, noise and wear of both the rails and wheels. The forces, which develop at the rail–wheel contact patch, influence the stability of the trains, the wear of the rails, the wear of the wheels and lastly the cost incurred in maintenance of the railway network. An optimized rail–wheel interaction permits a better quality of rides with less noise and vibrations and longer service life for both components [2–4]. In addition to that, rail–wheel interaction is also a factor of energy consumption. Due to this, rolling resistance is minimized while promoting traction such that lesser energy is utilized and operational costs are lower. In light of the growing need for high-speed and heavy-haul trains, these factors become increasingly important in operating modern railway systems.

3.1.2 Historical Development

The history of studying rail–wheel interaction has surely changed since the inception of railway transport. At an early age, most engineers were preoccupied with ensuring the rails and wheels had enough strength to bear the weight of the trains. But then it became clear that the physics of the interaction between wheel and rail is of importance. In the beginning of the 19th century, when railway systems developed, it based on observation and practical experience. Revolutionizing theoretical mechanics and the development of material science in the late 19th and early 20th centuries witnessed the beginning of a scientific approach towards rail–wheel interaction. Earlier researchers made paradigm-shifting contributions towards knowledge concerning the moving parts of a rail vehicle [5].

DOI: 10.1201/9781003659983-3

In the second half of the 20th century, high-speed trains like those of Shinkansen in Japan and TGV in France called for a new approach to study rail–wheel interaction. During this period, contact mechanics and vibration and wear studies were dramatically improved. Also, Kalker's linear and nonlinear contact mechanics theory and rolling contact fatigue proved to be important in forecasting and controlling the rail-wheel dynamics under certain operating conditions.

3.1.3 Basic Principles of Rail-Wheel Dynamics

The contact mechanism between the wheel and rail is also aided by the contact mechanics principles which look and explain how contact occurs between the wheel and rail surfaces while the train moves along the railways. It is noted that the region where the wheel meets the rail also known as contact patch is the centre of this interaction. In this small region, axial spring forces in addition to shear forces in the tangential directions on the rail–wheel interface are transmitted into the zone where the rail and wheel come into contact [6, 7]. The distribution of these forces has multiple reasons such as the profile and characteristics of the wheel and rail as well as the load being transported by the train and its speed. The relationships established by modifying these parameters result in various types of contact conditions, conformal and non-conformal for instance, that influence the wear and fatigue of the rail-wheel system.

3.1.4 Dynamics of Rail–Wheel Interaction

The dynamics of rail–wheel interaction is defined by several core parameters:

Normal Forces: The normal forces are those perpendicular to rail's surfaces. They could be battering forces standing to the track due to the weight of the train. Alteration of normal forces distribution also includes changes of geometrical track features like curves or gradients and such changes impact stability and wear on wheels and rails. Tangential Forces: This type of force acts in the plane of the contact surface and is usually associated with traction and braking effects. Factors like wheel-rail adhesion, train speed and the braking or accelerating force also affect these forces. The tangential forces are responsible for introducing slip or creep at the contact patch, which will contribute to wear and in extreme cases excavate wear of wheel or rail.

Creep and Traction: Creep is defined as the relative displacement of the wheel to the rail in the contact patch. It is caused by the difference in speed of the wheel surface to the rail. Creep is also a highly important element in traction because it permits the transmission of shear forces in the tangential direction with minimal sliding. Creep in moderation is advantageous, but too much of it will accelerate wear and increase energy losses [8, 9]. Impact of Track Geometry: Track geometry features such as curves radius super-elevation and alignment play a critical role in rail–wheel interactions. For curves for instance, the outer rail tends to bear significant normal loads then under such circumstances there is a risk of high wear and potential derailment if no corrective measures are taken. Table 3.1 provides an overview of rail–wheel interaction concept.

TABLE 3.1
Overview of Rail–Wheel Interaction Concepts

Aspect	Description
Significance	Importance of rail–wheel interaction in safety, ride quality, noise reduction, and maintenance cost management.
Historical Development	Evolution from empirical methods to scientific analysis, particularly with the rise of high-speed trains requiring sophisticated understanding of rail-wheel dynamics.
Basic principles	Concepts of contact mechanics, distribution of forces at the contact patch, and influence on wear and stability.
Dynamics	Interaction of forces like normal, tangential, and lateral forces affecting train stability and track wear.

3.1.5 Evolution of Analytical and Computational Methods

As the years went by, changes came about in the analytical methods that were used to study the rail–wheel interactions. Early methods adopted the use of simplified models whereby rail and wheel system was regarded as a rigid-bodied complex system. These models however were lacking in their prediction of the true behaviours which were observed in the real world. Towards the end of the century, more advanced models were coming up due to the introduction of computational technologies. Finite element analysis (FEA) and multi-body dynamic (MBD) were also utilized from this point amongst other simulation technologies in the study of rail–wheel interaction. These methods assist the engineers in creating simulations whereby the rail and the wheel are stiffer bodies instead of being flexible with the assistance of geometric properties stresses, contact conditions and dynamic loads.

Computational modelling and simulation have been embedded in the design and the modification of designs of railway systems so as to incorporate its use. It also brings useful predictions of the rails-wheels' operational behaviour under various adversative conditions minimizing any rail failure and streamlining maintenance problems. All in all, the interaction between the wheel and rail comprises many aspects of railway engineering that have grown tremendously for the last 100 years. Its investigation and control are needed for or rail systems to be used in a safe and efficient manner, and more so in a sustainable manner. Further, as technology progresses in rail transport, further research in this area will continue to be relevant to future transportation needs.

3.2 FORCES AND DYNAMICS OF RAIL-WHEEL CONTACT

The rail–wheel interaction pertains to a variety of forces and dynamics that are important in the functioning of any railway system. It is necessary to comprehend the interplay of these forces and their action on the rail and the wheel for making the maximal engineering development's periphery and their servicing and efficiency. This section addresses the classification of forces acting in the rail-wheel contact, the process of creep and traction, as well as the dependence of these forces on speed, load and geometry of the track.

3.2.1 Normal and Tangential Forces at the Contact Patch

Normal and tangential forces that are active in the contact patch are the primary constituent that contributes to the rail–wheel interface. The rail-wheel contact patch is the section that is made when the wheel rotates against the rail. This area however minute is the centre of all the interaction and exchange forces between the rails and the wheels.

Normal Forces: Such forces are always normal to the point of contact surface and are mainly concerned with bearing the load of the train. The direction and the intensity of these forces, however, depend on certain parameters such as weight of the train, the geometry of the track and the situation of the rail and wheel surfaces. Such variations in these forces may result in nonuniform wear and tearing, deformation of materials and even structural failure in extreme cases.

Tangential Forces: These, directed tangentially to the contact surface with the rolling stock, are closely associated with its tractive and retarding actions as well as side sway. They have great bearing on the stability and handling of the train whenever it is increasing its speed, reducing it, or going round a bend. The tangential forces are a function of roll adhesion which interlinks the rail and the wheel and therefore depends on the geometry of the rail, prevailing weather conditions such as ambient moisture, ambient temperature and the material used in building the wheel and rail.

3.2.2 Longitudinal, Lateral and Vertical Forces

The various forces present at the interface between the rail and wheel are divided into longitudinal, lateral and vertical forces.

Longitudinal Forces: These forces are the ones apparent in either traction or in braking. They are there to propel the train forwards or to stop it. Longitudinal forces are determined by the level of adhesion achieved between the rail and the wheel and if the conditions are too severely exceeded, it may cause the wheel skidding or slipping. Proper care for these forces is exceedingly vital in ensuring adhesion is achieved even in adverse scenarios such as wet or oily floors [10].

Lateral Forces: Lateral forces are those which are at right angles to the movement of the train and are most active for example when a train is taking a turn A particular case is with lateral forces and any inclination of the track relative to the train movement. When lateral forces are extreme, the wheel flanges can come in contact with the rail leading to derailment and massive wastage of the contact areas.

Vertical Forces: These forces are static in nature and are vertical in action hence they are the load-bearing forces. These forces vary as per the weight of the train and the contour of the designed track. Since vertical forces may be complicated by sudden deviations in rails geometry, which are rather frequent at changes in track alignment, track switches or road crossings, such forces are also expected to contribute to the wear and damage of the rail as well as the wheel, especially when present. Distribution of forces in rail–wheel interaction is shown in Figure 3.1.

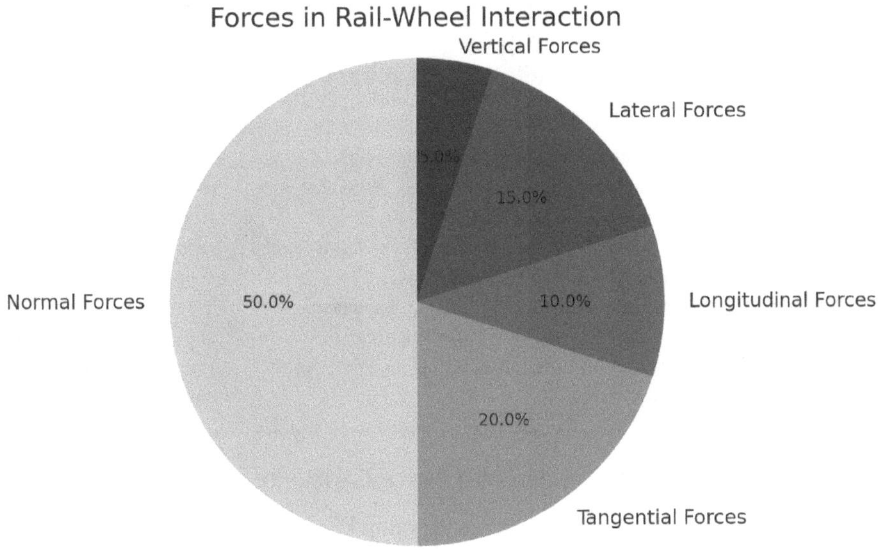

FIGURE 3.1 Distribution of forces in rail–wheel interaction.

3.2.3 Creep and Traction: Concept an Idea

Creep is a universal and useful term when it comes to rail–wheel interaction. It describes the slippage of the wheels in the contact patch in the direction of the friction force, which results from the difference between the actual rolling velocity of the wheels and that of rolling without slipping.

Longitudinal Creep: This is the phenomenon that occurs when the forward motion of the vehicle is not equal to the rotational motion of the wheel. Longitudinal creep has a fundamental role during traction and braking as it facilitates the relationship of tangential forces between the wheel and the rail. On the other hand, very high longitudinal creep promotes slippage, which diminishes the efficiency of traction, and wear is increased.

Lateral Creep: Relative lateral creep can take place as the wheel and rail have some degree of relative motion in the transverse direction. Most engineers will experience lateral creep during curve radius snow and the lateral creep affects the generation of lateral forces and the stability and steering of the train. Engineers will have an interest in lateral creep because of derailment risk and lateral friction forces induced wear of the rail and wheel.

The idea of traction relates to creep since it is the degree to which the wheel can generate forces against the rail system without skidding. The degree of traction that can be achieved is a function of the friction that exists between the surfaces of the rail and the wheel. Traction is also affected by other external factors, for example maintenance of rail surface, wheel profile and weather conditions: rain or leaves on the rails or frosting. Table 3.2 shows the key forces and dynamics of rail-wheel contact.

TABLE 3.2
Key Forces and Dynamics of Rail–Wheel Contact

Force Type	Description	Factors Affecting
Normal forces	Perpendicular forces supporting train weight.	Train load, track geometry.
Tangential forces	Forces parallel to the contact surface, associated with traction and braking.	Adhesion, speed, braking force.
Longitudinal forces	Forces acting along the direction of travel, impacting acceleration and braking efficiency.	Traction, load, track conditions.
Lateral forces	Perpendicular to the travel direction, influencing stability during curves and affecting wear.	Curvature, super-elevation, alignment.
Vertical forces	Primary load-bearing forces acting perpendicular to the rail surface.	Train weight, speed, track condition.

3.2.4 Influence of Speed, Load and Track Geometry on Forces

At the rail–wheel interface, the forces that develop keep changing with the variations in speed, load and some geometrical aspects of the track.

Speed: Increasing the speed impacts more dynamic and increased force that would cause more stressing of variation in the geometric/asphalt profiles of the track system. At great speeds, because of the interactions between the wheel and rail dynamics, excessive vibration and noise emission, greater wear and tear and instability may occur.

Load: Heavier trains will exert greater normal forces which will increase the contact stresses at the rail–wheel interface and result in normal contact. This will wear out faster and increase the chances of rolling contact fatigue (RCF) and plastic deformation of the rail and wheel active surfaces. Proper load distribution and balancing are very important in how these conclusions will be constrained.

Track Geometry: The curvature, super-elevated structures and alignment of the track contribute mightily to rail–wheel interaction. In curved sections, the outer rail bears most of the normal force Perfect Evidence While the inner rail suffers reduced forces and this may lead to what is called differential wear and formation of rail defects such as buckled gauge corner. To accomplish effective interaction of wheel-rail system design, track design, as well as track maintenance is very important to alleviate excessive and unwanted wear and the chances for derailment. The wheel track contact dynamics and forces are elementary in the daily operation and management of railway systems. The knowledge of these forces and how they act explains the reasons for having different rail and wheel configurations for satisfying maximum traction braking forces and less abrasion. Linear, transverse and vertical loads, combined and separately with the concepts of creeps and traction, complicate the problem at a rail and wheel interface. Putting a good strategy for handling these issues is what is going to enhance the safety, reliability and efficiency of today's railways operation.

3.3 WEAR AND FATIGUE MECHANISMS

Possessed and endured at the rail–wheel interface wear and fatigue are a natural result in the interaction of these two components as long as this interaction persists. They are among the concerning issues in the maintenance and operational aspects of the railway systems owing to the influence on both wheels and rails lifespan. For optimal design and selection of materials, effective maintenance and management of machinery during railway operation, it is necessary to understand the mechanisms of wear and fatigue [11–13]. In this section, basic concepts of different types of wear are reported, followed by RCF and factors affecting these forms of wear. Table 3.3 shows the types of wear and fatigue mechanisms. Wear and fatigue progression over time is shown in Figure 3.2.

TABLE 3.3
Types of Wear and Fatigue Mechanisms

Wear Type	Description	Influencing Factors	Consequences
Adhesive wear	Material transfers due to frictional forces between wheel and rail surfaces.	High contact pressure, friction conditions.	Surface roughness, noise, increased wear.
Abrasive wear	Removal of material due to hard particles between rail and wheel.	Presence of contaminants, hardness differences.	Rapid material loss, increased maintenance.
Fatigue wear	Crack initiation and propagation due to cyclic loading.	High stresses, material properties, load cycles.	Rolling contact fatigue, structural failure.

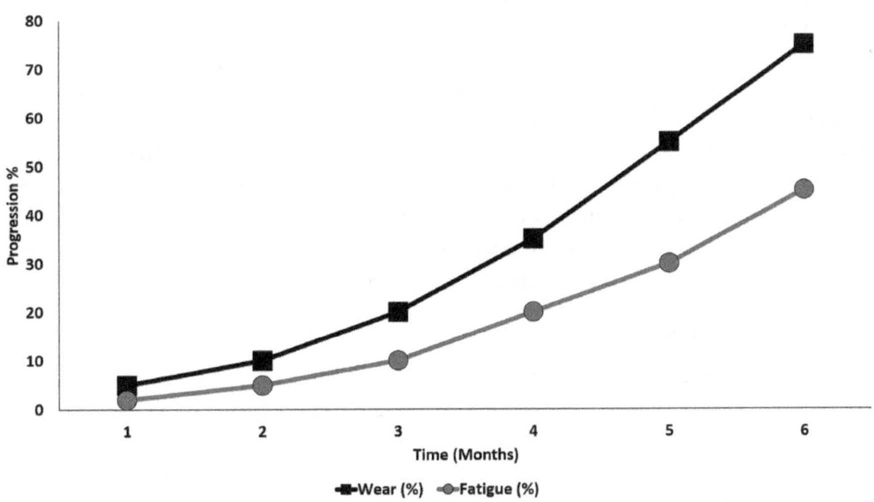

FIGURE 3.2 Wear and fatigue progression over time.

3.3.1 Types of Wear

For the rail-wheel mechanism wear is divided into three major types: adhesive wear, abrasive wear and fatigue wear. Each type of wear occurs due to different interaction mechanisms at the rail–wheel interface and is related to phenomena, such as contact pressure, material properties and surroundings.

3.3.1.1 Adhesive Wear

Adhesive wear is one type of wear part that lands itself as two solid surfaces experiencing a linear motion through the application of normal load, leading to a transfer of material from one surface to another. In the rail–wheel interface, the frictional forces between a wheel and a rail contribute to the adhesive wear. It occurs when the pulling force of the join is less than that of the joining materials and very small bits of material come off the surface, thus causing wear. This type of wear is also suffered most in situations of high contact pressure and/or high frictional forces like while towing or braking. The increased wear by adhesive mechanism can lead to the production of roughened rail and wheel surfaces which has been reported to amplify the noise and vibrations levels of the vehicle. Severe adhesive wear, which usually develops as time passes, will lead to crack formation. This development will further promote the deteriorating state of the rail–wheel interface.

3.3.1.2 Abrasive Wear

According to ASTM G40, abrasive wear takes place when a harder surface or a harder particle is sliding against a softer surface with cutting or ploughing action that results in the removal of to cutting action, micro-cutting. In the rail-wheel system, abrasive wear is mainly attributed to foreign materials like sand dirt or metal parts that are caught in between the rail and the wheel. Such particles are eventually embedded in friction surfaces and work as abrasives to scratch and gouge the surfaces, facilitating wear.

Abrasive wear affects more so the users operating in a debris-ridden environment, such as in industrial or mining railway system. It can cause problems for the material because of the fast loss if contaminated with foreign particles which are hard and sharp. The thickness of this type of wear varies because of the difference in the hardness of wheel and rail materials. In conclusion, the most exposed parts to abrasive wear are the wheels and rails of rolling stock, whose surfaces are made of hard materials. But the use of harder materials for wheel rail can base the abrasive wear but this has to be balanced with other issues such as RCF.

3.3.1.3 Fatigue Wear

RCF is the wear that results from the cyclic loads acting on the wheel and the rail surfaces' contacts. This wear pattern is associated with the process of crack initiation and subsequent increase in size up to the point of the surface detachment, otherwise called spalling. Wear of parts due to RCF is the predominant form of damage in rails and wheels, which have considerably lower performance of these parts.

The influence of RCF is affected by a number of factors such as the magnitude and distribution of the contact stresses, the properties of the rail and wheel materials

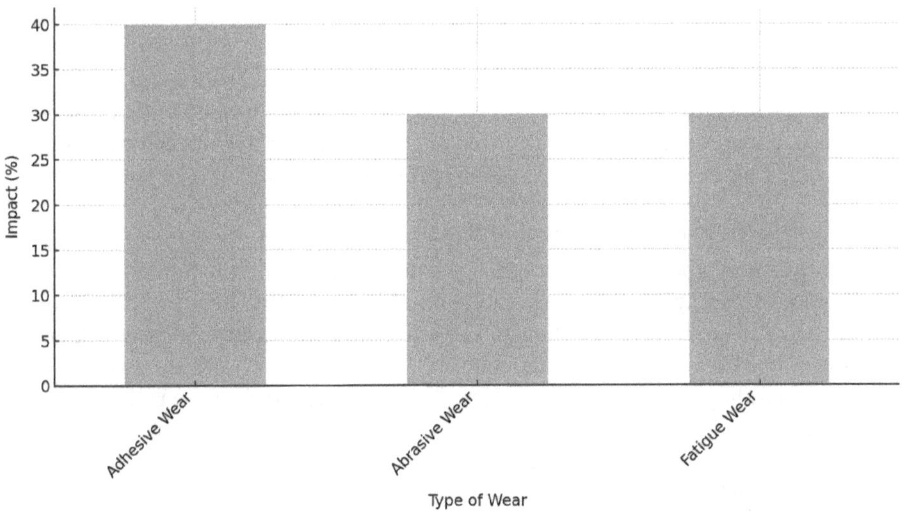

FIGURE 3.3 Impact of different types of wear on rail–wheel interaction.

and also on the working conditions such as speed and load. Against global high contact stresses there is considerable cracking at some point (rail joints or switches) throughout the period. When such damage is involved, controlling its potential effects in the design phase will normally be effective. If not confined and controlled at the early stages then RCF can lead to adverse destructive axle bore failure and wheel fractures. Figure 3.3 shows the impact of different types of wear on rail–wheel interaction.

3.3.2 RCF and Its Effects

The RCF can be defined as damage caused in operation which is a result of the cyclic force interactions in a certain contact area within the wheel and the rail. If this repeated loading continues, this results in the occurrence of micro-cracks on the material's surface or within its subsurface layers [14–17]. Such defects are likely to grow and extend with each load sequence making it impossible to contain material spalling or complete breakdown of the component.

3.3.2.1 Mechanisms of RCF

RCF is enabled through the accumulation of plastic strain and cyclic stress in the contact region which causes plastic deformation. The mechanisms of RCF arises majorly include:

Surface-Initiated RCF: This is RCF where surface defects such as pitting, roughness or embedded wear debris create stress-risers at the surface. The RCF crack emanating from the vertex surface i.e., surface-initiated cracks generally extend on the surfaces, or below, causes peeling or flaking off.

Subsurface Initiated RCF: This type of RCF occurs whenever shear stresses are induced on and into the rail or wheel wear surface. These stresses are contributed by the congestion in Hertzian contacts while rolling takes place Wanahayi Getting of ceramic socket implants.pdf. For cracks initiated subsurface, the direction of crack extension is usually to the surface resulting in cracking or in severe cases, total rupture of the body.

3.3.2.2 Effects of RCF on Wheel and Rail Life

Both the rails and wheels endure a RCF in order to perform their functions and this inflicts a lot of damage to RCF. The head checks, squats and shelling that develop around the rail due to RCF applies as defects on the rail and tend to weaken the rail's structure. However, if such defects are not remedied by grinding or replacement or some other maintenance activities, they are likely to result in rail breaks and derailment. With regard to wheels spalling, shelling and out round are some of the RCF-related defects that can be caused. These defects are destructive with regard to the life of the wheel and also the quality and safety of the ride. It has been seen that under serious RCF conditions, excessive damage to the wheel will lead to excessive vibrations and noise emissions from the wheels which will cause further harm rolling stocks and track systems.

3.3.3 Factors Influencing Wear and Fatigue

A number of variables including material properties, lubrication, environmental environment and operational parameters determine the rate and intensity of rail-wheel wear and fatigue at the interface.

3.3.3.1 Material Properties

There is no doubt that material properties of the rail and wheel determine the extent of wear and fatigue each component will experience. Hardness, toughness and microstructure are a few examples of such properties.

Hardness: Hardness contributes positively in terms of improving abrasive wear resistance although it exposes the material to cracking and fatigue Ngugi prisons get cheap homework help. Provide scientists as well discuss tradeoffs between hardness and toughness that is if hardness is. A toughened material is likely to be brittle and survive shock failure. Materials defined as durable: sacrifice impact on achieved structure or.

Toughness: Toughness is defined as the capacity of a material to withstand traumatic conditions in order not to fail completely, a characteristic known as plastic deformation. Cyclic loading is withstood more successfully by those with extremely sturdy constituents who have a lower tendency to develop cracks.

Microstructure: The microstructure, length scale, raging from phase distribution to the grain size of the material played an important role in determining fabricated components' wear system and estimating fatigue life. Grain structure finer produced more wear and toughness properties than corse.

3.3.3.2 Lubrication

In the end, lubrication is the most decisive among these factors, which reduces wear and fatigue by restriction of contact temperatures and stresses at the rubbing zones. The application of an effective adhesion wear protection technique would lead to the great enhancement of service life of headers and axes. There are different means of achieving lubrication for example by use of fixed lubricators mounted on trains and trackside lubricators for direct application of the lubricant to the rail. The application of lubricant as well as the type of lubricant used varies from track to track due to track and environmental conditions and the wear and fatigue encountered. Table 3.4 shows rail–wheel interface management techniques

3.3.3.3 Environmental Conditions

Environmental conditions include temperature, humidity and the presence of contaminants, which can have a critical influence on wear and fatigue. For example:

Temperature: The wear and RCF resistance of rail and wheel materials might be compromised as temperature increases due to the loss of hardness and strength of the materials. On the contrary, low temperatures may lead to increased brittleness in the materials which would thus become easily prone to fracture.

Humidity and Moisture: Moisture can be a lowering factor in adhesion at the rail–wheel perimeter which can cause slippage and increased damage. Moisture fosters corrosion that accelerates the rate of material degradation and reduces the fatigue life.

Contaminants: The presence of sand, dirt and other particles constitutes contaminants and increases abrasive wear and scratch. In places where there is a high level of industrial activity, there are chemical contaminants that contribute to corrosion and wear.

TABLE 3.4
Rail–Wheel Interface Management Techniques

Technique	Description	Benefits
Lubrication	Reduces friction and wear at the contact interface.	Lower wear rates, reduced maintenance costs.
Rail grinding	Restores rail profile, removes surface defects.	Improved ride quality, reduced noise.
Wheel re-profiling	Restores wheel shape to maintain optimal contact with the rail.	Enhanced stability, reduced wear and tear.
Laser cladding	Applies protective coatings to rail surfaces to enhance wear resistance.	Increased lifespan, improved surface durability.
Coatings	Surface treatments to reduce wear and improve fatigue resistance.	Enhanced material performance, longer service life.

3.3.3.4 Operational Parameters

Such operational parameters as the speed, load and braking intensity can exert direct influence on the dealt forces and stresses at the rail–wheel interface, hence surface wear and fatigue.

Speed: Faster speed means that there are more forces captured within the contact patch, this might increase the magnitude of wear and fatigue. In regions of very high speeds, even small imperfections are very dangerous as very high kinetic energy is present.

Load: For increased loads, normal forces will increase further leading to increased contact pressure and the tendency to enhance plastic deformation and RCF. Improper load mechanics and distribution will lead to excessive wear and fatigue.

Braking Intensity: Harsh conditions of the brakes, high torque and rapid movement can produce reliant tangential forces at the patch of contact increasing the chances of adhesive wear on the surface of materials. Alternatively, the high intensity of repeated brakes impacted on the same spots can lead to thermal damage thus cracks and degradation of the material.

3.3.4 Mitigation Strategies

Different measures are put in place so as to overcome the difficulties caused by wear and fatigue in the design and maintenance of railway systems.

3.3.4.1 Material Selection and Treatment

Material composition involving the correct combination of hardness, toughness and fatigue resistance is necessary for controlling wear and fatigue. Heat treatment processes like induction hardening or bainitic transformation do increase the wear resistance and fatigue life of rail and wheel materials [18].

3.3.4.2 Rail Grinding and Wheel Re-Profiling

Scheduled rail grinding and engineered wheel profiles are some of the most crucial maintenance practices that enable axes and tracks to revert to their original profiles thus preventing the occurrence of disharmony, and hence wear and fatigue in components. These procedures remove operational defects and restore the flat contact area, improving the comfort of the operation of the system and durability of parts.

3.3.4.3 Lubrication and Friction Management

Good lubrication methods and friction strategies help to a great extent in eliminating unnecessary wear and fatigue. Elaborate creams of lubrication like high-pressure lubricators and top-of-the-rail friction modifiers assist in sustaining the levels of frictional contact and helping reduce levels of surface damage. The problems of wear and fatigue in the rail–wheel interface are of interest in design, safety performance and maintenance of the railway system, as they are inevitable. Engineers must learn the processes and causative factors of these processes in order to devise measures that will help contain these factors. Technologies in the new materials, lubrication and monitoring systems are improving the durability and performance of rail and wheel componentry and hence their safe and efficient operation of railway.

3.4 RAIL–WHEEL INTERFACE MANAGEMENT

Effective management of the rail–wheel interface is critical in augmenting the safety, efficiency and durability of the railway systems. It encompasses a series of operations and methods to be applied to the rail-wheel contact in a bid to reduce wear, fatigue and noise without compromising performance. Good management of the interfaces cannot only reduce costs related to maintenance but also enhance operations reliability as well as the lifespan of the rail and the wheel. This section looks at some of these methods such as lubrication, rail grinding, wheel re-profiling and advanced methods such as laser cladding and coatings ICS.

3.4.1 Lubrication

Lubrication has been highlighted as one of the best and most frequently applied techniques in the management of the rail–wheel interface. Surfaces in contact get into relative motion with reduced wear and rolling resistance which inversely enhances the consumption of energy and the life span of the interacting components. In railway systems there are some applied lubricants and ways of their application:

Grease-Based Lubricants: This class of lubricants is used for instance on the railhead or wheel flange to prevent erosion. Grease lubricants play a positive role in reducing the metal-to-metal contact hence the occurrence of adhesive wear. Under certain conditions such lubricants have their performance compromised for instance in terms of temperature and moisture.

Water-Based Lubricants: Such lubricants are mostly utilized during high-speed rail, where the friction levels have to be low. They perform their function well and do not allow excessive residues which in some cases can be disturbing.

Top-of-Rail (TOR) Friction Modifiers: Such lubes are used on the upper part of the rail. Raisins are utilized in TOR framers, but unlike the rest, these do not smear the cleaning process of raisin swathers, instead they are used to steer with finessing friction. Here, wear and noise can be decreased while the traction and braking action remain untouched.

Application Methods: Lubricants can either be applied by a trackside system, which dispenses lubricant on the rail at prescribed positions, or an onboard system, by which lubricant is directly sprayed onto the wheel flange while the train is in motion. Trackside systems are primarily utilized on predominantly wear arme circumstances such as bends, meanwhile, onboard systems are not limited to a single point of use and may be adopted to rectify a particular wear problem.

3.4.2 Rail Grinding

Rail grinding is a process of maintenance to a track that defines the contours, eliminates imperfections on the surface and extends the life of the rail surface. Doing this is very critical in the management of the rail–wheel interface as it affects the contact and wear of the rail-wheel directly.

Profile Restoration: Over periods of time, the rail head is bound to have irregular wear patterns due to incessant trains running on it. Rail grinding is done in order to bring back the required rail sections in line with the contact between the wheel and rail to reduce failings such as head checks and gauge corner cracks.

Surface Defect Removal: Rather than preventing or bridging cracks, rail grinding results in the removal of bettering the surface of a defect filled with noise such as corrugations, squats and spalls. Smoothing out the rail surface reduces dynamic forces at the rail–wheel interface yielding better ride quality and lower damage on the rail and the wheels.

Preventive Grinding: More commonly, light periodic grinding is used as a routine preventer of further mothballing or future wear. It is useful in restoring the rail surface conditions preventing the formation of defects including cracks that would translate to greater damages if the cracks are not dealt with.

3.4.3 Wheel Re-Profiling

Wheel, which is also referred to as camber, re-profiling is also referred to as cambering is the process in which a worn or damaged machine wheel is artificially shaped by restoring or even recontouring. It is relevant as it helps optimize wheel-rail contact geometry and wear and noise reduction.

Causes of Wheel Profile Degradation: Due to different factors such as heat or friction, poor track conditions, or intense braking, the wheels can sustain damage and wear out. In turn, this may bring flat spots, out of roundness and uneven wear, which impact negatively the ride quality and raise the load stress on the rail and any track's components.

Re-Profiling Techniques: There are specialized turning lathes or grinding machines that are designed to rotate a wheel with the tread, and or, flange against cutting instruments to achieve the desired profiles. By using the wheel re-profiling processes, the wheel not only regains its original form but by so doing makes the wheel sets more balanced, hence less vibration and improved stability.

Benefits of Re-Profiling: The practice of regular wheel re-profiling reduces the chances of derailment and improves the ride quality and the lasting chances of both the contact surfaces of the wheel and the rail are increased in instances where the contact geometry is maintained in a set perimeter. It also helps in reducing localized rubbing of the wheel and rail surfaces by facilitating better distribution of the means of wear and tear.

3.4.4 Advanced Techniques for Wear Resistance

Apart from these conventional approaches lubrication and grinding, there will be some more improvements. This will focus on developing advanced techniques to further improve the abrasion resistance and durability of wheels and rail material.

Laser Cladding: Laser cladding is the application of a protective layer onto the surface of rail or wheel by the use of a high-powered laser. This process enables the deposition of wear protective coatings in such quantities and shapes that the surface hardness and wear as well as fatigue resistance is increased significantly. Most often due to the laser cladding technique, the operating life of the elements, which experience increased loading in areas like curves and turnouts, can be extended [19–23].

Specialists Equipment Enclosures: Placement of sensitive equipment such as switch machines or electrical components within an enclosure aids in containing the operational noise that would have escaped from the enclosure. These enclosures are engineers with sound insulation to a cellular frame of the structure.

3.4.5 Advanced Monitoring and Predictive Maintenance

Incorporating special advanced monitoring systems that are equipped with sensors, data analytics which will be used to monitor wheels and rails in order to reduce vibrations and noise is paramount. It makes use of predictive maintenance where the problems are dealt with before they are experienced.

Acoustic Emission Monitoring: Acoustic emission sensors counting on the absence of visible external sources are based on detecting acoustic emissions originating from obscure internal sources; this is in turn understood as high peak vibrations that occur as a result of defects like cracks or spalling. As such emissions are monitored by operators, such defects can be rectified even before they become serious and end up causing excess noise [24–28].

Vibration Monitoring: It helps strategic sensoring of acceleration values especially gimbaled devices or accelerometers or other locations on the track or hinged rolling stock which provides information about vibration. Such data when analysed helps towards an indication of the problem areas and assists in the decision-making on the maintenance that needs to be done to reduce the noise and vibrations [29, 30], which discuss advancements in rail vehicle stability, ride performance, crashworthiness and vehicle dynamics.

Vibrations and noise created as a result of interaction between rail and wheel are great limitations to railway operations, as they affect both passenger satisfaction and the offender's health per the railway structure. It is crucial to define the cause of these phenomena and identify their effects to be able to address them comprehensively. By making use of regular maintenance, advanced monitoring and additional innovative methods, it is possible to alleviate the adverse effects of vibration and noise hence

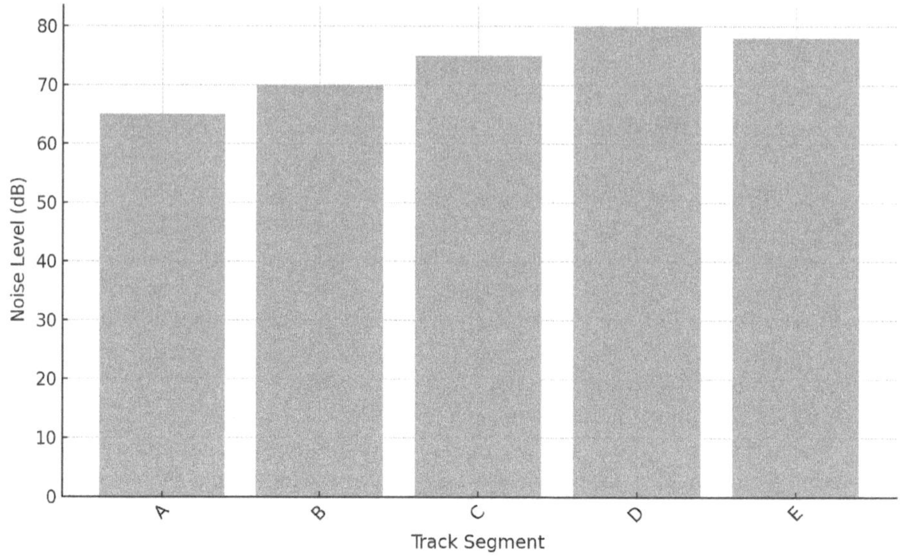

FIGURE 3.4 Noise level over different track segments.

transforming the railway systems into a safer, more effective and greener alternative means of transport. Figure 3.4 shows the noise level over different track segments

3.5 INNOVATIONS IN RAIL-WHEEL MATERIALS AND DESIGN

The modification of materials and the design of rail-wheel systems have led to a notable improvement in the performance, durability and safety of railway operations. Scientific and engineering advancements have contributed to the making of more robust materials and efficient designs that are more resistant to the conditions at the rail wagon wheel such as very high forces, temperatures and erosive conditions. This section reviews some of the current trends in rail-wheel materials and design and provides evidence of the practical application of these technologies with case studies.

3.5.1 Advanced Materials

The contribution of advanced materials has particular on the rail and wheel wear resistance, fatigue life and other relevant components of durability.

Bainitic and Pearlitic Steels: Traditionally pearlitic steels have served as the rail steel due to their properties of toughness and strength. However, bainitic steels are able to prevent RCF and wear thanks to their fine microstructure. These steels have also better resistance to impact, thus making them more suitable for high speed and heavy haul.

Composite Materials: These materials also include polymer composites, which are typically reinforced by fibres. These are currently used in non-structural components such as wheel centres or rail fasteners that add weight to the structure but do not contribute much to the overall sturdiness of the design. They are still being developed for use in non-primary structural components of wheeled vehicles, as research is still being conducted to bring them to use in rail and wheel making.

High-Strength Alloy Steels: Other novel high-strength steels with improved wear resistance or toughness are being designed. Such materials are preferably applied in high contact stress environments, which are usually the situations for heavy freight railroads and places with steep curves. They aid in reducing maintenance intensity and enhancing service life relative to other steel grades.

3.5.2 Innovative Design Approaches

Along with these changes in the materials employed, new design changes have also been undertaken with a view of improving rail-wheel interaction and performance. Table 3.5 shows Innovations in Rail-Wheel Materials and Design.

Steps in wheel profiles design: New wheel profiles have been developed which improve significantly the stability and wear characteristics of railway wheels, for instance, the 'CONTOUR' and the 'S1002' profiles. The characteristics of these profiles are such that contact stresses are reduced and the chances of RCF and wear of the flanges are minimized which increases the service life of the wheel in question.

Rail new profiles and its effectiveness: New rail profiles provide for the more even distribution of contact stresses and lessen the risk of stress problems. For example, atypical rail profiles have been found to lessen the lateral and wear forces that are normally found on the curvature rail. This leads to uniform wear of the main contact surfaces and less wear of the machine.

Effective wheels and ribbed wheels. New concepts such as elastic wheels composed of rubber or polymer layers interposed between the tread and the hub of the

TABLE 3.5
Innovations in Rail–Wheel Materials and Design

Innovation	Description	Benefits
Bainitic steels	Advanced steel grades with improved wear and fatigue resistance.	Longer service life, reduced maintenance.
Composite materials	Lightweight, durable materials for non-structural components.	Reduced weight, enhanced environmental resilience.
Optimized wheel profiles	New profiles designed to minimize contact stress and reduce wear.	Improved stability, reduced rolling contact fatigue.
Resilient wheels	Incorporate damping elements to reduce noise and vibrations.	Enhanced ride comfort, reduced noise pollution.

wheel have been developed to control noise and or vibration. These types of wheels are primarily used in urban and light rail transit systems that require attenuating noise as well as comfort during rides.

3.5.3 Case Studies of Successful Innovations

Many rail networks from different countries have successfully adopted these advanced materials and designs with improvements in performance and reductions in costs.

Bainitic Rail Steels in High-Speed Lines: There has been the introduction of bainitic rail steels for high-speed rail lines such as in China and Germany which has contributed to a reduction in the total life cycle cost and lesser occurrences of RCF compared to the pearlitic steel. Areas with a high density of traffic and constrained operational time schedules have greatly benefited from the advanced characteristics of the bainitic steels.

Composite Brake Blocks in Freight Operations: The employment of headache-resilient composite brake blocks on North American freight trains has increased its braking efficiency due to optimal wheel braking and decreased wheel abrasion. These blocks have been shown to perform more efficiently than traditional cast iron blocks in that they have enhanced thermal characteristics and lower noise emissions.

Resilient Wheels in Urban Rail Systems: Other cities such as London and Paris have incorporated the use of resilient wheel design in their metro systems to help alleviate the problems of noise and vibrations. Outfitting the rail systems with these wheels has caused a reduction in noise levels which enhances passengers' comfort as well as mitigating the noise impacts to the community.

3.5.4 Future Directions

Continuing to improve the performance of the rail-wheel materials and designs occupies a significant place in ongoing research, engineering and further development. Among these is the production of nanostructured steels containing less construction material so as to enhance wear-resisting ability, composing variable materials that will respond to change in the environment, integrating the moving surfaces of the rail and wheel with online monitoring systems for predictive maintenance to decrease the likelihood of surprises. The further improvement of rail-wheel material and design has played an important role in enhancing the evolution of railway systems and rendered improved efficiency, security and longevity of the systems. Further study and application of these new technologies will become inevitable in addressing the increasing requirements of modern-day rail network configuration and operation for their viability and effectiveness in serving the current and future needs. Table 3.6 shows future trends and research directions

TABLE 3.6
Future Trends and Research Directions

Trend	Description	Potential Impact
Smart materials	Materials that adapt to changing conditions, such as load and temperature.	Improved performance, reduced wear and maintenance.
Real-time monitoring	Use of sensors and data analytics to monitor rail-wheel conditions in real-time.	Early detection of defects, predictive maintenance.
AI and machine learning	Advanced algorithms for analysing data and optimizing maintenance schedules.	Enhanced efficiency, reduced operational costs.
Sustainable solutions	Development of eco-friendly materials and energy-efficient designs.	Reduced environmental impact, lower costs.

3.6 SUMMARY AND FUTURE DIRECTIONS

Ongoing development concerning rail–wheel interaction would not be possible without the impact of technology, material sciences and computational modelling. Whenever appropriate, these new ideas are intended to solve problems like: wear, fatigue, noise and vibrations while augmenting safer and greener railway practices in the future. This section covers trends and research areas that are expected to influence the development of rail-wheel systems.

3.6.1 New Technologies for the Rail–Wheel Interaction

There are still many a maturing technology, which improves the functionality and reliability of the rail-wheel systems. These include:

Smart Materials: Smart materials such as shape memory alloys and magnetorheological fluids are also being investigated for their use in developing an adaptive rail-wheel system. They respond to changes in external load or temperature by changing their properties, thus allowing a constant adaptation of the optimal contact and reducing wear to the components.

Condition Monitoring in Real Time: Improved condition monitoring technologies of rails and wheels, including real-time monitoring due to various sensor interfaces and analytics, are becoming widespread. Systems containing vibration sensors or acoustic emission detectors as well as infrared imagers can be used to monitor the operational health states of rail-wheel components. Furthermore, such data can also be beneficial to predict upcoming failures and assist in planning maintenance schedules before any major problems occur, thus improving downtime and maintenance cost efficiency.

Analytical methods such as Artificial Intelligence and Machine Learning: Large amounts of operational data are being gathered, and there is an increasing application of machine learning methodologies on it. Such technologies have the capability to avail the early detection of abnormal patterns and needs for maintenance, thus optimizing the requirement for the maintenance schedule in the systems and enhancing their efficiency.

3.6.2 Computational Modelling and Simulation Techniques

Computational modelling and simulation techniques are becoming more relevant in the design and optimization of rail–wheel interaction. Notable improvements include:

Finite Element Analysis (FEA): FEA is a powerful tool as it can fully replicate the stress and strain distribution of rails and wheels under various load conditions. Much stone has been cast in this arena in the recent past, that is aimed at improving the timeframe required to carry out such simulations, making it possible to assess more models within shorter durations [25].

Multi-Body Dynamics (MBD) Simulation: MBD simulations are applied to an entire train system to understand the motion response of train isolators, particularly the motion responses in contact with the wheels, rail and isolation systems. Progress in this specialty is enhancing the understanding and modelling of complex issues such as hunting instability, the dynamics of derailment and the interaction of vehicles with high-speed tracks [3].

Integrated Simulation Platforms: Advanced technologies are now emerging that allow for the consolidation of FEA, MBD and other modelling methods into a single system. These platforms provide a full analysis of rail-wheel systems, including their mechanics, thermodynamics, materials, etc. Such a systemic view enables optimizing designs and enhancing their maintenance [1, 9, 31].

3.6.3 Prospects of Smart Materials and Adaptive Systems

Developing smart materials and adaptive systems is one of the most promising areas for development. These innovations may lead to rail-wheel systems that self-adjust to optimal operation conditions making them safer and more efficient.

Self-Lubricating Materials: There are also efforts on materials that can be activated to function under lubricate deprivation or wear, lubricate itself due to frictional heat generated by movement. These types of lubricants may be effective, especially in high motion or hard-to-access regions where maintenance of equipment is minimal. Due to this precision down time can be exercised.

Adaptive Suspension Systems: Automation of the suspension system that employs sensors and actuators to change suspension attributes according to the characteristics of the track is on course. The successful implementation of these systems will alleviate the moving comfort level of a passenger often improving bouncing motion stability and enhancing safety by cutting down the level of bouncing motion instability even at maximum wheel-rail contact performance.

Embedded Sensor Technology: Another area of research is embedding the sensors in the rail and wheel materials. Such sensors can be used to collect data on internal stresses and even the early formation of cracks and other imperfections which aid diagnostical purposes.

3.6.4 Sustainable and Environmentally Friendly Solutions

With more researchers entering the field due to the rise of global warming, efforts are being made to build more environmentally friendly and sustainable rail-wheel systems.

Eco-Friendly Materials: There is increasing interest in the creation of eco-friendly materials for wheels and rails like the use of recyclable composites and even the use of soy-based, bio-degradible lubricants. Such materials are meant to minimize the ecological impact of the railway operation whenever possible, without affecting the efficiency and durability.

Noise and Vibration Reduction Technologies: New devices like rail dampers, sonic barriers and ground-hugging locomotives are under development in order to reduce the extent of noise and vibration produced by rail transport. These kinds of solutions are quite vital, especially for urban and suburbanite settings where noise pollution is a huge problem.

Energy-Efficient Designs: Studies have also been aimed at enhancing the profiles of wheels and the rails to optimize rail vehicle propulsion, traction means and lightweight materials. Such innovative ideas also cut down the running cost as well as carbon emissions arising from the operations of railway systems.

3.6.5 Collaborative Research and Industry Partnerships

The interaction between the rail and the wheel involves complex issues that warrant different expertise from academia, industry and governmental agencies. This way, the three stakeholders form partnerships that are important for the application of research findings and in devising several solutions within the railways industry.

Industry-Academia Collaborations: A number of these Institutes of Technology are collaborating with railway organizations for instance on material design, fatigue analysis and predictive maintenance research. Sponsored collaborations carry the scope of knowledge transfer and supporting creativity.

International Research Initiatives: European Shift2Rail initiative and research programmes of the International Union of Railways (UIC) are other examples of international research collaboration, that seeks appropriate uniform practices for effective management of rail–wheel interaction across countries and railway systems. It is an exciting time in the field of rail–wheel interaction research with the increasing integration of new technologies and materials that we will see more and more emerging. Railway industry transformation is becoming inevitable, with such trends as smart materials, in situ monitoring and predictive maintenance associated with artificial intelligence (AI). With continued advancement in research, the next generation of rail-wheel systems will not only be but more effective, safer and eco-friendly which will guarantee the place of rail transport as part of world transport infrastructure for years to come.

REFERENCES

1. Badgujar, P., S. Sharma, D. Rahul, S. K. Sharma, and R. K. Upadhyay. "Feasibility Study of Mumbai-Pune Hyperloop Corridor: Challenges and Implementation." *Research in Transportation Business & Management* 52 (2024): 101089.
2. Banerjee, D., S. K. Chakraborty, S. Bhattacharyya, and A. Gangopadhyay. "Appraisal and Mapping the Spatial-Temporal Distribution of Urban Road Traffic Noise." *International Journal of Environmental Science and Technology* 6 (2009): 325–335.
3. Bokaeian, V., M. A. Rezvani, and R. Arcos. "The Coupled Effects of Bending and Torsional Flexural Modes of a High-Speed Train Car Body on Its Vertical Ride Quality." *Proceedings of the Institution of Mechanical Engineers, Part K: Journal of Multi-Body Dynamics* 233, no. 4 (2019): 979–993.
4. Chen, X., L. Shen, X. Hu, G. Li, and Y. Yao. "Suspension Parameter Optimal Design to Enhance Stability and Wheel Wear in High-Speed Trains." *Vehicle System Dynamics* 62, no. 3 (2023): 1–23.
5. De Filippi, P., M. Corno, M. Tanelli, and S. M. Savaresi. "Single-Sensor Control Strategies for Semi-Active Steering Damper Control in Two-Wheeled Vehicles." *IEEE Transactions on Vehicular Technology* 61, no. 2 (2012): 813–820.
6. Dotoli, M., A. Fay, M. Miśkowicz, and C. Seatzu. "Advanced Control in Factory Automation: A Survey." *International Journal of Production Research* 55, no. 5 (2017): 1243–1259.
7. EN 16286-1:2013. *Railway Applications - Gangway Systems between Vehicles - Part 1:* Main Applications, 2013.
8. Fu, D., W. Wang, and L. Dong. "Analysis on the Fatigue Cracks in the Bogie Frame." *Engineering Failure Analysis* 58 (2015): 307–319.
9. Gao, S., L. Liu, Y. Liu, H. Liu, and Y. Wang. "API Recommendation for the Development of Android App Features Based on the Knowledge Mined from App Stores." *Science of Computer Programming* 202 (February 2021): 102556.
10. Garg, V. K., and R. V. Dukkipati, eds. "Tangent Track Response of Railway Vehicles." In *Dynamics of Railway Vehicle Systems*, Academic Press, Toronto, ON, pp. 177–224, 1984.
11. Günay, M., M. E. Korkmaz, and R. Özmen. "An Investigation on Braking Systems Used in Railway Vehicles." *Engineering Science and Technology, an International Journal* 23, no. 2 (2020): 421–431.
12. Han, J.-W., J.-D. Kim, and S.-Y. Song. "Fatigue Strength Evaluation of a Bogie Frame for Urban Maglev Train with Fatigue Test on Full-Scale Test Rig." *Engineering Failure Analysis* 31 (2013): 412–420.
13. Han, J., S. Q. Zhong, X. Zhou, X. B. Xiao, G. T. Zhao, and X. S. Jin. "Time-Domain Model for Wheel-Rail Noise Analysis at High Operation Speed." *Journal of Zhejiang University: Science A* 18, no. 8 (2017): 593–602.
14. Huang, Q., C. Zhang, Y. Jin, C. Yuan, and X. Yan. "Vibration Analysis of Marine Propulsion Shafting by the Coupled Finite Element Method." *Journal of Vibroengineering* 17, no. 7 (2015): 3392–3403.

15. Iwata, K., M. Onosato, K. Teramoto, and S. Osaki. "Virtual Manufacturing Systems as Advanced Information Infrastructure for Integrating Manufacturing Resources and Activities." *CIRP Annals - Manufacturing Technology* 46, no. 1 (1997): 335–338.
16. Iwnicki, S., M. Spiryagin, C. Cole, and T. McSweeney, eds. *Handbook of Railway Vehicle Dynamics*. CRC Press, Boca Raton, FL, 2019.
17. Kopp, G., E. Beeh, R. Schöll, A. Kobilke, P. Straßburger, and M. Kriescher. "New Lightweight Structures for Advanced Automotive Vehicles—Safe and Modular." *Procedia - Social and Behavioral Sciences* 48 (2012): 350–362.
18. Kumar, U., R. Kasvekar, S. K. Sharma, and R. K. Upadhyay eds. Wear of wheels and axle in locomotive and measures taken by Indian railway. In *Advances in Engine Tribology. Energy, Environment, and Sustainability*. Springer, Singapore, 2022. https://doi.org/10.1007/978-981-16-8337-4_5
19. Liu, X.-A., D.-Q. Yang, and Q. Li. "The Load Criteria for Ship Mechanical Noise Prediction in Low Frequencies and Experimental Validation." *Journal of Ocean Engineering and Science* 8, no. 3 (2023): 278–293.
20. Mei, T. X., H. Li, R. M. Goodall, and A. H. Wickens. "Dynamics and Control Assessment of Rail Vehicles Using Permanent Magnet Wheel Motors." *Vehicle System Dynamics* 37, no. supl (2002): 326–337.
21. Miao, B. R., Y. X. Luo, Q. M. Peng, Y. Z. Qiu, H. Chen, and Z. K. Yang. "Multidisciplinary Design Optimization of Lightweight Carbody for Fatigue Assessment." *Materials and Design* 194 (2020): 108910.
22. Nishimura, Y., M. Endo, K. Yanase, Y. Ikeda, S. Miyakawa, and N. Miyamoto. "High Cycle Fatigue Strength of Spring Steel with Small Scratches." In P. Raghu, J. Vikram, S. Ashok eds. *Advances in Structural Integrity*, pp. 541–548. Springer, Singapore, 2018.
23. Ohta, K., K. Ikeda, G. Zheng, T. Okimoto, and I. Honda. "Vibration Response and Noise Radiation of Engine Block Coupled with the Rotating Crankshaft and Gear Train." *Journal of Environment and Engineering* 6, no. 4 (2011): 765–777.
24. Okokon, E. O., T. Yli-Tuomi, A. W. Turunen, P. Tiittanen, J. Juutilainen, and T. Lanki. "Traffic Noise, Noise Annoyance and Psychotropic Medication Use." *Environment International* 119 (2018): 287–294.
25. Qiu, N., Y. Gao, J. Fang, Z. Feng, G. Sun, and Q. Li. "Crashworthiness Analysis and Design of Multi-Cell Hexagonal Columns Under Multiple Loading Cases." *Finite Elements in Analysis and Design* 104 (2015): 89–101.
26. Rallabandi, S. R., L. Srinivas, S. Palli, R. C. Sharma, N. Sharma, A. Sharma, and S. K. Sharma. "A Contrastive Characterization of Pure Mg and AZ91D Alloy Based on the Testing of Mechanical, Corrosion, Wear, and Erosion Properties." *Engineering Research Express* 6, no. 1 (2024): 015017.
27. Sharma, R. C., and M. Dhingra. "Braking Systems in Railway Vehicles." *International Journal of Engineering Research & Technology* 4, no. 1 (2015): 206–211.
28. Sharma, R. C., S. Palli, L. Dharmana, S. K. Sharma, A. Sharma, A. Kharola, and N. Sharma. "Simulation of Transient Dynamic Response of a Locomotive Considering Its Flexibility and Rigidity." *Noise & Vibration Worldwide* 55, no. 1–2 (2024): 33–47.
29. Sharma, R. C., N. Sharma, S. Palli, S. Sharma, S. Joshi, S. Rallabandi, and A. Sharma. "Dynamic Behavior of Three-Wheel Vehicle with Inerter-Based Dynamic Absorber Suspension System." *Noise & Vibration Worldwide* 55, no. 4–5 (2024): 253–263.
30. Sharma, S. K., R. C. Sharma, and J. Lee. "Propelling Precision of Longitudinal Vibration Mitigation in Ship Propeller Shafts Through Advanced Nonlinear Intelligent Semi-Active Control Leveraging Adaptive Neuro-Fuzzy Inference System with Linear Quadratic Regulator." *Journal of Vibration and Control* 31, no. 7–8 (2024): 1472–1484. https://doi.org/10.1177/10775463241244836.
31. Song, Y., J. W. Lee, and J. Lee. "A Study on Novel Filtering and Relationship between Input-Features and Target-Vectors in a Deep Learning Model for Stock Price Prediction." *Applied Intelligence* 49, no. 3 (2019): 897–911.

4 Digital Twins in Railway Systems

Srihari Palli, Sivasankara Raju Rallabandi, Samson Abiodun Aasa, Seepana Praveenkumar, Pavankumar Rejeti, Azad Duppala, and Abdulsamad Ibrahim

4.1 INTRODUCTION

Industrial revolutions (IR) in the last two centuries have changed the world around us and shifted technology to higher levels. With each IR there were better products with superior quality raising the process standards and making the human lives less laborious. By the end of the third IR, product design and manufacturing have been assisted mostly by digitization and advanced computing. Digital manufacturing (DM) has taken over from the conventional design and manufacturing sequences where product and process design happen simultaneously. Digital manufacturing or digital Factory integrates, 3D CAD visual graphics, simulation, analytics, and collaborative tools which enable industries to design the entire manufacturing process sequence on the computer and upon digital verification and validation only, the first physical product is rolled out with almost zero scrap. DM uses simulation and modelling to optimize processes, which can help reduce operating costs, speed up production, and improve cohesion between stages of the manufacturing process.

On the other hand, railways have been rapidly growing over the past 50 years taking big strides in terms of train speed, covering longer distances than early-aged rail vehicles, isolating and absorbing the vibrations, lighter vehicles, passenger ride comfort and more importantly the most preferred means of transport all over the world. The demand for sophisticated seats/berths with better aesthetics and the best ride feel for passengers is inviting the use of advanced technologies and superior or premium services. Improved train and track components design with the minimum failure probability is becoming the need of the hour for railway operators. The use of semi-active and active suspensions, shift from diesel to electric and electric to hybrid locomotives, magnetic levitation, etc. are some of the major changes the railway vehicles have seen in the modern era [1]. However, these features are expensive, and it is important for the manufacturers to develop the rail and train-related products with a chance of low failure rate, lesser maintenance cost and repairing at a faster rate [2–11]. The fundamental components of digital twin (DT) system is shown in Figure 4.1.

To improve product design and innovation, reduce downtime, optimize supply chains, and enhance operational efficiency virtual replicas of objects which are updated in real time with data from the physical world have been developed in the

 DOI: 10.1201/9781003659983-4

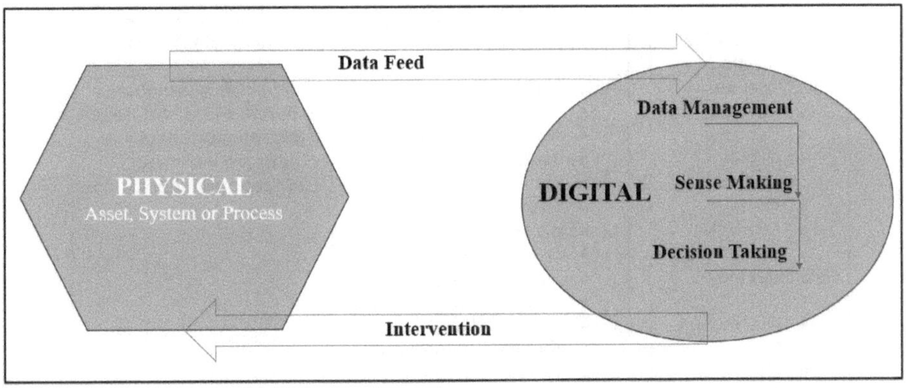

FIGURE 4.1 The fundamental components of a digital twin system.

past decade known in the technology world as DTs. DTs can also help with predictive maintenance, safety and risk management, and customer and client engagement. DTs can be used in any industry that manufactures items, including automotive, agriculture, aerospace, consumer goods, and healthcare [12].

4.1.1 DT versus Simulation Model: A Comparative Analysis

There is a differentiation regarding the parameters and the functional aspect between the simulation model of a product and a DT since both are virtual representations. Simulations rely on true parameters and design features [13, 14]. After the creation of the digital model, the parameters must be changed if the designer wants to incorporate new parameters. The picture with those specific parameters can only give details in regard to that particular design [15]. New designs require engineers to build another simulation from scratch. DT technology, on the other hand, starts off the same way as a simulation. However, since it uses real-time data, it will change the simulation automatically. It will keep looking for ways to improve the product through active simulation. For example, it can simulate the entire product lifecycle over time, allowing designers to get information a simple simulation can't provide.

Since simulations accurately replicate real-world products in a digital environment, they can only simulate what could happen with the product after extended use. The key technologies used for DTs are shown in Figure 4.2. Designers don't have the option to change the settings and use different setups, so they can't test the product in various environments. DT wins the race here as well. It's not only able to replicate the product but the entire environment and different use cases. Simulations depend on the designer's ability to enter accurate parameters to change the outcome. The data used by DT is generated in the real world, allowing designers to test product performance using accurate information. That way, they can simulate various use cases and find better design solutions based on real-world applications. In the same manner, real-world data may be used to conduct a test of elaborate manufacturing operations within a changing context. The crux is that in simulations, everything is conceptual while DT is real and correct. Table 4.1 outlines the main differences between simulation and DTs.

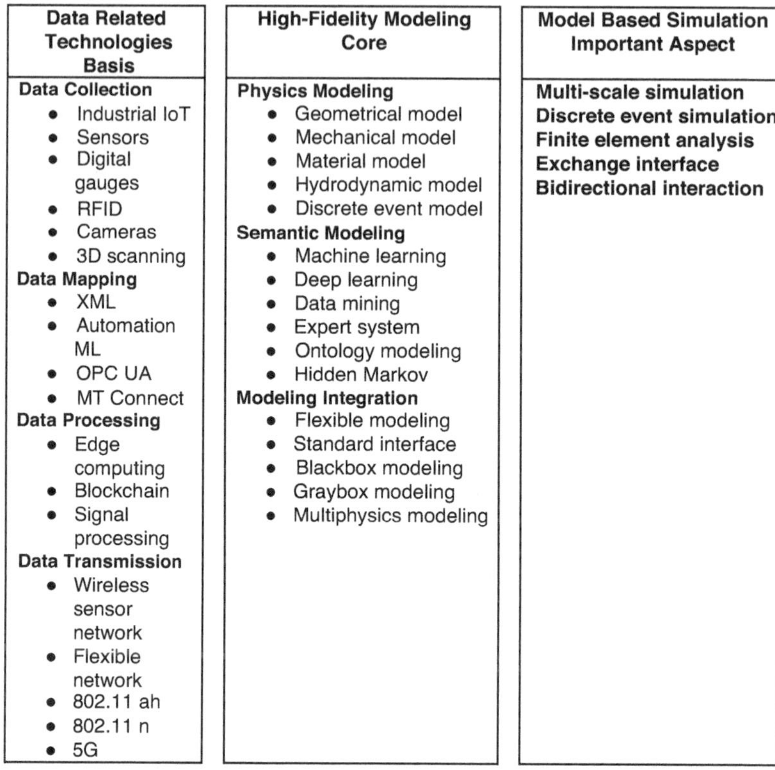

FIGURE 4.2 Technology architecture for a digital twin. (From Ref. [16].)

TABLE 4.1
Comparison between Simulation and Digital Twin

	Simulation Basis	Data Elements and Interaction	Scope
Digital twin	Real world feedback from a specific product/process	Active	Wide – cross business
Simulated model	Potential parameters input for testing	Static	Narrow – primarily design

4.1.2 DTs in Railways

For the modernization of their operations, and the re-engineering that comes with the quest for a competitive advantage in efficiency, safety, and sustainability, rail operators are looking at DTs as a key ally that has been gaining importance. DTs of the railways are the graphical representations of existing railway systems and enable the performance of simulation, monitoring, and control of operations in real time. These three-dimensional models are built by combining various data from Internet of Things (IoT) sensors, other devices, and historical data allowing the railway proprietors to enhance operations, and the safety of their infrastructures.

4.2 ROLE OF DTS IN RAILWAY VEHICLES AND RAILWAY SYSTEMS

There are several important railway system parameters such as signal processing, route and traffic control, vehicle and passenger safety, vehicle and infrastructure status, etc. in which DTs on the ever-evolving technologies have an effect. Furthermore, there are so many domains whereby for example a DT can be applied and that simulates the whole business which if done successfully also predicts all technical, social, and economic impacts [17, 18]. It may utilize these various systems to build the model more accurately and evaluate the results of operations [19, 20]. The catch to this technology is that it is more costly than what the simplistically termed as simulations are. This also has its own drawbacks whereby it requires long periods to be kind of integrated and engaged with the existing data analytics solutions.

DTs in railway vehicles open noteworthy that assist maintenance improvement, safety, and operational efficiency that will help to push the future of rail transport towards more reliable and sustainable systems. These models are a composite of information intakes: sensor data, history of maintenance tasks, and certain operational measurements.

4.3 CRITICAL FEATURES OF DTS IN RAILWAYS

Digital twins, or DTs as are commonly referred to, are one of the latest technologies that can enhance the efficiency, safety and overall performance within the railway industry. DTs are defined as the theoretical copies of real-world assets processes or systems in time that can be used for monitoring, maintenance, editing, and most importantly for making a decision. These technologies do not only provide an all-round solution to vehicle design or manufacture but also many aspects of the whole railway industry such as track components, signalling devices, energy control systems, and passenger experience factors. With time progresses in technology this will not only be of outstanding importance as it shall highly endorse various areas in railways ranging from ride comfort to operational efficiency.

4.3.1 Real-Time Monitoring

One of the primary advantages of DTs, as applicable to the railway industry, is real-time monitoring. Notably, it is achieved through the integration of numerous sensors and data acquisition systems that target key observations from railway resources. Such sensors are capable of tracking temperature or noise or speed or wear levels and other such information and continuously sending it to the twin model. This helps in ensuring that a physical system can be graphically represented at any time and accurate decisions are made instantly.

Data Integration: The integration of data is fundamental to real-time monitoring. On the other hand, railway vehicles, tracks and infrastructure use sensors for a variety of applications some of which can be incorporated into the DT platform. These

TABLE 4.2
Common Sensors Used in Railway Digital Twins for Real-Time Monitoring

Sensor Type	Data Collected	Deployment Location
Vibration sensor	Vibration levels	Tracks, rolling stock
Temperature sensor	Temperature readings	Locomotive engines, track components
Wear level sensor	Material wear levels	Tracks, train wheels
Pressure sensor	Track and bridge strain	Track, bridges
Speed sensor	Train speed	Locomotive, track

include data about the environmental conditions, mechanical wear, and performance counters. As these data streams are processed in these systems, rail operators are provided with the capability of monitoring the performance and health of rail systems without manual checking.

Performance Tracking: After being incorporated into the DT, this information is incredibly useful as regards the asset's performance. For example, a system that measures vibration in vehicles can determine the wearing out level of components such as wheels, axles, and suspension systems through the reading of abnormal values. Continuous performance monitoring will enhance the capacity of maintenance teams by enabling them to fix problems before they occur and are extremely expensive as well as cause unanticipated downtimes. Furthermore, pre-emptive action can be taken by operators in scheduling maintenance procedures before the entire operation is forced to shut down due to catastrophic failure of some components. The capability to witness in a real-time perspective the state of entire fleets and infrastructure system components is very critical in enhancing seamless operations and the safety of rail services. With real-time information about asset conditions, railway operators can take the necessary steps in a timely manner thereby increasing efficiency as well as safety throughout the network. An overview of key railways-related sensors most commonly used as part of DTs for railway infrastructure monitoring is presented in Table 4.2. To avoid data loss, these sensors are embedded into the tracks, the rolling stocks, and many other important parts of the system for real-time monitoring and instant analytics.

4.3.2 Predictive Maintenance

One of the important benefits of the introduction of the DTs is predictive maintenance in the railways. With this method, one can utilize historical facts and condition records to estimate the period when an object, be it a train, or part of a track, is to be serviced. The models of a DT can predict how given assets will behave in telemetry images of different situations, therefore predicting the failure of the asset even before it happens.

Failure Prediction: By making use of a twin, it is possible to analyse big datasets containing past maintenance data as well as real-time sensor data, in order to detect trends that signify the likelihood of failure. For example, a specific component may be expected to fail within a particular range

TABLE 4.3
Predictive Maintenance for Rolling Stock Components Using Digital Twins

Component	Predicted Maintenance Interval (Days)	Prediction Method
Train wheels	50	Vibration and wear analysis
Braking systems	75	Temperature and stress data
Axles	120	Strain and load analysis
Suspension systems	90	Vibration and temperature analysis

because of numerous factors such as increased vibration and high temperature in conjunction with a component's other worn-out parts. These prognostications draw upon the sophisticated machine learning techniques and statistical models provided for within the DT environment.

Less Downtime: One of the key benefits of predictive maintenance is the reduction of downtime thanks to the more effective scheduling of maintenance. Rather than waiting for a breakdown to occur, which results in long outages of services, railway operators can schedule their maintenance operations based on the anticipated occurrence of failures. This minimizes the chance of unforeseen breakdowns, which occur more frequently due to inefficient operations with idle trains. In addition, the predictive nature of predictive maintenance is more economical in the allocation of maintenance resources as funds are only utilized for assets that are in need of maintenance as opposed to time-allocated schedules. With the employment of predictive maintenance practices, railway companies stand to gain high operational reliability and low operational expenditure while increasing the life span of their assets. The maintenance intervals for some critical rolling stock components appear in Table 4.3. These estimations are mostly attributable to the effective utilization of vibration, temperature, and strain sensors in planning timely interventions.

4.3.3 Enhanced Design and Testing

The implementation of railways DTs can be extremely helpful for accelerating the railway project development lifecycle during design and testing. According to a common practice in design processes, engineers construct a physical model or a prototype and carry out real-life tests to evaluate the performance of newly introduced designs/intended modifications. In contrast, with DTs, they are capable of altering the design or performing a wide range of simulations without any real physical construction.

Simulation of Scenarios: Different operational modifications are possible in respect of a specific DT. For instance, changes in train schedules, new track designs, or new train structures can be attempted and planned module over a specific system. This offers itself to virtual deployment and testing to see the effect different changes will have on the system. For instance, if the system installs a new rail route, the system predicts how it will impact overall operations, traffic distribution, and wear rates on various forms of assets. This method reduces the potential of making expensive mistakes because it addresses the major concerns in an accurate manner prior to the implementation of the plan.

Design Optimization: Apart from mere visualization of operations, DTs can also be applied to improving designs. For instance, one can construct models for railway vehicles, stations, or components of the rail track and test them for their strength, load and energy efficiency before production. In a virtual setting, engineers can programme different constructions of the project in order to see which would give the best performance at the lowest cost. These photorealistic models of the infrastructures developed through DTM are crucial to the planning and decision-making process for railway companies, bringing about more efficient designs and cost-effective operational systems. Moreover, DTs can facilitate inter-team cooperation in the completion of a project. The obstacles caused by dissimilar spatial representation concerning ArchiCAD and Revit will be eliminated, as all three parties will have the same digital model, and all improvements in the design will relate directly to the same up-to-date data.

4.3.4 Operational Efficiency

The implementation of railways DTs can be extremely helpful for accelerating the railway project development lifecycle during design and testing. According to a common practice in design processes, engineers construct a physical model or a prototype and carry out real-life tests to evaluate the performance of newly introduced designs/intended modifications. In contrast, with DTs, they are capable of altering the design or performing a wide range of simulations without any real physical construction.

Fleet Management: Fleet management is the process of scheduling trains, allocating train cars and avoiding delays in train service operations. With the help of DTs, fleet managers can remotely examine the performance of particular units and fine-tune operational programmes. Employment of such an approach allows for improvement in utilization of trains, reduction of downtimes and better management of resources. For example, a DT system can help operators decide the best possible routes for a given train by considering the current traffic, weather, and the state of the infrastructure.

Energy Management: Energy management is one of the challenges in the railway field, particularly with respect to the reduction of operational costs for fuel and protection of the environment. As DTs have the capability of monitoring energy usage and allowing for efficacious use of energy, they are helpful in these cases. Recommended changes by DTs which include reduction in train speeds, enhancing acceleration and deceleration profiles, and optimizing train schedules can save energy by reducing train speeds, and improving braking and fuel efficiency. Over time, these changes will have cost implications and lead to reduced emissions and fuel savings.

DTs permit rail operators to achieve high performance in their operations without adverse effects on the environment through constant monitoring and changes.

4.3.5 Safety and Compliance

Safety and compliance with regulations are two of the most important areas in the railway industry and the use of DTs is pivotal in making sure that they are adhered to. DTs enable the prevention of incidents and the observance of safety requirements through the provision of continuous monitoring and predictive capabilities.

Regulatory Adherence: DTs can ensure that railway vehicles comply with safety standards and regulations by continuously monitoring key parameters such as braking distances, load limits, and system health. If an asset is at risk of deviating from these safety standards, the DT can trigger an alert, prompting the necessary inspections or corrective actions. This helps railway operators maintain compliance with industry regulations and avoid potential penalties or accidents caused by non-compliance.

Incident Analysis: DTs are capable of determining the factors that make a system fail especially when all the real-time activities that happened prior to an incident are available. In that case, analysis of failed processes in real time allows the operators to understand what went wrong and develop a plan to ensure that such systemic faults do not exist. Decisions are made in such a way that contained incidents will not occur again. Such decisions are aided by the fact that it is also possible to create and analyse system failures which help significantly in planning for an emergency such as communicating to all the relevant departments about their assigned duties or testing what would happen in the event of a given failure.

4.3.6 Passenger Experience

Passenger experience is an increasingly important focus in the railway sector, and DTs are being used to improve safety, comfort, and service delivery for passengers.

Passenger Safety: It can be noted that DTs focus on congestion management by monitoring the passenger flows in real time. As an example, if a certain percentage of train or station occupancy is reached, the operators can be notified to either suppress the running of the train to that station or to regulate access to the station to avoid overcrowding, thus enhancing comfort and safety. In addition, risks or dangerous aspects can be identified and early models can be developed to provide a sheath or protection to the hazards.

Passenger Comfort: Passengers' comfort can be increased through DT technology not only by heat or position of the seats but also by air quality sensors which can monitor and regulate temperature or ventilation automatically. This would enable the travel system to comply with many comfort standards automatically. DTs can also increase comfort by determining the maximum load for lighting or air conditioning systems.

Digital Shaping of Services: Using information on passenger behaviour, DTs help adapt services. For instance, peak travel times could be applied in making adjustments to the train schedule, or more directed services can be provided in accordance with the preferences of the target audience regarding entertainment preferences, food and drink choices or space between the seats. This not only enhances passenger satisfaction but also supplements the system's operational effectiveness as it is possible to match resources with demand in an efficient manner.

4.4 APPLICATIONS OF DTS IN RAILWAYS

The integration of DT technologies in the railway market can transform a great deal of operational and engineering processes across the board. At first, DTs were limited to a few, specific and hi-tech applications but now, many are being spread across quite a large number of areas of the railway system such as infrastructure management to rolling stocks and autonomous. Virtual duplicates of the physical site allow operators in the railway business to observe the site, simulate scenarios, foresee developments and optimize business activities with a new level of accuracy. The present section is intended to address the emergence of DTs around three themes: smart railway systems, rolling stock, and autonomous rail systems. Table 4.4 presents key advantages associated with DT technology in railway systems such as cost reduction, improved safety, and improved capacity.

4.4.1 Smart Railway Infrastructure

There are substantial applications for DTs in the administration as well as in the upkeep of the railway infrastructure. There is therefore a distinctive advantage of operating with DTs which is the ability to constantly track the operational integrity of key assets, such as tracks, bridges, tunnels, and others. Structures that contain embedded sensors supply data into the digital model which is being analysed to understand the problem before it escalates.

4.4.2 Tracking Physical Health of Railway Elements

Many components of railroads, including rail lines, bridges, and tunnels, can be continuously monitored with the use of sensors. For example, strain and vibration

TABLE 4.4
Key Benefits of Digital Twin Applications in Railway Systems

Benefit	Description
Cost efficiency	Reduces downtime, minimizes unscheduled repairs, optimizes resource usage.
Improved safety	Enhances proactive monitoring and failure prediction for critical systems.
Increased capacity	Optimizes train scheduling, tracks, and infrastructure utilization.
Energy efficiency	Optimizes energy consumption by adjusting train speed and operations.
Passenger experience	Improves comfort and service delivery through real-time data management.

TABLE 4.5
Data-Driven Scheduling Optimization in Fleet Management Using Digital Twins

Parameter	Data Collected	Use Case
Train speed	Speed, acceleration, deceleration	Optimizing fuel consumption
Track conditions	Track wear, vibration	Schedule maintenance based on track wear
Train occupancy	Passenger count	Adjust train capacity during peak hours
Fuel efficiency	Fuel consumption	Optimize energy usage during travel
Environmental data	Weather, temperature	Adjust schedules to avoid delays due to weather

sensors positioned on track and bridge structures offer an ever-present perspective on mechanical stress and vibrations that may cause damage or preclude failure. This information is collected into the DT framework and used to forecast and simulate real-time virtual representations of real assets.

4.4.3 Predictive Repair Actions

As far as predictive maintenance is concerned, one of the greatest advantages of managing infrastructure with the help of the DTs is the ability to determine the time and location of repairs. Instead of waiting for a scheduled maintenance inspection, which is unable to detect a developing problem, DTs apply predictive analytics to estimate and avert operational failures. For example, depending on temperature, humidity, and wear characteristics analysis, DTs can estimate track defects or the level of bridge deterioration and maintain the structures in advance in order to avoid progressive failure. Sections of Table 4.5 explain the use of aggregate sensor data on systems for the control and maintenance of the fleet composition in railway systems, improving the effectiveness of its operational scheduling.

4.5 ROLLING STOCK MANAGEMENT

Rolling stock (locomotives and carriages in operation within the railway network), on the other hand, can drastically be improved with the application of DTs. With the help of technology, it is possible to manage the entire life cycle of the vehicles from the design stage through the stages of manufacture, operation, and ultimately, decommissioning [21–28], which elaborate on the improvement of stability, ride comfort, crashworthiness and dynamics of rail vehicles.

4.5.1 Management of Life Cycles of Trains and their Core Components

DTs allow for tracking and control over the entire life cycle of rolling stock assets and separate parts such as engines, wheels, and braking systems. The data received from sensors that are installed in these components, is processed in the DT system which utilizes them to consistently refresh the virtual model with the current health state of each component. This helps the operators assess the performance of individual parts and provides assurance that such components are working efficiently and safely.

4.5.2 Performance Prediction and Maintenance of the Systems

Besides performance assessment of the rolling stock, DTs do support the optimal fleet and maintenance scheduling. In tandem with real-time and past operational parameters, DTs can also assess when a particular train part will need servicing or replacement. This cuts back on unexpected slack time and slashes the costs of repairs by tackling concerns before the situation is critical. In this case, the DT model would help to detect and evaluate vibration frequency levels and the thresholds at which wheel bearings will require lubrication or delineation, hence permitting engineers to remedy problems at the planned maintenance instead of unanticipated failures.

4.5.3 Autonomous Rail Systems

The presence of DTs becomes crucial in designing, testing, and implementing autonomous trains as railways begin automating their systems. Autonomous systems depend on accurate representations of their operating environment and the current conditions in order to make decisions. DTs offer these representations by reproducing actual operating situations in the simulated.

4.5.4 The Development and Availing of Tests for Automated Trains

DTs rely on a unified data model that supports the planning, operation, maintenance, and early failure prediction of railway systems integrated with intelligence and automation. A DT is a replicative model that enables engineers to create virtual conditions like the necessity for emergency braking, or if rerouting is possible, without risking actual trains that rely on costly physical tests.

4.5.5 Modelling Operating Conditions

The autonomous train must respond adaptively to the surrounding conditions including the state of the rail where it is on, climatic as well as other moving objects within the line. Creating autonomous parts without the DT findings would be irresponsible since other moving objects as cars can interfere in the realistic world where everything is constantly changing. The DT is therefore a crucial useful tool that helps predict the functioning of any autonomous part of the train system.

4.6 BENEFITS OF DTS IN RAILWAYS

All economies are advancing towards the use of advanced DT technologies in their physical railway systems. It stems from a number of specific economic factors such as efficiency, improved safety, growth in capacity, and operational optimization. These are through real-time data, predictive visualization, and advanced computing technologies such as artificial intelligence (AI) and machine learning (ML), IoT, and Cloud Computing. The more the railways incorporate DTs, the greater the efficiency, safety, and sustainability of the entire system.

4.7 COST EFFICIENCY

Cost efficiency is perhaps the biggest benefit offered by the integration of the DT technology in the operations of a railway company. In the previous maintenance and operating model, maintenance and operating works were performed after something had already gone wrong which came with high repair costs and unplanned downtimes. This is not the case with DTs as they create a 3D or virtual model of tangible assets which allows for predictive maintenance, condition monitoring, and resourcing. This enables a huge reduction in operational costs.

4.7.1 Reducing Cases of Unscheduled Repairs

The case of high operational costs within the railway industry is because of too many frequent unscheduled repairs. Conventional maintenance typically operates on time intervals or is dependent on occasional visual inspections, which may fail to notice problems in the making. The application of DTs enables continuous monitoring of railway assets such as tracks, trains, or stations as it integrates real-time data from sensors installed in these assets. Information of this nature is beneficial in estimating the breakdown time of specific components so that maintenance activities can be planned to eliminate unplanned maintenance.

For example, sensors on train wheels determine friction, heat, and vibrations. These parameters, within the DT paradigm, make it possible to predict wheel failures in advance, thereby allowing maintenance crews to change or repair the wheels before any unplanned failure occurs. The predictive method is so advantageous since it eliminates the need for costly emergency repairs, which may take a long period of time while the equipment remains idle and a lot of money in terms of labour costs.

4.7.2 Reinforced Asset Integrity

To extend the overall life cycle of the assets, DTs ensure optimum asset performance of all railway components – from the trains to the tracks. This becomes possible because of the continuous monitoring of parameters like temperature, tilt, pressure, and load giving the operators a chance to carry out maintenance on an actual wear-and-tear basis during scheduled maintenance instead of waiting for the recommended maintenance schedule by the manufacturer or recommending time-based limits. Not only does this make sense from a cost standpoint but also constitutes an in-depth strategic rationale as asset replacement or repairs will only be performed when absolutely necessary to prevent equipment deterioration from over-maintenance.

For instance, tracks are usually subjected to inspection in a periodic manner. On the other hand, DTs can continually monitor the condition of the track and as well, incorporate temperature changes, load gradients, and wear, to help establish "in future, how much repair will be needed". This ensures that maintenance of the tracks and other elements of the infrastructure is done at the appropriate times alleviating the need to undertake such activities at unnecessary durations.

4.7.3 Optimization of Spare Parts Inventory

The implementation of DTs also retains the spare parts cost as those components are only ordered when required. Since the circumstances of certain specific parts are monitored in real time, they need not have to accumulate a lot of spare parts but only order them bearing in mind the usage trends as well as the predictive maintenance hope. This results in improved inventory management and working capital costs to be incurred on base-level parts storage to be minimal. For instance, Benbam provided an example where a DT could estimate when a specific rail or train component would need replacement, therefore, providing the operators with the necessary time to order for replacement components. This approach reduces the need for spare parts stock levels and the frequency of purchasing them since both functions are carried out on time.

4.8 IMPROVED SAFETY

The safety of both the passengers and the railway's personnel is critical in the industry. DTs help to boost railway safety with two major functions: monitoring potential hazards and emergencies, therefore, improving the anticipation and mitigation of accidents and their impacts when they do happen.

4.8.1 Failure Prediction without Actually Failing

Safety incidents are usually caused by unpredicted breakdowns of the devices. Such incidents are avoided by using DTs that keep acquiring and processing real-time data from sensors on various assets, such as trains, tracks, and signalling systems. The data is then incorporated into the DT model for the purpose of monitoring the condition of the asset and anticipating failures in advance.

Taking for example the case of train brakes, they cannot be overlooked as they are vital in safe train operations. Failure in the brake system would have disastrous effects on train control. Therefore, with the use of A DT embedded into the brake system, sensors can record the wear and tear of braking force and temperature. Thus, the system optimizes the replacement or maintenance of the brakes saving unexpected brake failures.

4.8.2 Modelling Emergency Situations

The use of a DT model facilitates the testing of emergency events and simulating those instances. Such tests allow operators to develop a feel of how the system reacts during emergency events such as collision of trains, blockage of tracks, or adverse weather conditions. For example, varying simulations of the braking system under specific load conditions can result in systems designed to function efficiently in critical scenarios. With such simulations performed, the DT lowers safety practices by running scenarios of anticipated emergencies and coming to the solutions almost instantly.

For example, when experimenting with new autonomous train systems, a DT can recreate emergency situations including emergency brake application, track failure or any other scenario that could be a risk to the system, thus enabling the engineers to test the system in safe conditions.

4.8.3 Real-time Monitoring and Diagnosis of Incidents

Apart from having capabilities of failure predictions, DTs er also useful for real-time incident diagnosis. In the event of an accident such as track and structure failures or derailment, the system assists in determining the cause of the incident as it immediately synthesizes information from relevant sensors of the DT technology. This helps operators gain the understanding of what the issue is to the best so as to avoid occurrences of similar issues in the future. As an illustration, a DT can address why an incident happened by assessing train speed, track condition, and the signal system used during a derailment. Such real-time evaluation will be useful in directing decisions in all crucial situations thus reducing the effectiveness of the safety response.

4.8.4 Documented Improvement in Capacity Management

Capacity management has never been less of a concern as railways seek to increase the transport service. DTs structure resource allocation, real-time scheduling, and traffic management to increase system capacity.

4.8.5 Data-Driven Scheduling and Traffic Management

DTs promote the efficient use of railway assets through real-time data information for scheduling and traffic management. Most traditional schedules rely on the assumption that time intervals are constant. However, with the digitized decision-support tools, wait schedules can be optimized by taking into consideration traffic, occupancy, weather, and track clearances. For example, if a train is running late through no fault of a dispatcher or scheduler, the twin may, in real-time, shift the timing of all subsequent trains in sequence, divert upcoming trains to avoid system congestion and elevate the entire system throughput. This approach mitigates choke points and allows the capacity of delays in the railway system to increase the number of trains within the system.

4.8.6 Optimizing Fleet Usage

The operation of rolling stock (trains) is another section in which the capacity is enhanced by DTs. By effectively leveraging on the data available on train breakdowns, wear of parts, and efficiency in use of components, the DT allows fleet managers to streamline the usage of every single unit of train. For instance, in the event of a mechanical breakdown, the DT may provisionally set repair times and alert the fleet operator to adjust schedules to avoid under-utilization of the entire fleet.

DTs also enable the operators to determine the most appropriate train type for specific routes. In order to maximize efficiency for a given route, for example, a train system with a higher capacity can be used in some sections where the rails tend to wear out easily or the passenger traffic is quite high.

4.8.7 Infrastructure Optimization

Thanks to DTs, there is a functional model of the complete railway systems such as tracks, stations, signals, and stations. The preceding image can serve a purpose to reveal the network's deficiencies and help in its reconceptualization. For example, DTs could identify regions that are slow due to inordinate downstream traffic and congestion and facilitate the improvement of such infrastructures. Such improvements might include the reconstruction of stations to accommodate more passengers or upgrading track systems to reduce traffic delay.

In particular, the portion of the line that cuts across an area with notorious snail speeds may have its DT employ different routes or in extreme cases replace the stretch to resolve the problems.

4.8.8 Surrounding Environment Supporting DT Technologies

DTs sit within a larger system where AI, ML, IoT, and Cloud Computing make it possible to provide real-time views, predictions, and improvements. The integration of these technologies leads to the development of an automated system whose operations are dependent on the most recent data available.

4.8.9 Cloud Integration and Real-Time Data Collection

Data comes in from sensors and cameras in real-time which makes up the basic components of the DT. Sensors installed on trains, tracks, and other objects of interest continuously monitor temperature, vibration, wear, load, and speed. This information is relayed to the cloud which processes and analyses the data. With the use of cloud technology, enormous amounts of information can be kept and processed from a single site. DTs combine information sourced from multiple locations which enables them to accurately reflect the current status of the physical asset in question making it possible for managers to take action based on real-time conditions.

4.8.10 AI and ML as Predictive Tools

AI and ML algorithms are required in order to manage the excess of data included in the simulations of the DT. These technologies build predictive models predicting when particular equipment is expected to require servicing, which track infrastructures would need repairs, or when the network would be prone to delays. AI and ML develop the expectation of prediction accuracy when the DT system is reinforced with more historical instances combined with around-the-clock real-time data analysis and experiencing the past. Here, for example, the ML models may examine the rate at which wheels of trains are worn out in order to forecast the remaining useful life (RUL) of the wheels. This makes it possible for operators to make convenient and appropriate maintenance schedules and avert untimely breakdowns. Data acquisition and data fusion as a Basis of future accurate predictions integration of multiple datasets aids in providing a better structure of the railway system. Therefore, with the help of DTs, forecasting about the performance of the system, likely failures and ability to optimize capacity, by combining information of trains, tracks, weather, stations, and maintenance activities, would be very possible. Furthermore, the multiplicity of data obtained from other vehicles and infrastructural components ensures that firm conclusions are drawn from a well-considered view of the system's condition.

For example, the integration of track sensors, train sensors, and the weather can enhance the predictions on how much track will wear out in the future along with the maintenance needed for specific weather conditions. Table 4.6 mentions the main AI and the ML models which are embedded in the DT ecosystems for the predictive maintenance of railway systems. Such models use historical data which helps predict when a failure might occur and what preventative measures should be taken.

4.9 SUMMARY

With the emergence of DT technology, there have been several changes across various industries with the railway industry being among the leading beneficiaries. A new dimension to monitoring, simulating, and optimizing the causes and effects of railway systems is made available by DTs. As a result of DTs, the railways have benefitted from coordinating real-time data and leverage predictive analytics, AI, ML, IoT, and Cloud Computing which are technologies of the current era. Such systems are also applicable in the growth stage of designing rail infrastructure and rolling stock in addition to the rest of the life cycle of the railway system with the aim of improving all the later stages of operations.

TABLE 4.6
AI and ML Models Used for Predictive Maintenance in Digital Twin Ecosystems for Railways

AI/ML Model	Application	Outcome
Decision trees	Predicting component failures	Identifies failure modes based on historical data
Neural networks	Track degradation prediction	Predicts long-term wear and tear based on usage patterns
Random forest	Predicting service interruptions	Estimates failure likelihood across multiple assets
Support vector machines	Fault detection in rolling stock systems	Classifies abnormal operating conditions in real time
Clustering algorithms	Segmenting infrastructure types	Helps categorize different track and station conditions for tailored maintenance schedules

4.9.1 The Revolution of DTs in Railways

However, in contrast to the conventional models or static simulations where parameters are fixed, in railways, there is a dynamic dimension of DTs. They develop and change over time, considering the information flow from sensors installed in various elements of the railway system, which includes trains, tracks, signals and stations. This frequent feedback provision allows digital clones to approximate real-world conditions very quickly. They can be defined as active virtual models based on physical objects. Such models also have the ability to "learn" over time from the data and enhance their forecasting and modelling capabilities, thereby "shifting the centre of gravity" to the more active management of rail operations.

4.9.2 Real-Time Supervising and Pre-Emptive Action

A major advantage that comes with the incorporation of DTs in railway systems is the capability to observe the systems in real time monitoring. DTs are real-time figurative images of the entire railway system encasing continuous data obtained from numerous sensors on trains, rails, stations, etc. This enables the system to consistently track operational parameters such as the speed of the trains, the position of the trains on the track, the health of the rails, and the reliability of the system as a whole. It is possible to monitor any resource that operates within a designated normal range of parameters and is capable of quick action to rectify any problem before it develops to a complete failure. For example, abnormal wear on train wheels can be signalled by abnormal train vibration, while rail failure due to expansion may be monitored by temperature sensors on the tracks. In addition to the problem of damage, these insights provide information on when specific components will be repaired or replaced, saving a lot of time and costs associated with unscheduled downtimes and emergency repairs.

4.9.3 Improved Operating Efficiency and Safety

The DT technology enhances performance in the operations of the railway systems. Scheduling, fleet management and resource allocation are made more efficient due to decision-making based on live data. Therefore, if tracking train conditions and performance monitoring were done at the same time, the operators would anticipate bottlenecks and suggest alternate routes or reschedule them so that losses are minimized and throughput is increased. Since his DTs allow for predictive maintenance, they assist in mitigating the risks of unexpected interruptions in service thus enhancing the reliability of the railway network. In addition, DTs allow for improved safety in railway operations. Safety is an aspect where railways also face a number of challenges such as operating fast trains, locatability of track or systems failures and human factors especially error. Once more, DTs permit operators to design potential failures and prepare how to handle emergency situations such as loss of signals, track failures, or storms. This function enhances safety measures since operators can ensure that once a risk is identified, all control procedures are put in place before the risk is realized. For example, reproducing scenarios wherein an emergency action is executed within a digital twinning environment may allow for the practical training of operators and system engineers on the potential braking capabilities and the need for adjustments should a need to implement protocols or reduce train speeds is necessary. Additionally, the presence of such a monitoring system allows for abnormal situations such as failures in important control systems, for example train signalling and or its control, hence safety concerns are efficiently addressed.

4.9.4 Passenger Comfort and Satisfaction

DTs have far reaching applications by also enhancing passengers experience providing not only operational efficiency but also increased comfort and improved service quality. A systematic approach to managing passenger flows, seat allocation, and environmental control is made possible through DTs in real time. To maintain ideal conditions on board trains and at stations, ensure the minimum number of overcrowded situations and improve the level of comfort as much as possible. For example, and thanks to the sensors installed on board trains or at train stations, occupancy figures can be collated to estimate and forecast the number of passengers expected, enhancing managerial efforts, for instance with regards to how many seats are available and the most popular times to run certain trains. Also, even with systems that maintain train and passenger conditions such as temperature, humidity and air quality the rail operators through the aid of DT systems can change heating, cooling and ventilation conditions to the required levels. There is also the possibility of offering individual services based on a passenger targeted thanks to the use of DTs. The analysis of such travellings makes it possible for the operators to recommend specific services, such as the best routes to take or what to provide onboard, which contributes to customer satisfaction.

4.9.5 Enhanced Operational Efficiency and Capacity

Using DTs, it becomes possible to not only increase the capacity but also enhance the efficiency of the utilization of the railway assets. A key issue for contemporary railways is about providing high capacity to meet the demand with limited amount of infrastructure and other resources. DTs through their advanced capabilities of real-time analytics and predictive analytics, allow the operators to optimize the use of the existing resources, be it rolling stock, track or station facilities, all while prioritizing safety and minimizing the operational expenditures. DTs are effective in scheduling and ensuring that the trains are placed into service as and when required taking into consideration the current trends and expected traffic levels. For instance, during peak periods, DTs may propose the optimal use of longer trains or more cars for higher travel load. In the same way, train-operating mode managed by automated scheduling systems, based on DTs, can change the steams and times for train departures in real time, preventing overcrowding of specific lines and thus increasing their efficiency without considerable investment in new infrastructure. Moreover, DTs add value in the area of energy consumption optimization as well. In this case, the energy optimization is achieved using intelligent DTs that utilize data on train speed, acceleration, and break forces to suggest more energy-efficient procedures such as decreasing speed or modifying bracing order. It not only reduces operational costs but also helps environmental sustainability of the railway network.

4.9.6 Contribution of AI, ML, IoT and Cloud Computing to DT Ecosystems

The information twins are not stand-alone technologies; rather, they are complementing components of an umbrella that has AI, ML, IoT, and Cloud Computing. Such technologies allow the DTs to aggregate and interpret millions of data from the railway system for enhanced understanding and faster decisions. AI and ML are also used to detect trends in the data obtained from the sensors fixtures on the railway infrastructure. Such algorithms are used to forecast possible breakdowns, maintenance workloads and operational requirements. Their prediction and simulation functions improve as the DT receives more information over time. Real-time data is constantly delivered to the DT model by IoT devices integrated into railway infrastructure, including onboard and track sensors and monitoring devices at stations. This connectivity ensures that every element of the railway system is kept track of and maintained in a continuously updated instantaneous version. Cloud Computing enables the big data sets produced by the DTs to be stored, processed, and shared. Such cloud integration ensures direct and indirect factors of the entire railway network such as the train schedules; weather conditions and track conditions can be combined and be used to make quicker and better decisions throughout the railway system.

4.9.7 Long-Term Impact on Future-Focused Rail Systems

In the future, the application of an increasing number of DTs is anticipated to help steer the development of advanced railway systems worldwide. Since rail stakeholders across the globe are under increasing strain to update their systems, streamline processes, and limit their carbon footprints, digital solutions will be an integral part of such initiatives. The ability to scope out probable enhancements, simulate system interactions and modify modes of operations in real time will enhance the robustness of rail systems to events that cannot be anticipated such as adverse weather, operational failures, and even operational strangleholds. This kind of base flexibility will be critical as the advancements in technologies enable the shift towards intelligent urban centres and further, greener forms of transport. And with governments and businesses insisting on 'greener' transport solutions, the development of DTs will assist in optimizing oil consumption and cutting down on energy use, as well as reducing waste through better-designed and managed railway systems.

4.9.8 Data-Driven Smart Rail Systems

The knowledge obtained from DTs is vital in creating smart rail systems which are enhanced, more reliable and efficient system. With the further development of digitization, automation, and sustainability of railways, the adoption of DTs will not only increase the efficiency of operations but also foster the development of advanced railway systems designed to cope with expanding population in a more ecologically friendly manner. This would allow countries to grow their rail networks and create autonomous technologies like self-driving trains and low-energy rail systems while laying the groundwork for tomorrow's rail sector.

REFERENCES

1. Sharma, R. C., Palli, S., and Sharma, S. K. "Ride Analysis of Railway Vehicle Considering Rigidity and Flexibility of the Carbody." *Journal of the Chinese Institute of Engineers, Transactions of the Chinese Institute of Engineers, Series A* 46, no. 4 (2023): 355–366.
2. Adeoye, A. O., Quadri, R. O., and Lawal, O. S. "Assessment of Biofuel Potential of Tenera Palm Kernel Shell via Fixed Bed Pyrolysis and Thermal Characterization." *Results in Surfaces and Interfaces* 9 (2022): 100091.
3. Adeoye, A. O., Quadri, R. O., and Lawal, O. S. "Biomass Waste Pyrolysis in the Context of Nigeria's Oil and Gas Industry: An Overview." *International Journal of Ambient Energy* 44 (2023): 2525–2541.
4. Adeoye, A. O., Quadri, R. O., Lawal, O. S., and Emojevu, E. O. "Physicochemical Characterization, Valorization of Lignocellulosic Waste (Kola Nut Seed Shell) via Pyrolysis, and Ultrasonication of Its Crude Bio-Oil for Biofuel Production." *Cleaner Waste Systems* 7 (2024): 100138.
5. Adeoye, A. O., Quadri, R. O., Lawal, O. S., Malomo, D., Emojevu, E. O., Omonije, O. O., Odeniyi, O. K., Fadahunsi, M. O., Yelwa, M. J., Aasa, S. A., Onakpa, A. E., Omoniyi, B. P., Mark, I. N., Usman, J., Ismaila, A. M., and Saidu, A. U. "Advanced Techniques in Upgrading Crude Bio-Oil to Biofuel." In *Energy, Environment, and Sustainability*, edited by Sharma, S. K., Upadhyay, R. K., and Kumar, V., pp. 321–353. Springer, Singapore, 2024.
6. Adeoye, A. O., Yelwa, J. M., Imam, N., Quadri, R. O., Lawal, O. S., Malomo, D., Aasa, S. A., Onakpa, A. E., Dorgbaa, R. K., Omoniyi, B. P., Ahmad, N., and Garba, H. A. "Pyrolysis of Plastic Wastes Towards Achieving a Circular Economy: An Advanced Chemistry and Technical Approach." In *Energy, Environment, and Sustainability*, edited by Sharma, S. K., Upadhyay, R. K., and Kumar, V., pp. 215–259. Springer, Singapore, 2024.
7. Adeoye, A. O., Quadri, R. O., Lawal, O. S., Malomo, D., Aliyu, M. T., Dang, G. E., Emojevu, E. O., Maikato, M. J., Yahaya, M. G., Omonije, O. O., Edidem, V. G., Abubakar, Y. K., Offor, O. F., Sochima, E. H., Peter, B. E., and Hikon, B. N. "Sustainable Energy via Thermochemical and Biochemical Conversion of Biomass Wastes for Biofuel Production." In *Energy, Environment, and Sustainability*, edited by Sharma, S. K., Upadhyay, R. K., and Kumar, V., pp. 245–306. Springer, Singapore, 2023.
8. Adeoye, A. O., Quadri, R. O., and Lawal, O. S. "Wet Synthesis, Characterization of Goethite Nanoparticles and Its Application in Catalytic Pyrolysis of Palm Kernel Shell in TGA." *Results in Surfaces and Interfaces* 11 (2023): 100118.
9. Avesh, M., Imam, N., and Adeoye, A. O. "Assessing Environmental Impacts and Mitigation Strategies for the DFCCIL Rail Projects." In *Energy, Environment, and Sustainability*, edited by Sharma, S. K., Upadhyay, R. K., and Kumar, V., pp. 333–353. Springer, Singapore, 2024.
10. Momodu, D. U., Adetuyi, A. O., Adeoye, A. O., Akinlolu, O. S., and Apara, E. O. "Surface Functionalization of Linen Fiber and Its Dye-Fastness Assessment Using (4-[(E)-(4-Amino-2-Hydroxyphenyl) Diazenyl] Benzene-1-Sulphonic Acid and 4-[(E)-(4-{(Z)-[4-(Dimethylamino)Phenyl] Diazenyl}-2-Hydroxyphenyl) Diazenyl] Benzenesulfonic Acid)." *Suranaree Journal of Science and Technology* 30 (2023): 030118.

11. Quadri, R. O., Adeoye, A. O., Ayanda, O. S., and Lawal, O. S. "Conversion of Palm Kernel Shell to Sustainable Energy and the Effect of Wet Synthesized Nanoparticles of Iron on Its Thermal Degradation Kinetics." *Bioresource Technology Reports* 27 (2024): 101933.
12. Liu, M., Fang, S., Dong, H., and Xu, C. "Review of Digital Twin about Concepts, Technologies, and Industrial Application." *Journal of Manufacturing Systems* 58, Part B (2021): 346–361.
13. Sharma, R. C., Palli, S., Dharmana, L., Sharma, S. K., Sharma, A., Kharola, A., and Sharma, N. "Simulation of Transient Dynamic Response of a Locomotive Considering Its Flexibility and Rigidity." *Noise and Vibration Worldwide* 55, no. 1–2 (2024): 33–47.
14. Palli, S., and Koona, R. "Analyses of Dynamic Response of a Railway Bogie." *International Journal of Vehicle Noise and Vibration* 11, no. 2 (2015): 103–113.
15. Wright, L., and Davidson, S. "How to Tell the Difference Between a Model and Digital Twin." *Advanced Modeling and Simulation in Engineering Science* 7 (2020): 13.
16. Qi, Q., Tao, F., Hu, T., Anwer, N., Liu, A., Wei, Y., Wang, L., and Nee, A.Y.C. "Enabling Technologies and Tools for Digital Twin." *Journal of Manufacturing Systems* 58 (2021): 3–21.
17. Bhatti, G., Mohan, H., and Singh, R. R. "Towards the Future of Smart Electric Vehicles: Digital Twin Technology." *Renewable and Sustainable Energy Reviews* 141 (2021): 110801.
18. Madubuike, O. C., Anumba, C. J., and Khallaf, R. "A Review of Digital Twin Applications in Construction." *Journal of Information Technology in Construction* 27 (2022): 145–172.
19. Boschert, S., Heinrich, C., and Rosen, R. "Next Generation Digital Twin." In *Proceedings of TMCE*, edited by Horvath, I., Suarez Rivero, J. P., and Hernandez Castellano, P. M., vol. 6. Las Palmas de Gran Canaria, Spain, 2018.
20. Boschert, S., and Rosen, R. "Digital Twin—The Simulation Aspect." In *Mechatronic Futures*, edited by Hehenberger, P., and Bradley, D., pp. 59–74. Springer, Cham, 2016.
21. Bhardawaj, S., Sharma, R. C., and Sharma, S. K. "Development and Advancement in the Wheel-Rail Rolling Contact Mechanics." *IOP Conference Series: Materials Science and Engineering* 691, no. 1 (2019): 012034.
22. Goswami, B., Rathi, A., Sayeed, S., Das, P., Sharma, R. C., and Sharma, S. K. "Optimization Design for Aerodynamic Elements of Indian Locomotive of Passenger Train." In *Advances in Engineering Design: Select Proceedings of FLAME 2018*, edited by Prasad, A., Gupta, S., and Tyagi, R., pp. 663–673, Springer, Singapore, 2019.
23. Sharma, R. C., Sharma, S. K., and Palli, S. "Linear and Non-Linear Stability Analysis of a Constrained Railway Wheelaxle." *International Journal of Vehicle Structures and Systems* 12, no. 2 (2020): 1.
24. Sharma, R. C., Sharma, S. K., Sharma, N., and Sharma, S. "Analysis of Ride and Stability of an ICF Railway Coach." *International Journal of Vehicle Noise and Vibration* 16, no. 3–4 (2020): 127–144.
25. Sharma, S. K., and Lee, J. "Crashworthiness Analysis for Structural Stability and Dynamics." *International Journal of Structural Stability and Dynamics* 21, no. 4 (2021): 2150039.
26. Sharma, R. C., Sharma, S., Sharma, N., and Sharma, S. K. "Linear and Nonlinear Analysis of Ride and Stability of a Three-Wheeled Vehicle Subjected to Random and Bump Inputs Using Bond Graph and Simulink Methodology." *SAE International Journal of Commercial Vehicles* 15 (2021): 65–79.
27. Sharma, S. K., Sharma, R. C., Lee, J., and Jang, H.-L. "Numerical and Experimental Analysis of DVA on the Flexible-Rigid Rail Vehicle Carbody Resonant Vibration." *Sensors* 22, no. 5 (2022): 1992.
28. Vishwakarma, P. N., Mishra, P., and Sharma, S. K. "Characterization of a Magnetorheological Fluid Damper: A Review." *Materials Today: Proceedings* 56 (2022): 2988–2994.

5 Regenerative Braking System for Railways

Sai Sarath Gonuguntla and Samson Abiodun Aasa

5.1 INTRODUCTION

Regenerative braking, activated in the form of energy when a train slows down, is a certain technology aimed at capturing and converting the kinetic energy of the moving train into electric energy. Conventional braking systems simply dissipate losses in the form of heat as the wheel brakes are used. In the case of regenerative braking systems, such energy is not wasted, but rather collected and returned to the power system or else utilized at a later time. This helps not only to save energy that would have been used otherwise in mechanical systems but also to minimize damage to the finesse of mechanical parts. In railway applications, this recovery of energy using regenerative braking systems is critical for improved operational performance, decreased energy costs, and environmental protection [1, 2].

In regenerative braking, the traction motors which are often used to propel the train instead serve as generators when the braking takes place. It permits to invert the flow of electrical energy and renew the energy of the system. It is useful in rail systems that operate mainly on start and stop modes as in urban transit and suburban systems, energy recovery is easily maximized.

5.1.1 Importance and Benefits of Regenerative Braking

When discussing the advantages of incorporating braking systems into the train units this system should not be left out since it possesses a number of advantages.

a. Improving Energy Efficiency: Regenerative braking systems can convert up to 30%–50% of the kinetic energy into electrical energy during braking which gets lost usually in the form of heat during the regenerative braking of the systems. The energy recovered can also be used for auxiliary systems on the train, for accelerating the train with less energy or electricity, and even swept back to the grid thereby massively reducing the overall energy which totals up [3, 4].
b. Reduced Operating Expenses: The amount of energy that would have obnoxiously been consumed in the consumption of the power system in place is alleviated by the energy recovered through regenerative braking. This would help investors to incur less energy expenses and as such, at that particular time invest less funds operating the railways. Further, the introduction of regenerative braking also goes beyond reducing the energy recovered but

DOI: 10.1201/9781003659983-5

mechanical braking components which require action to bring about changes are also conserved, making maintenance to be low [5–7].

c. Benefits accrued to the Environment: More energy not used means less emissions produced and this is indeed why regenerative braking features systems which have the ability of energy optimization. In consequence, by adopting such systems the operational activities of the railways reduce carbon dioxide emissions and energy consumption.
d. Better Stopping Ability: Regenerative braking systems have a considerable effect on train braking abilities, due to the one acting as an auxiliary they reduce the input needed from conventional friction brakes which in turn softens and better controls the rate of descent. This is especially crucial in the urban mass transit systems where large numbers of stops are the norm [8, 9].
e. Linking with the Smart Grid: Today's technology has made it feasible to link modern regenerative braking systems with the smart alternative energy grid system. The stored additional energy during excessive energy generation due to braking can be supplied to the grid giving energy to other trains or for other uses than trains. This interconnectivity leads to improved efficiency in the use of the energy grid.

Table 5.1 summarizes the primary benefits of implementing regenerative braking systems in railway operations, highlighting improvements in energy efficiency, cost savings, environmental impact, and braking performance.

5.1.2 The Historical Evolution of Regenerative Braking and its Adoption in Railways

The modern regenerative braking system usage in rail transport in comparison with the last century has witnessed a considerable shift. Major events include:

a. Early Attempts (1900s–1950s): The concept of regenerative braking as a means of returning energy to the electric grid was further developed in electric traction-powered locomotives. However, such early efforts suffered the

TABLE 5.1
Key Benefits of Regenerative Braking in Railways

Benefit	Description
Energy efficiency	Recovers and reuses kinetic energy, reducing overall energy consumption.
Reduced operating costs	Lowers energy costs and reduces wear on mechanical braking components.
Environmental impact	Decreases greenhouse gas emissions, contributing to a more sustainable railway system.
Improved braking performance	Enhances braking force and reliability, particularly in urban transit systems.
Integration with smart grids	Allows for efficient energy distribution and management within the railway network.

challenges of lack of advanced power conversion requirements and energy storage components.

b. Technological Advancements (1960s–1980s): Improvements in power electronics and control systems made the application of regenerative braking for railways more practical. These innovations also brought better control of the electrical energy, and these linear electric motors became more reliable and efficient. The realization of this technology in the first place was made within suburban rail systems, being later introduced in high-speed trains.
c. Modern Era (1990s-Present): Regenerative braking has been redefined due to the emergence of new and advanced energy storage such as batteries and supercapacitors which are integrated with railways. This period has witnessed the application of regenerative brakes in urban public transport systems and also in high-speed rail system. Hybrid systems combining regenerative and conventional brakes have effectively gone beyond the two systems and achieved optimal braking performance with high energy recovery [10, 11].
d. Current Trends and Future Directions: Currently, the efficiency and capacity of the regenerative braking system remain the major focus of research and development initiatives. Advancements in energy devices, digital controls, and grid dynamics are contributing to the establishment of more sophisticated and efficient brake systems. Additional features are expected to include advanced smart braking systems, on-the-spot energy management, and the use of regenerative braking together with clean power sources [12, 13].

5.2 REGENERATIVE BRAKING SYSTEM TECHNOLOGY

5.2.1 Basic Principles of Regenerative Braking

In a moving train, regenerative braking system transforms the kinetic energy into electrical energy at the time of braking. The main idea here is to use the train's already-installed traction motors (which are used to push the train forward under normal circumstances) in the case when the speeds are reduced and make it behave as a generator [14–23]. This can be further described in a number of sequential operations: It is the process of converting the kinetic motion into electrical energy. Figure 5.1 depicts the fundamental mechanism of regenerative braking in railways [24]. When a train decelerates, the traction motors switch to generator mode, converting kinetic energy into electrical energy. If recovery is impossible, this energy is either returned to the power system, or it is kept in on-board energy storage systems, or it is lost as heat. The diagram covers how energy flows during braking while also detailing the main elements of the system including the traction motor, energy storage, and the braking resistor.

FIGURE 5.1 Basic working principle of regenerative braking system.

As the wheels of the train are decelerated, they are still rotating and thrusting the traction motors in the opposite direction. Instead of using current to provide motion to the motors, the motors instead, take the kinetic energy from the train and make it into electricity.

Electricity Generation: The power from the traction motors is turned into mechanical energy. Mechanical energy in the form of electric energy depends on what type of motor is used, either DC or AC.

Energy Management: The electricity that has been produced may either be returned into the power supply system or be used for the operation of on-board systems, or may be stored in energy storage devices such as batteries or supercapacitors.

Braking Force Sharing: The braking force has been achieved by instantly switching over from a regenerative brake to a conventional friction brake or vice versa.

Such a technology not only contributes to the enhancement of the energy efficiency but is also an enhancement of the braking ability, hence making it mandatory in modern railway systems. Table 5.2 provides an overview of different types of regenerative braking systems, their basic working principles, and typical applications in railway networks.

Table 5.2 provides an overview of different types of regenerative braking systems, their basic working principles, and typical applications in railway networks.

5.2.2 Components and Working Mechanism

Several ancillary elements are critical for the performance of a regenerative braking system, and these components work in a few also work together in that they convert, handle as well as use the recovered energy such as

1. Traction Motors: These are devices, which are important in the application of regenerative braking. They assist in kinetic energy recovery by changing from moving the train to generating current when braking takes place.

TABLE 5.2
Types of Regenerative Braking Systems

System Type	Description	Applications
Dynamic braking	Converts kinetic energy into electrical energy, dissipated as heat through resistors.	High-speed trains, freight trains
Plug braking	Reverses motor polarity to oppose motion, effective for low-speed braking.	Low-speed operations, emergency braking
Regenerative with energy storage	Recovers energy into storage systems like batteries or supercapacitors.	Urban transit, suburban rail
Hybrid regenerative systems	Combines regenerative and traditional mechanical braking for optimal performance.	Mixed-use networks, variable terrain conditions

Whether these motors perform well determines how effective the regenerative braking system will be.

2. Power Electronics: The energy is either returned to the power system, kept in the on-board storage systems, or converted to heat when recovery is not possible. The schematic illustrates the sequence of energy recovery during deceleration of the machine and the elements that participate in the process, such as the traction motor, energy storage, and the braking resistor.
3. Energy Storage Devices: Energy generated on braking can be stored in various systems which include the following:

   Batteries: Long-term storage of energy is done using such means whereby after breaking the train is able to draw upon that energy at a further point for accelerating the train or through auxiliary systems.

   Supercapacitors: Mainly used in delivering energy for a short time and from the supercap, which quickly recharges and discharges, mostly useful instant power as during starting steps and step in sprawling movement in cities.

   Flywheels: They retain energy in the form of rotational energy and can convert it back into electrical energy rapidly whenever warranted.
4. Braking Control System: This system integrates the regenerative and mechanical braking systems with each other. It makes sure that the maximum amount of energy is recovered while bringing the vehicle to a standstill in a safe manner by preventing abrupt braking and enhances the movement efficiently by dynamically controlling the amount of brake force to the speed, load, and track conditions of the train.
5. Resistive Braking System: If devices are fully charged and it is not required to energize storage, control devices use a considerable amount of energy and source of system generation to produce excessive heat energy. This protects the machine from injury and contributes to the effective and fail-safe braking of the system even when regenerative braked systems are common and pervaded throughout the environment. When there is no need for energizing storage and the devices are fully charged, the excess energy is used by the system to generate heat. This saves the machine from damage and assists in the safe and dependable braking of the system, even in circumstances where regenerative braking is prohibited.

5.2.3 Regenerative Braking Systems in Railways

As far as railways are concerned, regenerative braking systems can be classified by the particular operational mode and specific configuration of the system which includes the following types:

1. Dynamic Braking: In this case, traction motors are running under a large voltage, which makes it possible to transform kinetic energy into electricity that can be converted to heat energy in the resistors. It does not harness energy for reuse, but it does serve to reduce the strain on the mechanical system and enhance the efficiency of brake mechanisms in a moral sense.

2. Plug Braking: This technique is based on reversing the traction motors so they oppose the motion of the train causing a braking effect on the train. This is more often done at low speeds or in emergencies, and while it is an effective means of stopping the locomotive, it is not very energy efficient compared with other systems.
3. Braking Control System: This system provides interconnection between the two braking systems: regenerative and mechanical. It ensures that the maximum energy is conserved in the process of bringing the vehicles to a halt and encourages the movements by controlling the amount of braking force to the speed, load and track conditions of the train so as to avoid sudden stops.
4. Resistive Braking System: The system can heat up the storage devices without charging them and still function. This saves the machine from damage or assists in safe and dependable system braking, even in conditions when regenerative braking is not permitted.

5.2.4 Technological Advancements and Innovations

Improvements in regenerative braking recently have been placed on thrusting efficiency, design, whether it can withstand harsh environment and how well it can integrate with other systems: Recent progress in this field centres on:

1. Advanced Power Electronics: The evolution of semiconductor materials has resulted in the mass production of more compact and efficient power electronics hence better energy management control. This includes the evolution of high-power inverters and converters, which can operate at even higher voltages and currents.
2. Enhanced Energy Storage Solutions: Considering the recuperative friction braking practice, the most significant advancement is the improvement of batteries and supercapacitors in their ability to store and reuse braking energy. Hybrid energy storage systems composed of battery and supercapacitor elements offer the protection of both elements and provide very high volumetric energy density as well as quick response time.
3. Smart Braking Control Systems: The efficient management of regenerative braking systems has been improved due to the use of advanced techniques which include computer algorithms as well as real-time stress conditions. Adaptive systems can change depending on such features as: train velocity, train load, track gradient etc. in order to achieve stability and safety in train systems.
4. Integration with Smart Grids: In the same fashion, regenerative braking systems are more often now integrated with smart grids which assist in effective distribution and consumption of the recovered energy. This facilitates the management of power transfer within the railway network more effectively and also makes it possible to supply other users outside the rail network.

5. Digitalization and Automation: The improved design and reliability of regenerative braking systems stand out due to the application of digital twins, machine learning, and automated control systems. These kinds of technologies allow for conducting predictive maintenance, optimizing operational performance, and managing energy in a more effective way.

5.3 OPTIMUM UTILIZATION OF REGENERATIVE BRAKING ENERGY

5.3.1 Energy Recovery and Conversion

Energy recovery and conversion are imperative for the efficient design and control of regenerative braking systems, especially in railways. As a train slows down its movement, the electrifying current created by the traction motors converts the mechanical energy into electrical energy. This process can be optimized through several methods:

1. Real-Time Monitoring and Control: High-performance control systems keep track of speed, brake application, and energy generated efficiently and in real time. This information enables appropriate modulation of braking force so as to utilize as much energy as possible when decelerating the motion without causing any vibrations.
2. Adaptive Braking Algorithms: These determine how the electrical energy developed through regenerative braking will be utilized and how much of it will be released to the mechanical brakes making the electrical, mechanical and track regenerative braking systems work effectively under different operating conditions like train load, speed and gradient. They make sure that safe limits of the brake performance are ensured and the available energy is used properly.
3. Efficient Power Conversion: It is important to also unmake any transport energy that has been transformed during the propelling break which is then stored or used again. The use of modern power converters and inverters during regenerative braking makes this process efficient by reducing the energy losses experienced during energy use.
4. Regeneration Efficiency Monitoring: There are modern systems with sensors and monitoring tools that measure the efficiency of the energy recovery process. They can be useful in improving the working of the energy recovery system over an extended period so that the system reliably recovers the energy lost in braking. Figure 5.2 presents a bar chart comparing the energy recovery efficiency of different braking systems used in railways. Regenerative braking systems exhibit the highest efficiency, recovering up to 50% of the energy, while dynamic braking systems recover less due to energy dissipation as heat. Mechanical braking systems show no energy recovery, as all energy is dissipated through friction. This comparison highlights the superior efficiency of regenerative braking in energy conservation [25–31], which discusses advancements in rail vehicle stability, ride performance, crashworthiness, and vehicle dynamics.

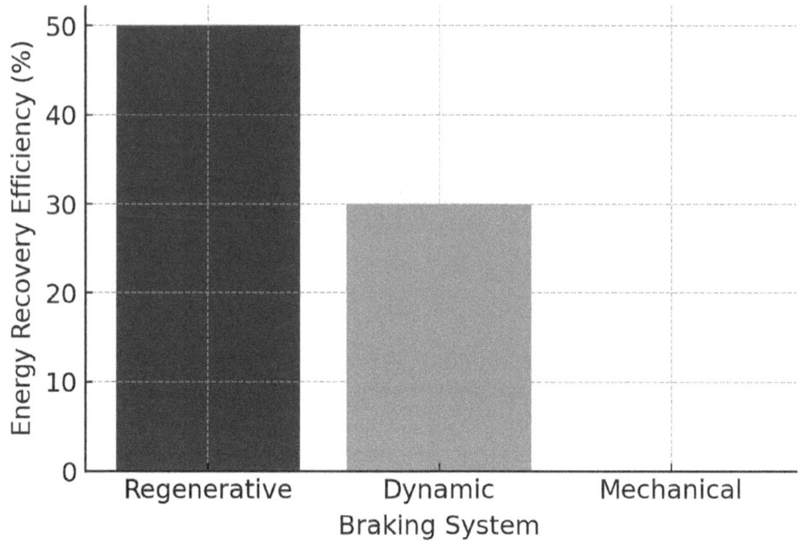

FIGURE 5.2 Comparison of energy recovery efficiency across different braking systems.

5.3.2 Energy Management Strategies

One of the challenges with the regenerative braking system is the quick release of recoverable energy. A number of techniques are countered to avoid the waste of any usable energy and its dispersing:

1. Facilities within the Carriage for the Purposes of Energy Recapture: Energy that has been recovered may be stored in the train using, for example, batteries, supercapacitors, or flywheels. This energy can then be used to run the train's auxiliary system or assist in the train's acceleration thus, lessening the burden to the principal power source.
2. Grid Integration: In the event of the rail system being tied up with the power grid, then the excess reacquired energy may be fed back to the grid for sale. This not only helps to avoid wastage of energy but also improves the efficiency of the electric energy system overall by catering to metered energy demand.
3. Energy Recovery Cross-Utilization: In the case where there are multiple train operations, the energy that is recovered by one train can be used in propelling another train. This is most advantageous in the case of metro and urban transit systems since energy resources may be maximized as several trains use similar infrastructure.
4. Effective Dynamic Load Management: In order to avoid over-stressing energy storage devices or the energy grid, dynamic load management is employed. These are normally placed so as to restrain the recovered energy

directed towards consumption or selling relative to the existing capacities or loads in order to preserve effective loads within the system.

5. Intelligent Energy Control Systems: There are incorporated intelligent energy management tools in most of today's regenerative braking systems that use predictive features to improve the system energy demands forecasting. This is beneficial in planning programmed energy uses and its storage and consumption in a systematic manner.

5.3.3 Integration with Railway Power Systems

Interfacing regenerative braking systems with the rest of the railway power system entails a number of technical as well as operational aspects which include the following:

1. Synchronization with Power Supply: It is necessary to ensure that the electrical energy produced during the braking process integrates into the railways electric network which may operate with varying (different) voltage levels, components as well as frequency ranges. Advanced converters and synchronization systems ensure seamless integration.
2. Energy Flow Management: The power system is required to have a dual capability of energy transfer which includes receiving energy from the train when braking and supplying energy while accelerating. This calls for advanced energy flow management technologies that are capable in forecasting both the supply and the demand in real time.
3. Infrastructure Upgrades: The deployment of regenerative braking on the larger grid may require the upgradation of certain other assets like substations, transformers, power lines, etc. so that they provide the additional electric overloads and enable appropriate power distributions.
4. Operational Coordination: The operations of trains, power supply systems, and energy storage facilities must all be conducted in a coordinated manner to ensure that energy losses are minimized during energy recovery processes. This involves system interaction and computation that relates to energy transfer in a more distributed way. An example of such energy flow is shown in Figure 5.3 for a regenerative braking system. The diagram explains the steps on how kinetic energy gets transformed into electrical energy and later on spreads it to a wide range of receivers which are mainly comprised of the power grid system, the on-board energy banks (batteries or supercapacitors), and auxiliary system. The importance of power electronics in control of energy flow and efficient conversion and storage is also addressed.

5.3.4 Optimization Algorithms for Energy Distribution

Modern optimization algorithms are based on the distribution and application of braking energy regeneration for a number of functions that include:

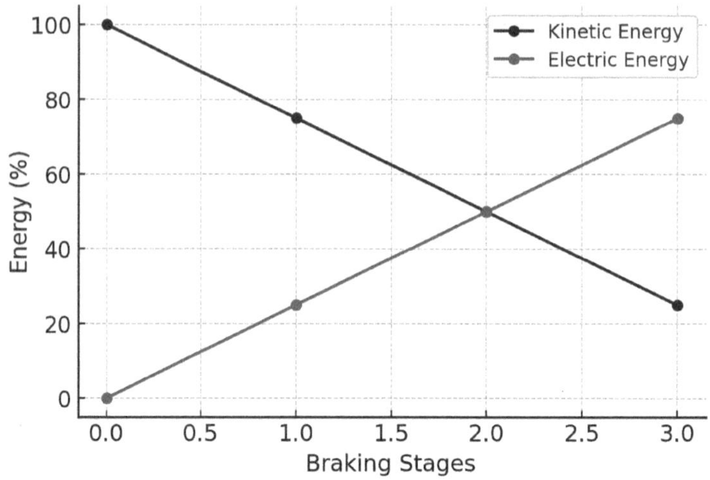

FIGURE 5.3 Energy flow in regenerative braking.

1. Predictive Energy Management: These algorithms are focused on the period of historical and real-time data to manage energy requirements and energy. In this case, it is possible to predict the acceleration and deceleration of the train, the amount of energy harvested, and the maximum load taken by power grid optimizing in this way the energy allocated over time to electricity consumption.
2. Load-Sharing Algorithms: In the case of multi-train operations, load-sharing algorithms allow for controllable parameters with which energy that is already recovered by trains may be supplied. This is important especially for systems like metro in which the energy demanded by and the energy that is expended by other trains might not be the same.
3. Real-Time Optimization: Real-time optimization algorithms manage the task of optimizing the energy of loads carried by trains in motion and the energy patterns acquired from movement with the trains for various operational parameters including train speed and track conditions. They assist in controlling optimal energy consumption and energy waste.
4. Machine Learning and AI Integration: Modern systems are now increasingly incorporating ML and AI approaches in their design for better performance in terms of efficiency in the management of energy systems. Such relations result in predictive maintenance, active energy delivery, or dynamic optimization of a system's performance.

In Figure 5.4, a flowchart depicting the energy management policy action in a regenerative braking system is presented. It demonstrates how, during braking, the energy produced is subsequently distributed to components such as energy storage units, auxiliary units, and the grid. Emphasized in the diagram are advanced control algorithms and predictive analytics in relation to energy optimization and overall system dynamics.

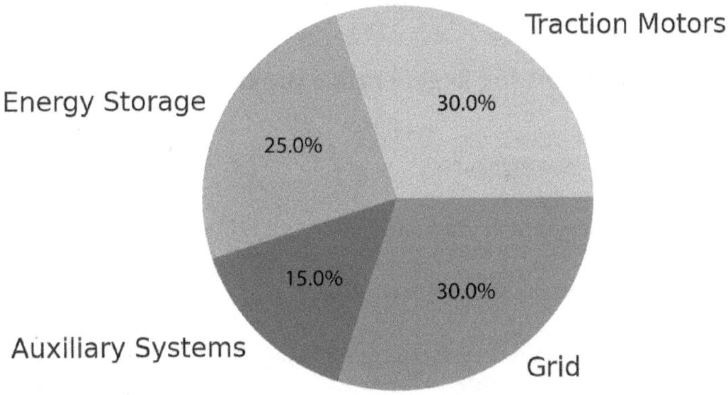

FIGURE 5.4 Real-time energy management in regenerative braking systems.

5.4 ENERGY STORAGE FOR REGENERATIVE BRAKING

This involves the brief recovery of energy usually lost as heat during deceleration processes.

5.4.1 Description of Energy Storage Devices and Systems

Energy storage is, probably, the most important part in regenerative braking systems, since the energy that can be recovered can also be stored and used when it is desired. Different technologies are employed having their pros and cons:

1. Batteries: Within the energy storage of railway systems, lithium-ion batteries and other variants are employed. These batteries can store tremendous amounts of energy over a long duration due to their high energy density. Their disadvantages are weight, price, and charge time.
2. Supercapacitors: Due to their high charging and discharging rates, supercapacitors can be utilized in rapid energy stores that will experience constant on and off loads such as quick accelerations and brakes within urban bus systems. Though they have lower energy density when compared to batteries, they have a much longer cycle life (Table 5.3).

Figure 5.5 illustrates the distribution of various energy storage technologies used in regenerative braking systems, including batteries, supercapacitors, and flywheels. The chart shows that batteries are the most commonly used storage solution, accounting for 50% of applications, followed by supercapacitors (30%) and flywheels (20%). This visual representation provides an overview of the prevalent storage technologies and their share in regenerative braking applications.

5.4.2 Battery Systems for Railways

Energy storage technologies, with batteries in regenerative braking, have the greatest use of battery energy systems. Major issues that dominate batteries include:

TABLE 5.3
Energy Storage Solutions for Regenerative Braking

Energy Storage Type	Advantages	Disadvantages
Batteries	High energy capacity, suitable for long-term storage.	High cost, weight, limited lifespan.
Supercapacitors	Rapid charging/discharging, long cycle life.	Lower energy density, suitable for short-term use.
Flywheels	High efficiency, rapid response.	Complex control systems, space requirements.
Hybrid systems	Combines benefits of multiple storage types.	Higher complexity, integration challenges.

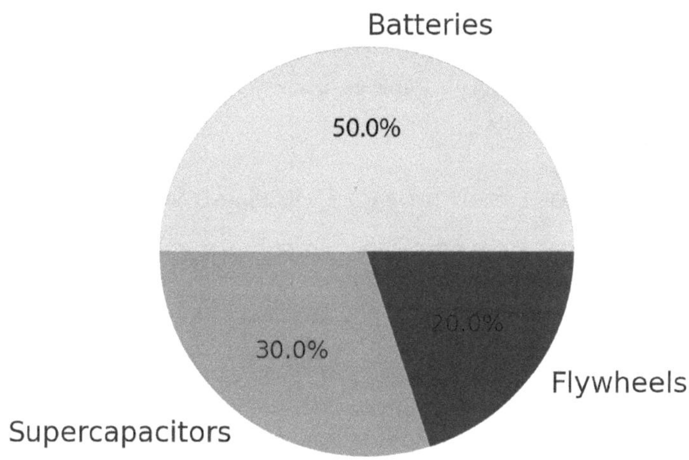

FIGURE 5.5 Types of energy storage systems for regenerative braking.

1. Types of Batteries: With regard to Li-ion batteries and their energy density and efficiency, this is the preferred type of battery. Others like nickel metal hydride and lead-acid batteries can also be used but will be determined by the nature of the system.
2. Energy Capacity and Density: Battery capacity is important in defining how much amount of energy can be stored and ultimately used. The higher the capacity of the batteries, the higher the energy recovery but the batteries become heavier and more expensive.
3. Charging and Discharging Cycles: The energy efficiency of a battery system also depends on how fast its stored energy can be charged and discharged. Batteries are constructed in such a way that many charging cycles, such as those that are standard in railway systems, can be carried out without significant battery deterioration.
4. Tube Diameter Trombone: Reduction in temperature is very important in preventing the performance from being affected by overheating. Maintaining the structure integrity of the system relies heavily on the use of these advanced

cooling systems and monitoring tools to enhance normal operating temperature and act as a preventive measure against thermal runaway.

5. What is the Overall System Design: Battery systems in an automobile can work effectively and efficiently alongside the regenerative braking mechanism. This requires advanced control systems that regulate and manage the energy flow into and out of the batteries when the braking system is in use.

5.4.3 Supercapacitors and Their Applications

Supercapacitors are fast charging and discharging storage technology is rapidly becoming integrated into regenerative braking. Supercapacitors are characterized by the following.

1. Power Density: Supercapacitors are self-energy dense devices, enabling rapid energy release compared with batteries making them most appropriate for applications that require fast energy burst for a short period, these include processes that require deceleration.
2. Number of Cycles and Life: Supercapacitors have cycles of charge and discharge in hundreds of thousands or even millions without any performance degradation, easily more than batteries are capable of. They are therefore cost-effective in the long run as they can be used universally with minimal environmental impact.
3. Response Time: For trains that are often started and stopped, supercapacitors can charge and discharge rapidly and are therefore ideal for many applications. They are capable of absorbing excessive braking and subsequently utilizing that energy while promoting efficiency in a sustainable manner.
4. Integration with Hybrid Systems: Supercapacitors are normally used in conjunction with a battery in hybrid energy storage systems such as these where a high energy capacity is required and a quick response is still a must.

5.4.4 Flywheel Energy Storage

Flywheel energy storage systems stand out in regenerative braking applications owing to the following advantages:

1. Mechanical Energy Storage: A flywheel is an energy device that stores energy in the form of motion and repurposes it electromagnetically usually as short pulse energy bursts. This makes the design beneficial in short energy-demanding applications.
2. High Efficiency: The efficiency of flywheels is quite high and on many occasions is above 90%, hence they are favourable concerning energy use, storage, and recovery.
3. Longevity and Reliability: Flywheel devices can last really long and charge and discharge quickly and thus are low risk in energy storage applications.
4. Application in High-Speed Rail: Flywheel systems can be useful in high-speed rail applications where scheduling requires high-power output and quick energy boost for fast acceleration.

5.4.5 Hybrid Energy Storage Solutions

Railway applications benefit greatly from the integration of hybrid energy storage systems which consist of batteries and supercapacitors. Such systems have significant energy capacity and quick response time. They dynamically balance surplus and capacitor shortages by using supercapacitors in high-power applications where acceleration/deceleration needs to be quick and batteries during prolonged power operations with low rates. This arrangement provides more efficient energy consumption and less stress on the single components which in return increases the reliability and general functioning of the regenerative braking system. Combination of various storage technologies gives hybrid systems the ability to work in a variety of operational regimes which can provide a reliability and resolution that no other single energy storage technology can. Crews report high satisfaction with the energy storage technology as it allows them to benefit from the advantages of both batteries and supercapacitors – achieving their common objective more robustly. Other areas of success include the diverse energy needs of various rail applications, ranging from local urban transit systems to high-speed rail systems.

The flexibility in hybrid systems aids in making application-specific solutions – from intracity public transport to high-speed intercity trains. Such versatility helps not only to increase the efficiency and reliability of the energy storage system but also allows for various optimizations in the performance of the rail mode operations. These energy storage and energy management solutions are critical in achieving the full capabilities of regenerative braking systems through increased energy recovery and use and therefore the railway dynamics. Figure 5.6 gives the configuration of hybrid energy storage for regenerative braking applications.

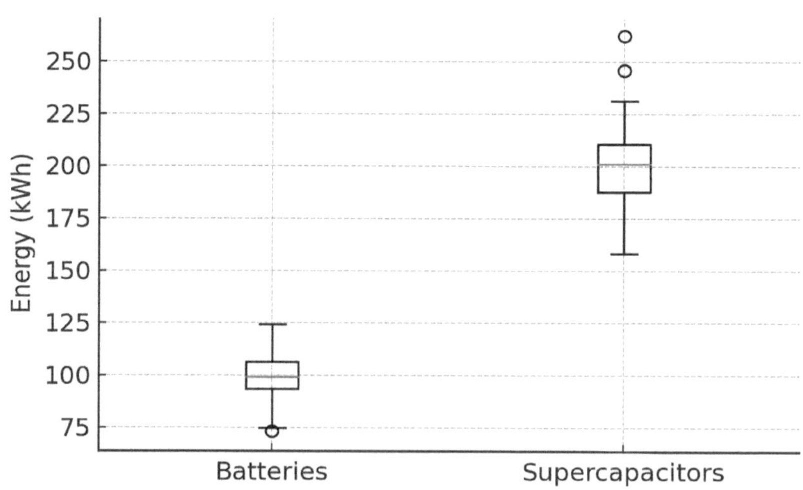

FIGURE 5.6 Hybrid energy storage configuration for regenerative braking.

5.5 CHALLENGES AND INNOVATIONS IN IMPLEMENTING REGENERATIVE BRAKING

5.5.1 Technical Challenges in System Integration

There are several technical barriers that need to be overcome while implementing regenerative braking systems in railways to make sure they work ER and reliably.

1. Compatibility with Existing Infrastructure: Retrofitting existing approaches in old rail networks that have not been originally built with these systems is not only highly expensive but hard to implement too. The power systems usually in place may not be bi-directional meaning the scope for power recovery is lost thus the system will call for great adjustments.
2. Energy Storage Limitations: Regenerative braking's foremost hurdle is the limitation posed by the energy storage unit available on-board. Barring batteries or supercapacitors, it is understood that energy storage devices have limitations. However, once such limitations are achieved, no extra energy can be stored and hence the efficiency of the system is reduced.
3. Power Quality and Stability: Injecting energy back into the grid system brings challenges of power quality and stability, particularly in situations where the system was not designed for such variations of output. This harkens back to voltage that has been high as well as power factors that one can scarcely comprehend owing to the power systems of the railway going on being adversely impacted upon.
4. Thermal Management: Energy recovery tends to produce a noticeable amount of heat which adversely affects both brake systems and the energy storage systems. Efficient thermal management is essential in an effort to avoid overheating and thus operational safety.
5. Regulation of Control System: There are real-time management software with advanced integration that can help achieve coordination of the regenerative and mechanical braking systems. Failures or latencies within the controller can lead to poor energy or braking operation recovery which can be termed as poor performance. Table 5.4 details some of the key difficulties in realizing the regenerative braking systems for rail transport and some of the possible measures that can assist in addressing the challenges such as technological intervention and economic measures.

5.5.2 Safety and Reliability Considerations

Safety is one of the major priorities for railway activities, and especially concerning the installation of regenerative braking systems brings new problems for resolution:

1. Distribution of Brake Forces: Deficient distribution in the share of the regenerative and mechanical systems during braking poses a danger to the motion of the train. Dependence on regenerative breaking may indeed affect the functioning of mechanical systems particularly when there are none or minimal emergencies.

TABLE 5.4
Challenges in Implementing Regenerative Braking Systems

Challenge	Description	Potential Solutions
Infrastructure compatibility	Retrofitting older networks to accommodate regenerative systems can be complex and costly.	Modular system design, phased upgrades
Energy storage limitations	Limited capacity of storage systems reduces the effectiveness of energy recovery.	Advanced storage solutions, hybrid systems
Power quality and stability	Feeding energy back into the grid can affect power quality and stability.	Improved synchronization systems, grid integration technology
Thermal management	Excessive heat generation can impact system performance and safety.	Advanced cooling systems, real-time temperature monitoring
Cost and economic feasibility	High initial investment and maintenance costs may be prohibitive for some railway operators.	Long-term cost-benefit analysis, government subsidies

2. System Redundancy: The principles of fail-safe applied to regenerative braking systems include the possibility to engage the mechanical brakes entirely in case of failure to other brakes. This is usually emphasized in high-speed rail applications where the braking system needs great attention and must perform effectively.
3. Reliability of Energy Storage Systems: The reliability of the supercapacitors and batteries greatly depends on temperature, ageing, or cycles of usage. To maintain them within their expectancy, consistent monitoring or maintenance ought to be done on them.
4. Emergency Scenarios: Factors such as emergency braking might limit the contributions of the regenerative system since it might not have enough time to recover any energy. Hence it is important to ensure the mechanical braking system takes care of the complete braking loads.

5.5.3 Cost and Economic Feasibility

When it comes to regenerative brake systems, the advantages will often come in the long run; however, the costs incurred in this case will be quite high at the beginning:

1. High Initial Investment: It creates disparity as the cost of implementing regenerative braking technology involves lots of capital acquisition on new technology, infrastructural enhancement as well as energy storage systems. This is bound to be a problem for railway managers especially those in areas where cash is a problem.
2. Cost of Energy Storage: High cost is incurred from using advanced energy storage systems such as batteries and supercapacitors which have a lot of energy density. The costs incurred on these systems should be compared to the amount of energy that will be saved and other operational gains that will be realized within the lifetime of that equipment.

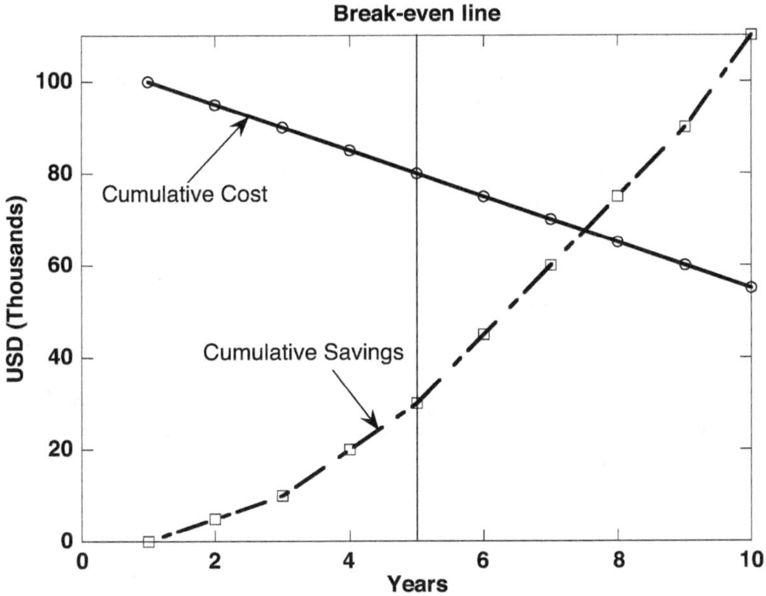

FIGURE 5.7 Cost-benefit analysis of regenerative braking systems.

3. Maintenance and Lifespan: Energy storage devices as well as power electronics are always in need of maintenance, in addition, in order for the systems to function effectively. Maintenance levels and the lifetime of batteries, being auxiliary components of a system, can create economic difficulties with respect to the total system.
4. Return on Investment (ROI): One of the setbacks in such systems is calculating the ROI of regenerative braking systems investment, as it is dependent on a number of factors, including energy costs, the way the train is used, and how much wear and tear will be saved on the mechanical parts. The line graph in Figure 5.7 takes a closer look at the cost versus benefits of the regenerative braking systems. It shows that initially high expenses arise due to investments in energy storage and the need for improved infrastructure, but after some years expenses on the operation and maintenance of the equipment fall for the reason of energy efficiency and decreased number of maintenance works required. The break-even point is identified, which states till the time savings are less than the investment made then why this technology should be used, emphasizing the long-term economic benefits of the technology.

5.5.4 Developments in Braking Systems and Their Control Systems

It is worth mentioning that, despite these challenges, there is a variety of innovations geared towards the improvement of the performance and efficiency of regenerative braking systems:

1. Advanced Energy Storage Solutions: The improved energy storage system by new battery or supercapacitor materials in terms of power, life, and efficiency is taking place. This involves solid-state batteries which have higher energy densities with enhanced safety prospects.
2. Predictive Maintenance Systems: Sensor and machine learning algorithms that prevent or forecast the failures of the regenerative braking systems, are highering the reliability and lowering the maintenance costs.
3. Intelligent Control Algorithms: AI applications and real-time data acquisition have advanced control systems of the architecture allowing for more recovery and better distribution of energy and overall system performance.
4. Joining the National Grids and Realizing Renewable Energy: The evolutionary developments recently witnessed under regenerative brake technology are linked to its marriage with smart grids and other energy sources. This also entails using the waste energy to help other sections of the railway system or pushing it to the national grid.

5.6 RECAPTURING ENERGY FROM REGENERATIVE BRAKING

5.6.1 Recovery Energy Usage in Railway Infrastructure

This is primarily because regenerative braking force which is its energy, can be used internally within the railway network itself:

1. Feeding on Auxiliary Systems: Energy that has been recovered can be utilized within the auxiliary systems of the train to power systems that include lighting, air conditioning, communication and thus lowering the strain placed on the general power supply.
2. Helping in Train Push-Off: Such energy can also help a train in push off after a standstill thus lessening the energy demand from the specific energy source. This is quite beneficial in urban transportation systems where a lot of train halting and pushing off is done per unit time.
3. Powering Different Trains: Such energy can also help a train push off after a standstill thus lessening the energy demand from the specific energy source. This is quite beneficial in urban transportation systems where a lot of train halting and pushing off is done per unit time.
4. Energy Storage at Rest: The energy that is recuperated by trains can always be kept in stationary batteries or supercapacitors which are contained in the depot. This energy can either be used to power station facilities or released to the electric grid when required.

   Table 5.5 illustrates various strategies for recycling regenerative braking energy within railway systems, including the use of recovered energy to power train and station facilities or feed into the public grid.

In the case where the current needs of the railway do not absorb the energy that has been recuperated from braking, users can redistribute it:

TABLE 5.5
Energy Recycling Strategies for Railways

Strategy	Description	Examples
Powering auxiliary systems	Recovered energy is used to power non-propulsion systems like lighting and HVAC.	London Underground, New York City Transit
Assisting train acceleration	Energy is stored and used to reduce power draw during train acceleration.	Tokyo Metro, Paris RATP
Grid feed-in	Excess energy is fed back into the public power grid, supporting overall energy stability.	Hong Kong MTR, Deutsche Bahn
Stationary energy storage	Energy is stored at stations for later use or grid support.	Siemens "Sitras SES" system in Berlin

1. Grid Feed-In Systems: In the case of excessive energy that enters the grid, systems with regenerative braking can offset the surplus capacity demand without sheet waste energy overload than anticipated in the system as energy is consumed only when necessary. There is high energy input demand at peak hours and thus support may be utilized to alleviate this demand.
2. Revenue Generation: For the railway industry, this brings about additional revenue as energy can be resold back to the grid, especially in areas with energy buy-back policies. Such revenues assist in offsetting the installation and operational lifetime gradually accumulating costs of regenerative braking systems.
3. Grid Stability and Support: As a result of low demand from the grid, energy can be pushed into the grid, and these systems provide support, especially during peak load periods. Such a situation is more experienced in the cities which consume a lot of energy.
4. Integration with Renewable Energy Sources: Available energy from electric braking can also be withdrawn for stabilization of solar or wind considerably intermittent energy generating systems when surplus renewable is not being harnessed.

5.6.2 Applications in Urban Transit Systems

Urban rail systems, in particular, have tremendous benefits from regenerative braking systems when the trains are resumed and stopped frequently:

1. Frequent Energy Recovery: While urban trains like metros and trams are in use, many stoppages are inevitable, which presents some opportunities to recover energy. This aids the regenerative braking system and makes it very effective in improving efficiency and reducing the costs of energy utilization.
2. Improved Operational Efficiency: or urban transit systems, part of the operating cost can be offset by harnessing energy and applying it for quick acceleration and operation of auxiliary systems at the same time.

3. Supporting Energy Infrastructure: In centrally located places, regenerative systems can cut down the heat demand on the local grid as the regained energy is used to propel the trains. This reduces the adverse influence of railways on the energy matrix of urban space.
4. Case Studies: Metro rail systems with provision for regenerative brakes have been in operation in cities such as New York, London, and Tokyo, making energy savings and operational costs much lower. Studies of this nature have been carried out and the findings reveal that regenerative braking can be incorporated into urban transit systems.

5.6.3 Case Studies of Effective Energy Recycling

Different railway systems across the globe have gone ahead and incorporated energy recovery with regenerative braking systems;

1. London Underground: In a number of lines within the London Underground, regenerative braking devices have been employed and every energy so produced is interconnected to the system for other loads within the stations, hence overall energy utilization will be decreased. This system contributed to cost and carbon emissions as the consumption of carbon emissions in the system was a result of electricity led to reduced carbon usage.
2. Tokyo Metro: The Tokyo underground system incorporates regenerative brakes. Energy from the braking units is also fed back into the system supporting other business functions including powering up trains and servicing the stations. This system of the buildings has been related to an intelligent energy management system that makes it cheaper to manage energy than if it was simply distributed in one.
3. Paris RATP: The sheer volume of surface railways and indeed even the Metro system enforced RATP to use regenerative braking systems on its rolling stock and tramway systems. In particular, energy storage devices were integrated at important junctions for storing and re-distributing regained energy. In this fashion, the operational efficiency of the system was also noticed to have been enhanced and energy bill concerns decreased.
4. New York City Transit: Regenerative braking schemes for the conductors in the New York City subway have been developed, with the potential charge to supply lighting and catering facilities in the stations by the recovered energy. Based on the results of these pilots, these measures are expected to be put into use in the entire network and the volume of the works undertaken based on the experience of the Wards. Figure 5.8 presents a flowchart detailing the real-time energy management process in a regenerative braking system. It demonstrates how braking energy is generated and how it is used or divided between parts like energy storage, auxiliary units, and the power grid. This picture illustrates how cutting-edge control algorithms and forecasting models function to enhance energy efficiency and the stability of the system.

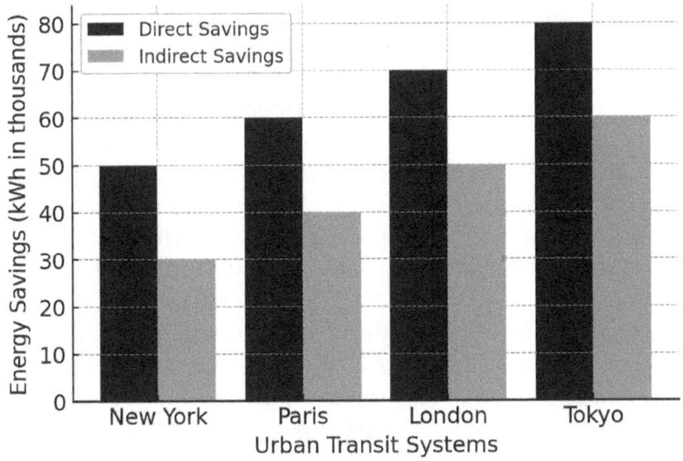

FIGURE 5.8 Case study: energy savings in urban transit systems.

TABLE 5.6
Case Studies of Regenerative Braking Systems

Railway Network	Implementation	Outcomes
London Underground	Integrated regenerative braking across multiple lines, energy used for station power.	Significant energy savings, reduced operating costs.
Tokyo Metro	Advanced energy storage and redistribution systems for urban transit efficiency.	Enhanced performance, reduced emissions, lower operational costs.
New York City Transit	Pilot projects using regenerative energy for station lighting and auxiliary systems.	Improved energy efficiency, potential for broader implementation.
Paris RATP	Regenerative braking with energy storage across metro and tram lines.	Increased network efficiency, optimized energy use.

Table 5.6 shows data related to functional regenerative braking systems that have been implemented in several railway networks covering the target areas such as energy savings, cost, and enhancement of systems performance.

5.7 SUMMARY

5.7.1 Key Findings and Insights

In this chapter regenerative braking systems on railways with emphasis on their principles, technologies, applications, and benefits of energy efficiency, cost savings, and sustainability improvement have been developed. They include:

1. Energy Recovery Efficiency: It has been estimated that up to approximately 70–90% of the energy used in regenerative braking systems is regained, thus contributing to on-site energy utilization and reduction of operational expenses.

2. Difficulties in Execution: To engage fully the advantages of regenerative braking systems, there are several issues of a technical, economic, and safety nature that must be solved. New development of energy storage systems, control architecture, and grid connection will be important for this.
3. Transformations of Urban Forms of Transportation: Obviously, the application of regenerative braking in urban transit systems is logical, since urban vehicles routinely stop to let passengers board/disembark and such technology minimizes energy loss during braking events. The booming expansion of metro systems in a number of countries should indicate that this technology can be successfully implemented in urban rail systems.
4. Anticipating the Future Innovations in Regenerative Braking Systems: Technology has improved the state of energy storage, integration, and control systems to the point where systems that feature reliable and efficient regenerative braking systems are in development. According to Table 5.7, the development of regenerative braking systems is also expected to keep pace with new challenges, such as the integration of advanced energy storage, smart grids, digital technologies, and international standardization to improve the performance and deployment of these systems in railway systems.

5.7.2 Benefits and Future Potential of Regenerative Braking

There are many advantages regenerative braking systems can bring to railway operation:

1. Reduced Energy Consumption: During train operations, most of the energy usage is put to work which ultimately converts to a pony saving on the overall energy usage target.
2. Cost Savings: Savings in energy and less strain on mechanical parts lead to cost benefits over the lifetime of the system.
3. Environmental Sustainability: This reduction in energy use is beneficial as it reflects a reduction in the emissions of greenhouse gases which makes regenerative braking systems vital for sustainability in railway operations.

TABLE 5.7
Future Research and Development Areas

Research Area	Description	Potential Impact
Advanced energy storage	Development of new battery chemistries and supercapacitor materials.	Improved storage capacity, efficiency, and safety.
Smart grid integration	Enhanced systems for integrating regenerative energy with smart grids.	Better energy management, support for renewable sources.
Digitalization and automation	Use of AI and machine learning for predictive maintenance and system optimization.	Increased reliability, reduced maintenance costs.
Global standards and best practices	Establishing guidelines and sharing best practices for regenerative braking systems.	Consistent implementation, improved global adoption.

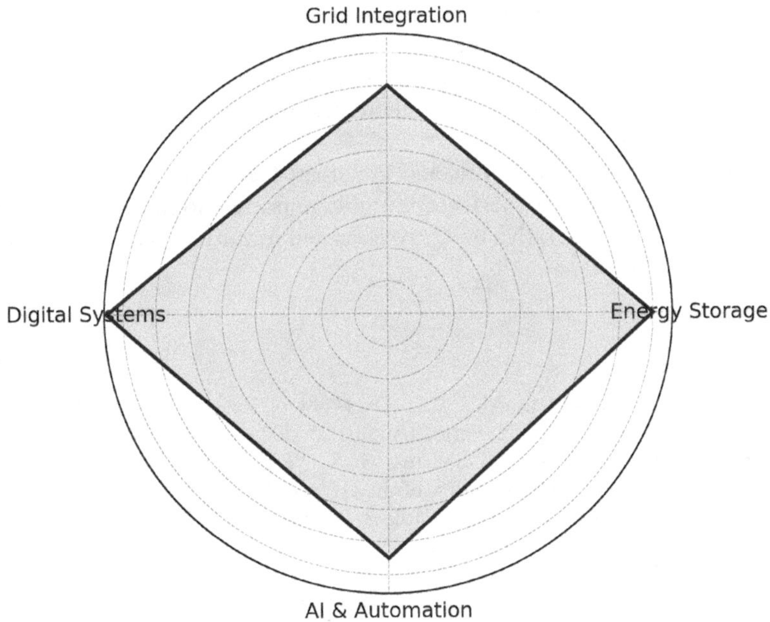

FIGURE 5.9 Future trends in regenerative braking technology.

4. Enhanced Performance: An operational benefit from the aforementioned engineering features is improved braking performance, less reliance on mechanical brakes, as well as delivery of energy which was put to rest in other parts of the railway network and is recovered from the use of brakes.

Developing plans for green grid energy management systems incorporate many components shown in Figure 5.9 like the use of advanced storage, smart grids, control systems, and predictive maintenance to regenerate braking systems. The motivation outlines goals that will influence innovative advances vital for the next generation's regenerative braking systems in railways.

5.7.3 Recommendations for Future Research and Development

To enhance performance and foster the integration of the regenerative braking systems, further research and development should be included:

1. Advanced Energy Storage Solutions: Other concepts that incorporate further energy storage enhancement such as new battery chemistry/supercapacitor materials or hybrid systems would be essential.
2. Smart Grid Interfacing: Implementing processes that would seek to interface efficiently with smart grids would allow for more versatile and effective energy management and would be beneficial for railroad systems and energy networks as a whole.

3. Globalization of Processes: Transforming high- and low-level structures by the incorporation of the technologies of t digital twins, machine learning, and robotic control of processes will improve the efficiency and reliability of industrial regenerative brakes due to real-time optimization and predictive maintenance.
4. Universalization of the Norms and Practices: Developing normative documents and sharing the experience is highly important for the effective introduction of the regenerative brake systems and rewarding their utilization in different railways.

REFERENCES

1. Hamada, A. T., and Orhan, M. F. "An Overview of Regenerative Braking Systems." *Journal of Energy Storage* 52, Part C (August 25, 2022): 105033.
2. Sharma, R. C., Palli, S., and Sharma, S. K. "Ride Analysis of Railway Vehicle Considering Rigidity and Flexibility of the Carbody." *Journal of the Chinese Institute of Engineers, Transactions of the Chinese Institute of Engineers, Series A* 46, no. 4 (2023): 355–366.
3. Oleksowicz, S. A., Burnham, K. J., Southgate, A., McCoy, C., Waite, G, Hardwick, G., and McMurran, R. "Regenerative Braking Strategies, Vehicle Safety and Stability Control Systems: Critical Use-Case Proposals." *Vehicle System Dynamics* 51, no. 5 (2013): 684–699.
4. Liu, M., Fang, S., Dong, H., and Xu, C. "Review of Digital Twin about Concepts, Technologies, and Industrial Application." *Journal of Manufacturing Systems* 58, Part B (2021): 346–361.
5. Sharma, R. C., Palli, S., Dharmana, L., Sharma, S. K., Sharma, A., Kharola, A., and Sharma, N. "Simulation of Transient Dynamic Response of a Locomotive Considering Its Flexibility and Rigidity." *Noise and Vibration Worldwide* 55, no. 1–2 (2024): 33–47.
6. Palli, S., and Koona, R. "Analyses of Dynamic Response of a Railway Bogie." *International Journal of Vehicle Noise and Vibration* 11, no. 2 (2015): 103–113.
7. Vasiljević, S., Aleksandrović, B., Glišić, J., and Maslać, M. "Regenerative Braking on Electric Vehicles: Working Principles and Benefits of Application." *IOP Conference Series: Materials Science and Engineering* 1271, no. 1 (2022): 012025
8. Yang, C., Sun, T., Wang, W., Li, Y., Zhang, Y., and Zha, M. "Regenerative Braking System Development and Perspectives for Electric Vehicles: An Overview." *Renewable and Sustainable Energy Reviews* 198 (July 2024): 114389.
9. Wright, L., and Davidson, S. "How to Tell the Difference Between a Model and Digital Twin." *Advanced Modeling and Simulation in Engineering Science* 7 (2020): 13.
10. Bhatti, G., Mohan, H., and Singh, R. R. "Towards the Future of Smart Electric Vehicles: Digital Twin Technology." *Renewable and Sustainable Energy Reviews* 141 (2021): 110801.
11. Madubuike, O. C., Anumba, C. J., and Khallaf, R. "A Review of Digital Twin Applications in Construction." *Journal of Information Technology in Construction* 27 (2022): 145–172.
12. Boschert, S., Heinrich, C., and Rosen, R. "Next Generation Digital Twin." In *Proceedings of TMCE*, edited by Horvath, I., Suarez Rivero, J. P., and Hernandez Castellano, P. M., p. 6. Las Palmas de Gran Canaria, Spain, 2018.
13. Boschert, S., and Rosen, R. "Digital Twin—The Simulation Aspect." In *Mechatronic Futures*, edited by Hehenberger, P., and Bradley, D., pp. 59–74. Springer, Cham, 2016.

14. Adeoye, A. O., Quadri, R. O., and Lawal, O. S. "Assessment of Biofuel Potential of Tenera Palm Kernel Shell via Fixed Bed Pyrolysis and Thermal Characterization." *Results in Surfaces and Interfaces* 9 (2022): 100091.
15. Adeoye, A. O., Quadri, R. O., and Lawal, O. S. "Biomass Waste Pyrolysis in the Context of Nigeria's Oil and Gas Industry: An Overview." *International Journal of Ambient Energy* 44 (2023): 2525–2541.
16. Adeoye, A. O., Quadri, R. O., Lawal, O. S., and Emojevu, E. O. "Physicochemical Characterization, Valorization of Lignocellulosic Waste (Kola Nut Seed Shell) via Pyrolysis, and Ultrasonication of Its Crude Bio-Oil for Biofuel Production." *Cleaner Waste Systems* 7 (2024): 100138.
17. Adeoye, A. O., Quadri, R. O., Lawal, O. S., Malomo, D., Emojevu, E. O., Omonije, O. O., Odeniyi, O. K., Fadahunsi, M. O., Yelwa, M. J., Aasa, S. A., Onakpa, A. E., Omoniyi, B. P., Mark, I. N., Usman, J., Ismaila, A. M., and Saidu, A. U. "Advanced Techniques in Upgrading Crude Bio-Oil to Biofuel." In *Energy, Environment, and Sustainability*, edited by Sharma, S. K., Upadhyay, R. K., and Kumar, V., pp. 321–353. Springer, Singapore, 2024.
18. Adeoye, A. O., Yelwa, J. M., Imam, N., Quadri, R. O., Lawal, O. S., Malomo, D., Aasa, S. A., Onakpa, A. E., Dorgbaa, R. K., Omoniyi, B. P., Ahmad, N., and Garba, H. A. "Pyrolysis of Plastic Wastes Towards Achieving a Circular Economy: An Advanced Chemistry and Technical Approach." In *Energy, Environment, and Sustainability*, edited by Sharma, S. K., Upadhyay, R. K., and Kumar, V., pp. 215–259. Springer, Singapore, 2024.
19. Adeoye, A. O., Quadri, R. O., Lawal, O. S., Malomo, D., Aliyu, M. T., Dang, G. E., Emojevu, E. O., Maikato, M. J., Yahaya, M. G., Omonije, O. O., Edidem, V. G., Abubakar, Y. K., Offor, O. F., Sochima, E. H., Peter, B. E., and Hikon, B. N. "Sustainable Energy via Thermochemical and Biochemical Conversion of Biomass Wastes for Biofuel Production." In *Energy, Environment, and Sustainability*, edited by Sharma, S. K., Upadhyay, R. K., and Kumar, V., pp. 245–306. Springer, Singapore, 2023.
20. Adeoye, A. O., Quadri, R. O., and Lawal, O. S. "Wet Synthesis, Characterization of Goethite Nanoparticles and Its Application in Catalytic Pyrolysis of Palm Kernel Shell in TGA." *Results in Surfaces and Interfaces* 11 (2023): 100118.
21. Avesh, M., Imam, N., and Adeoye, A. O. "Assessing Environmental Impacts and Mitigation Strategies for the DFCCIL Rail Projects." In *Energy, Environment, and Sustainability*, edited by Sharma, S. K., Upadhyay, R. K., and Kumar, V., pp. 333–353. Springer, Singapore, 2024.
22. Momodu, D. U., Adetuyi, A. O., Adeoye, A. O., Akinlolu, O. S., and Apara, E. O. "Surface Functionalization of Linen Fiber and Its Dye-Fastness Assessment Using (4-[(E)-(4-Amino-2-Hydroxyphenyl) Diazenyl] Benzene-1-Sulphonic Acid and 4-[(E)-(4-{(Z)-[4-(Dimethylamino)Phenyl] Diazenyl}-2-Hydroxyphenyl) Diazenyl] Benzenesulfonic Acid)." *Suranaree Journal of Science and Technology* 30 (2023): 030118.
23. Quadri, R. O., Adeoye, A. O., Ayanda, O. S., and Lawal, O. S. "Conversion of Palm Kernel Shell to Sustainable Energy and the Effect of Wet Synthesized Nanoparticles of Iron on Its Thermal Degradation Kinetics." *Bioresource Technology Reports* 27 (2024): 101933.
24. Bhardawaj, S., Sharma, R. C., and Sharma, S. K. "Development and Advancement in the Wheel-Rail Rolling Contact Mechanics." *IOP Conference Series: Materials Science and Engineering* 691, no. 1 (2019): 012034.
25. Goswami, B., Rathi, A., Sayeed, S., Das, P., Sharma, R. C., and Sharma, S. K. "Optimization Design for Aerodynamic Elements of Indian Locomotive of Passenger Train." In *Advances in Engineering Design: Select Proceedings of FLAME 2018*, edited by Prasad, A., Gupta, S., and Tyagi, R., pp. 663–673, Springer, Singapore, 2019.

26. Sharma, R. C., Sharma, S. K., and Palli, S. "Linear and Non-Linear Stability Analysis of a Constrained Railway Wheelaxle." *International Journal of Vehicle Structures and Systems* 12, no. 2 (2020): 1.
27. Sharma, R. C., Sharma, S. K., Sharma, N., and Sharma, S. "Analysis of Ride and Stability of an ICF Railway Coach." *International Journal of Vehicle Noise and Vibration* 16, no. 3–4 (2020): 127–144.
28. Sharma, S. K., and Lee, J. "Crashworthiness Analysis for Structural Stability and Dynamics." *International Journal of Structural Stability and Dynamics* 21, no. 4 (2021): 2150001.
29. Sharma, R. C., Sharma, S., Sharma, N., and Sharma, S. K. "Linear and Nonlinear Analysis of Ride and Stability of a Three-Wheeled Vehicle Subjected to Random and Bump Inputs Using Bond Graph and Simulink Methodology." *SAE International Journal of Commercial Vehicles* 15, (2021): 65–79.
30. Sharma, S. K., Sharma, R. C., Lee, J., and Jang, H.-L. "Numerical and Experimental Analysis of DVA on the Flexible-Rigid Rail Vehicle Carbody Resonant Vibration." *Sensors* 22, no. 5 (2022): 1992.
31. Vishwakarma, P. N., Mishra, P., and Sharma, S. K. "Characterization of a Magnetorheological Fluid Damper: A Review." *Materials Today: Proceedings* 56 (2022): 2988–2994.

6 Sustainability in Railway Engineering

Sivasankara Raju Rallabandi, Srihari Palli, Samson Abiodun Aasa, Seepana Praveenkumar, Badana Govindarajulu, and Bendi Venkataramana

6.1 INTRODUCTION

Development of sustainable railway engineering means taking social, economic, and environmental concerns into account when planning, constructing, operating, and maintaining the rail network. Besides taking into account social needs such as accessibility or safety, the aim is to reduce the environmental effects of trains, improve their energy efficiency, and stabilize the economy of the system in the long run. Broadly defined, sustainability means the actions aimed at the provision of a natural resource over time. Nonetheless, the term encompasses four separate domains, or the four pillars of sustainability: human, social, economic, and environmental [1, 2]. One benefit of electrification is the decrease of greenhouse gas emissions, which is a major concern when it comes to energy efficiency. A key component of making rail transit more sustainable and cost-effective is improving energy efficiency in railway engineering. For long-distance transport of heavy goods or many people, railways are already among the most energy-efficient options. Nevertheless, the railway industry may further improve its energy efficiency by implementing a number of tactics and technologies [3, 4].

When designing a track system (Figure 6.1), it is important to think about environmental impacts alongside technical, functional, and economic considerations. This is especially true when deciding between ballasted and ballast-less systems, as well as the materials and components that go into making these systems. The loads transfer from the rails to the sleepers via the ballast layer on ballasted tracks. On ballast-free or slab tracks, concrete slabs are placed on the ballast layer which serves as the support for the rails. The results of several researches show clearly that slab track environmental and construction costs are higher than those of ballasted track although the service life is longer for slab track – 60 against 40 years, the maintenance costs for slab track are 20%–30% lesser than that of ballasted track and the life cycle cost is lower after including all the environmental costs.

DOI: 10.1201/9781003659983-6

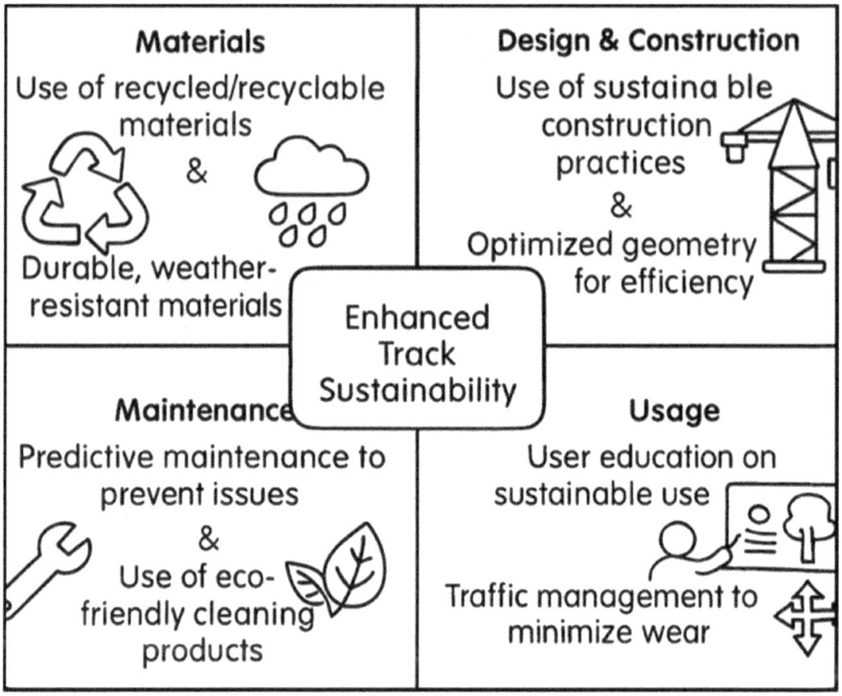

FIGURE 6.1 Potential enhancement areas for tracking sustainability over the life cycle [5].

6.2 CRITICAL APPROACHES TO RAILWAY ENERGY EFFICIENCY

6.2.1 Railways Transition to Electricity

The electrification of rail systems is one of the most preferred methods of reducing dependence on fossil fuels and hence reducing greenhouse gas emissions. Trains that are powered by electricity, particularly from renewable energy sources are far cleaner than diesel-powered trains. Rail systems can switch most of their energy from diesel to electricity and substantially lessen their emissions and operational expenses. The ability of modern power delivery systems such as smart grids and high voltage direct current systems (HVDC) to reduce energy losses in transmission and distribution is critical energy Energy-saving features make electric rail operations not only more effective but also much more environmentally friendly. Smart grids further present communication systems that capture valuable metrics that assist in the efficient distribution of electricity and many cost-effective renewable energy sources to the grid. The transition to electric trains therefore not only reduces the adverse effects on the ecosystems but also allows for wider improvements in energy efficiency [6].

6.2.2 Regenerative Braking via Energy Storage

Another important technology that assists in the energy efficiency of rail systems is the regenerative braking system. This permits the trains to make use of the energy that is produced during braking and convert it into electrical power that is stored in batteries or supercapacitors for later use. This technique minimizes the total energy expenditure of the train by enabling it to reuse the recuperated energy instead of dissipating it as heat. Furthermore, modernized train structures also assist a great deal in reducing the aerodynamic drag at higher speeds of operation. With the use of new composite materials or lightweight aluminium alloys, accelerating and maintaining high speeds requires less energy because trains are less resistant to motion. This includes both the recovery of energy and the strength of the construction which together allow avoid waste of energy and enable lower energy spend [7].

6.2.3 Economical Traction Systems

While attempting to improve the energy economy of rail systems the traction systems are vital and in particular these systems. Due to their better energy efficiency than induction motors, permanent magnet synchronous motors (PMSMs) are becoming favoured especially for rail applications as they maintain their efficacy in partial loads among other factors. PMSMs are particularly well suited for urban or suburban railway networks since they can operate on low energy and are frequent start-stop systems. There are also advanced control systems that distribute the essentials and traction that help with energy efficiency as energy is utilized in the most effective way. These systems are more efficient in the sense that they can easily change the amount of power supplied based on speed, load, and the conditions of the track which reduces energy wastage in addition to improving operational performance [8–17].

6.2.4 Energy-Efficient Operations

It is possible to make train operations more energy-efficient by managing the timetable, speed, and routing in a smart way through management systems. Smart train management system tries to achieve energy-efficient operations by utilizing up-to-the-minute data and advanced algorithms, facilitating the trains to move at the best energy-efficient speeds and routes. This prevents needless use of energy and improves efficiency on the operational level. Energy-efficient operations also depend on driver advisory system. They enable the operator to minimize excessive engined acceleration and braking by offering timely feedback to the train operator. Likewise, Automatic Train Operation (ATO) makes sure that trains are always operating in an energy-efficient manner. ATO systems are useful in densely populated urban areas where consistent speed and timing of the trains are necessary in order to prevent energy loss and achieve smooth and efficient operation of the network [18].

6.2.5 Enhancements to the Infrastructure

The basis of energy-efficient systems and processes is to support the reduction of total energy consumed by the rail systems. One such enhancement is the introduction of a modern hierarchy of energy-efficient signalling systems which allow for better spacing of the trains and less idle time at the stations which in turn helps in energy conservation. By optimizing train flow, these systems reduce the need for trains to idle or accelerate unnecessarily, which ultimately saves energy [19–26], which discuss advancements in rail vehicle stability, ride performance, crashworthiness, and vehicle dynamics. Another vital aspect is the need for smooth and well-curved tracks as they play a major role in increasing the rolling resistance which increases energy consumption. Track maintenance at all times and deployment of quality materials can help in lowering the friction tremendously and make sure that the trains are able to move with more ease. Further, green stations that incorporate renewable energy, energy-efficient lighting, and climate control also aid the railway network in terms of sustainability. Having said that, the actual carbon footprint of the rail infrastructure itself could be lowered by integrating such renewable energy technologies into the design of the stations, for example, solar panels.

The combination of hydrogen and biofuels with electric power, for instance, is a rising trend in the rail. Hybrid trains that can switch between electric power and alternative fuels offer flexibility in operation and reduce the environmental impact of rail transport, especially on lines that are not yet fully electrified. Battery-electric trains are particularly suitable for shorter distances or for operating on non-electrified tracks, offering an eco-friendly alternative to diesel-powered trains.

6.2.6 Railway Energy Management Systems

The focus for railway energy management systems (REMS) is the optimal use of energy throughout the railway network. These systems are capable of automatically and continuously measuring, monitoring, and controlling energy consumption, locating sources of inefficiencies, and assessing the effectiveness of energy corrective actions implemented. Monitoring systems coupled with energy assessment surveys allow rail operators to identify measures that can be undertaken to decrease the level of energy consumption and enhance the level of detail provided for the required improvements. They also enable better power distribution; prevent energy losses and enhance the efficiency of the system, hence reducing costs. These systems are very beneficial in the long run as they enable a significant reduction of the operational carbon footprint of the railways and ultimately contribute to the alleviation of climate change. Reduced consumption of energy by REMS also means reduced operating costs which in turn could mean reduced tariffs for passengers and more competitive freight services. In the long term, energy-efficient trains and energy-efficient systems lead to lower costs of maintenance and lower frequency of operating interruptions hence improvement of the reliability of the system [27].

6.3 UTILIZING ADVANCED COMPOSITES AND LIGHTWEIGHT METALS

When constructing trains, these materials can be of help in reducing energy requirements. To improve the efficiency, performance, and lifetime of rail systems, it is a must to incorporate lightweight materials by engineers. By reducing the weight of trains and other railway components, engineers can substantially improve fuel or energy consumption, operating costs, and ecological footprint. Less energy is required for acceleration and maintenance of speed of trains that are lighter use less energy to do these tasks, which means less fuel or electricity used. Composites are

TABLE 6.1
Mechanical Properties of FRP Material Types and Steel [28–30]

	Material Type				
Property	**CFRP**	**GFRP**	**AFRP**	**BFRP**	**Steel**
Density (gm/cm^3)	1.50–2.10	1.25–2.50	1.25–1.45	1.90–2.10	7.85
Tensile strength (MPa)	600–3,920	483–4,580	1,720–3,620	600–1,500	483–690
Young's modulus (GPa)	37–784	35–86	41–175	50–65	200
Elongation (%)	0.5–1.8	1.2–5.0	1.4–4.4	1.2–2.6	6.0–12.0
Coefficient of linear Expansion	–9.0 to 0.0	6.0–10.0	–6.0 to 2.0	9.0–12.0	11.7

created by mixing at least two materials with distinct properties that are not present in the combined materials. Fibres, resins, an interface, fillers, and additives are all parts of fibre-reinforced polymer (FRP) composites. Apart from that, the high elastic modulus of fibres is part of the mechanical properties of FRP material. Resins assist in stress transmission or distribution further preventing fibre breakage due to mechanical or environmental damage. Recognized to safely affect FRP composite behaviour is the contact between fibres and matrix. Aside from the aforementioned three primary ingredients, fillers have a role in lowering the cost and shrinkage. Composites can have their workability, physical and mechanical qualities, and more enhanced with the use of additives. The outstanding mechanical characteristics of fibre reinforced polymer composites mostly result from the incorporation of reinforcements inside the matrix, shown in Table 6.1. Principally used forms of reinforcement include natural and synthetic fibres.

6.3.1 Essential Lightweight Components for Railways Construction

Aluminium alloys are extensively utilized in train bodies, frames, and interior components, owing to their incredible corrosion resistance and strength-to-weight ratio. Aluminium's lightweight (approximately a third that of steel) makes it a great material for lightweight trains that don't sacrifice strength. One way that aluminium helps keep railways green is because it is easily recyclable [31]. It can also be suggested to use new materials such as FRPs in Train bodies and interior panels and other parts incorporating composites like carbon fiber reinforced polymer (CFRP) and glass fiber reinforced polymer (GFRP). These materials are of immense importance for structural as well as non-structural applications as they have high strength and low weight and are also corrosion-resistant. The composites are adaptable to unique design requirements which allows advanced and effective designs of trains [32, 33]. The earliest application of FRP was as reinforcing bars, primarily in Russia. Since 1990, FRP has been employed in the United States to reinforce compromised structures. Since then, other researchers have investigated FRP composites as a viable choice for strengthening, rehabilitation, and retrofitting systems. FRP is mostly composed of either externally bonded (EB) or interior reinforcement, such as rebar. These two methodologies are employed to enhance various forms of structures by utilizing distinct approaches. The tensile strength of FRP is approximately ten times higher than that of steel and it weights around 25% of the steel density, as shown in

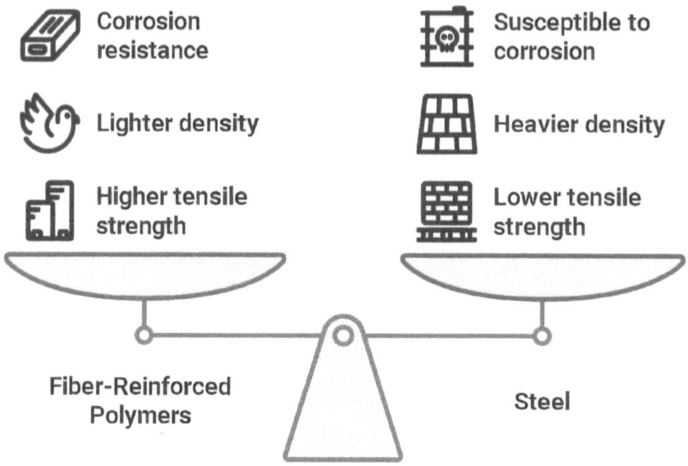

FIGURE 6.2 Comparative analysis of FRP and steel replicated.

Figure 6.2. In 1996, FRP was globally recognized as a structural support material. This material is now extensively used as an alternate reinforcement material for both new construction and the strengthening and rehabilitation of existing buildings [34].

Thirdly, high-strength steels have strength and lightweight are paramount, such bogies and structural components, high-strength low-alloy (HSLA), and ultra-high-strength steels (UHSS) are the way to go. These steels are lighter than regular steel but still provide enough strength, which helps to lower the train's overall weight [35, 36]. Titanium alloys are utilized in many high-performance components, including fasteners and structural pieces that necessitate a blend of low weight, high strength, and resistance to corrosion. Titanium has a few advantages over other materials, such as being strong, lightweight, and corrosion-resistant. However, its hefty price tag restricts its use to certain valuable applications [37]. It is critical for magnesium alloys to maintain weight as these alloys are employed in interior elements such as seat frames and other fasteners. Since they are one of the lightest structural metals available, magnesium alloys also have low density. Nevertheless, their production and material characteristics restrict their widespread use.

6.3.2 Application of Railway Engineering with Lightweight Materials

Many other sectors can benefit from the use of FRP composites and this is illustrated in the following Figure 6.3. Composites and aluminium are commonly used to build the shells of trains, which helps to reduce their weight and improve their fuel economy. Inside parts such as seats, panels, and flooring are all made of lightweight materials, which helps keep the interior fittings light without sacrificing comfort or durability.

Bogies are the train's undercarriage, which includes the wheels and axles, shown in Figure 6.4. By employing lightweight materials, the train's overall weight can be lowered, which in turn leads to less track wear and increased energy efficiency.

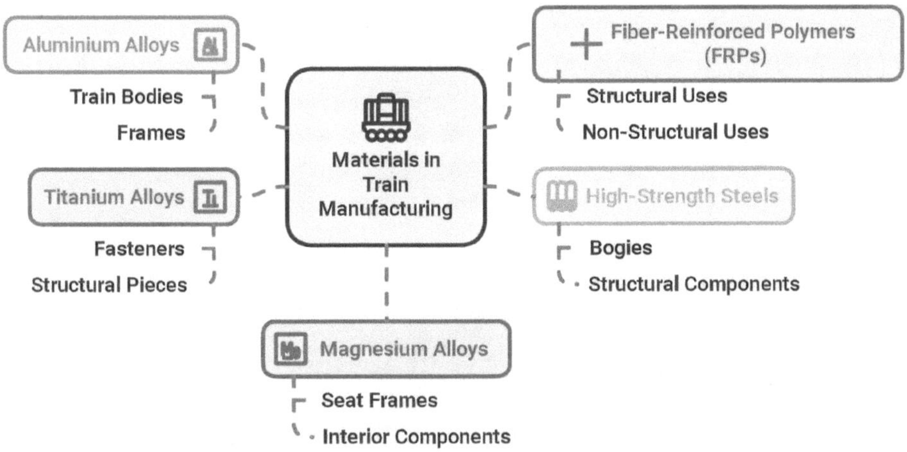

FIGURE 6.3 FRP Composites possess the potential for use in infrastructure, transport systems, renewable energy, and aircraft [38].

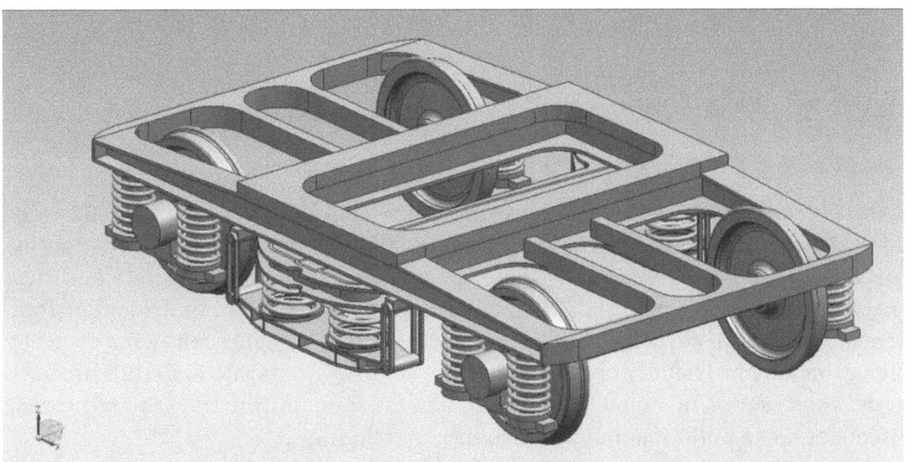

FIGURE 6.4 Construct lightweight and durable bogie frames [3].

Brake discs made of composite materials have less unsprung mass, which means they are more efficient and provide a better ride.

Overhead catenary systems are constructed using light materials that are stress-free on the support structures and are low on maintenance requirements (see Figure 6.5). It is possible to fabricate pantographs, the devices linking trains to the overhead electrical conductors, of lightweight materials so that stress wear is reduced and power transmission is made efficient. The inherent principle of line compaction is minimizing the side bearings movement of conventional suspension string. In this way, the right-of-way dimensions are reduced and line supports may be slender.

FIGURE 6.5 Railway insulator fitting [39].

Infrastructure of railway bridges and other such structures such as the windows, have lightweight, high-strength materials that minimize the construction costs and time while preserving the overall strong structure. Concrete and steel are said to be durable materials that will decrease the need for repairs and maintenance. The overall experience of train passengers can be significantly improved by the design of eco-friendly comfortable train cabins. Composite materials are used in the construction of sound barriers and fences along railway tracks to reduce the effects of noise pollution and improve the surroundings.

While lightweight materials have many advantages, their higher price tag compared to more conventional options may prevent them from being widely used. Composites and other lightweight materials may not be the most durable, impact-resistant, or repairable options, especially when faced with harsh environments. The railway systems will also need to be environmentally friendly, thus ensuring that lightweight materials should be successfully recycled or disposed of after the end of service. Train energy consumption is even further optimized because the heavier the train, the greater the rolling resistance of the train's wheels to the rails. Train structures that are not heavy can have higher acceleration and deceleration rates which improves the system's efficiency and decreases the duration of travel times. It can be seen that a weight-reduction structural design of the train helps to create a more efficient transport system as more areas are left for passengers or freight. Light trains exert less strain on the tracks thereby leading to low track maintenance costs and extended track lifespan.

6.4 EFFECTS ON THE ENVIRONMENT

Lightweight materials enable the conduct of train operations more efficiently, which results in reduced emissions of greenhouse gases. This is especially the case when electrification is applied in conjunction with renewable energy sources. The rail systems can also be made more sustainable through the utilization of recyclable and green materials. Many economic advantages are gained for the railway system throughout its lifetime because of a reduction in energy consumption and also in maintenance requirements. Trains and infrastructure using lightweight materials which are also corrosion-resistant and durable can be replaced after a much longer period. To some degree, peripheral environmental vibration and noise can be reduced by the use of technology. The preservation of biodiversity involves planning railways in a way that does not interfere with existing natural habitats, such as by incorporating wildlife corridors. The efficiency of drainage systems lies in their ability to control runoff and keep waterways clean. Building railways with recycled materials and keeping tracks in good repair to prolong their life are examples of sustainable building practices. Materials with a smaller amount of carbon embodied, including sustainable steel and concrete, are used. Reduce the footprint of railway infrastructure through efficient land use planning, which aims to minimize land use (Figure 6.6).

FIGURE 6.6 Comprehensive guide to life cycle assessment.

6.5 LIGHTWEIGHT MATERIALS AND EFFICIENCY

Aluminium alloys, carbon fibre composites, and high-strength steel are being used more often in railway operations for example. These materials have qualified for improving the general weight of the trains while also contributing to better performance that makes acceleration and cruising faster. With these materials onboard, rail systems turn out to be more non-carbon intensive which respectively contributes to lower production of carbon dioxide and other greenhouse gases emitting. In addition to that, by combining these materials with sunlight, wind, and hydropower, the effects are multiplied. Rail electrification, particularly using these renewable sources, has a much lesser environmental damage impact than that of diesel-powered systems. The greater move toward cleaner energy resources can shift together with the lightweight materials in railways for better configuration in reducing emissions in the entire transport sector. The significant elements of durable rail system materials with respect to the system's structure, weight, cost, and environmental impact have been compared using various materials in a similar table which is shown in Table 6.2. Parts made out of lightweight materials are not only cheaper to operate, but these also result in significantly lower fuel use as well as emissions of greenhouse gases throughout the period of operations of the trains. It is possible, however, that they would be more expensive or scarce. On the other hand, wear and tear performance of standard materials such as steel and concrete is significantly higher; however, this comes at a cost and the cost includes the environmental factors of mining and producing them. The table addresses the dilemma of durability versus cost versus environmental impact in the process of material choice for constructing environmentally friendly railway infrastructure.

6.6 SUSTAINABLE MATERIALS AND RECYCLING

Sustainable rail systems must also take into account the selection of materials. Recently, eco-friendly steel and concrete materials, which can be reutilized and have a lower ecological impact, have emerged as construction trends in railways. This is because these materials help reduce the carbon emissions of the railway sector due to lower energy requirements during production.

The waste diversion from landfills is also a result of the recycling of old tracks and used wagons, which is one of the recycling programmes in the railway sector. The new trains and facilities are expected to be long-lasting, corrosion-proof,

TABLE 6.2
Comparison of Energy Savings and Reduction of Emissions In Lightweight Rail System Materials

Material	Energy Savings (%)	Reduction in CO_2 Emissions (%)
Aluminium alloys	15%	10%
Carbon fibre composites	20%	15%
High-strength steel	10%	7%

TABLE 6.3
Environmental Impact of Sustainable Materials Used in Rail Infrastructure

Material	Carbon Footprint Reduction (%)	Recycling Potential
Sustainable steel	25%	100%
Recycled concrete	30%	80%
Recycled plastic composites	20%	90%

and recyclable as well at the end of their service. This strategy encourages emerging economies where it is always beneficial to reuse, repair, and recycle materials in order to minimize pollution and save resources. Key advantages of lightweight materials used in rail transportation systems are listed in Table 6.3. Because of their lower mass and clouds of energy needed to transfer them, these materials improve energy and greenhouse gas emissions during train operation. Furthermore, the use of lightweight materials improves the overall efficiency of rail operations such as track maintenance, which reduces damage to the rails and lowers maintenance costs. This table gives a comprehensive overview of benefits realized through the integration of lightweight materials in rail transport engineering, for example, direct and indirect carbon credit, lifecycle, and operational cost among others confirming the need for lightweight materials in sustainable railway design.

6.7 VIBRATIONS AND NOISE REDUCTION

Lightweight materials and the use of new technologies contribute to the reduction of energy needs and such engineering issues as vibrations and noise, which are an environmental nuisance, especially in cities, are also assisted. The noise and vibrations caused as a result of the running of railways are a cause of interference to the communities surrounding the railway lines. However, noise, vibrations, and movement of trains and structures can be hushed by the use of efficient suspension technology, sound, and vibration-dampening materials. In recent years, more designs with sound walls and acoustic treatment on the trains and also use of active sound control are in use. Such technologies can lower noise pollution on neighbouring residents to levels where healthy living can be possible and such cities can run on efficient rail systems.

6.8 PRESERVING BIODIVERSITY

The impact of railways on the local environment and natural diversity may be rather serious. To begin with, railway construction tends to be land hungry, and as a consequence, in most instances, destruction of natural ecosystem occurs and even animals can be endangered. Nevertheless, by accommodating animal migration, the design can assist in railway activities without adversely affecting the ecological biota. For example, wildlife corridors such as elevated structures and green tunnels allow animals to traverse the railroad tracks and not be at risk of harm. These steps guarantee that railway construction and operation do not negatively impact the ecological balance of the surrounding area or hamper animal migration processes.

TABLE 6.4
Wildlife-Friendly Design Features in Railway Infrastructure

Feature	Impact on Biodiversity
Wildlife corridors	Enables safe migration of animals across railways
Green tunnels	Reduces habitat fragmentation and preserves ecosystems
Elevated tracks	Allows animals to pass underneath without interference

Table 6.4 summarizes the environmental benefits of using recyclable materials in the construction and operation of rail systems. It juxtaposes the ensuing effect of materials such as steel, recycled concrete and plastics against materials that are not recyclable. Recyclable materials significantly reduce wastage and preservation of raw materials by making it possible to use the materials in any other intended construction or operational processes in the future. The table further observes that the application of recyclables reduces the environmental impact of railway infrastructure. This can lead to a reduction in the carbon embodied in construction materials which is an important aspect in attaining sustainable railway systems.

6.9 DRAINAGE SYSTEMS AND WATERWAY PROTECTION

They are crucial for protecting the surrounding environment. Efficient systems that prevent the pollution of water resources, as well as runoff control, must be incorporated when building railways. Runoff waters from railways can fetch domestic waste pollutants such as oil, metals, and chemicals among other dangerous wastes to the water bodies in vicinity. Such pollution response measures as effective drainage features that incorporate sediment and filtration systems are key in protecting the local water quality. Table 6.5 presents an overview of the four stages of lifecycle management (LCM) for rail systems, as defined in the life cycle assessment (LCA) framework. The stages include (i) defining objectives and scope, (ii) conducting a life cycle inventory analysis (LCI), (iii) evaluating the environmental impacts (LCIA), and (iv) drawing conclusions. Each stage is vital for understanding and minimizing the long-term environmental consequences of rail systems, from construction to decommissioning. The table highlights the importance of integrating sustainability at each stage of the rail system's lifecycle, ensuring that environmental impacts are considered at every phase, and demonstrating how LCM can lead to more sustainable rail infrastructure.

TABLE 6.5
Comparison of Drainage Systems for Railway Infrastructure

Drainage System	Effectiveness in Controlling Runoff	Waterway Protection
Traditional drainage systems	Moderate	Low
Eco-friendly drainage systems	High	High

6.10 LIFECYCLE MANAGEMENT

Lifecycle management is required for the furnished outcome of the railway systems, ensuring that the railways are operated and maintained in an environmentally sustainable way for the future. In this, attention is placed on the strength of construction materials, simplicity of maintenance, and final disassembly and recycling of elements after the end of service life.

6.10.1 Durability and Maintenance

The aspect of the durability of railway infrastructure is of the essence in enhancing the life cycle of the infrastructure while minimizing the adverse effects that are caused to the environment in the process. Employing long-lasting elements in making the trains and tracks minimizes the use of replacements and repairs and damage done. This will not only save resources but also limit the generation of wastes. Maintenance is also an important factor in the management of rail systems, as proper maintenance can aid in energy efficiency and decrease the amount of pollution produced by the systems. Managed assets that manage and record information concerning the state of the railway parts are of utmost importance since they assist in suspicious early signs of trouble and prevent expensive breakdowns, therefore extending the life of the system. Table 6.5 assesses the impact of railway infrastructure on local ecosystems, examining potential disruptions to wildlife habitats, biodiversity, and natural landscapes. These impacts are addressed through the provision of measures such as planting of wildlife corridors, retention of existing natural covers, and applying ecological design that protects critical areas. There are guidelines in Table 6.6 that indicate how railways can be built without significantly affecting local flora and fauna, but remain efficient in meeting transport requirements. This assessment helps underline the importance of environmental planning and impact mitigation in the design and construction of sustainable rail systems.

6.10.2 Recycling and Disposal

When a component of a railway becomes functional, it is vital to include in its design considerations for its eventual demolition and reprocessing. As a way of reducing wastes and the consumption of new raw materials, the metals and plastics and other materials used in highways and railways should be returned for reuse. Not only does this save materials, but it also reduces the number of resources that would have been used in manufacturing and transport activities.

TABLE 6.6
Impact of Maintenance on Railway Sustainability

Maintenance Activity	Impact on System Longevity	Environmental Benefit
Regular track inspections	Increases lifespan by 30%	Reduces waste and energy usage
Preventive maintenance of trains	Increases lifespan by 25%	Decreases the carbon footprint

6.10.3 Social and Economic Sustainability

The concept of sustainability is not limited only to the environment but also covers the socio-economic aspect of the environment. A fundamental aspect of achieving an equitable transport system is ensuring that the rail systems are available to every stratum of society, including physically challenged individuals. In addition, the construction and upkeep of the railway would also generate jobs within the area and thus boost the economies of the local communities.

6.11 TECHNOLOGIES AND INNOVATIONS

Apart from the implementation of reliability, availability, maintainability, and safety, which in its context means safety management, energy consumption, and mobility costs, the advancement of various technologies is crucial in rendering rail systems more efficient and environmentally friendly. Energy-saving features are being installed in different parts of the railway industry as well as developing them to ensure safety and improve the quality of the passengers' journey.

6.11.1 Smart Rail Systems

According to the definition presented earlier in this paper a "smart rail system" involves an integration of the rail within the transport system so as to enhance its operations with IoT, Big Data, and AI. Triage train operations and predictive maintenance via real-time control and monitoring systems which view the train conditions and schedules. This assures that trains are not going excessively fast or travelling faster than required and do not stop whenever it is not necessary to so do. Table 6.6 provides information on different available methods useful in vibration and noise reduction from rail transport, which impacts the communities residing close to the rail network. These technologies involve the use of engineering designs such as noise barriers, track vibration dampers, and adjusted train wheelsets aimed at reducing the transmission of vibrations. Table 6.7 also presents graphical records of the effectiveness of each technology in the fight against noise pollution and vibration particularly, which not only enhances the living standards of people surrounding the areas

TABLE 6.7
Benefits of Smart Rail Systems

Technology	Benefit	Environmental Impact
IoT-based monitoring	Optimizes train performance and energy usage	Reduces energy consumption
AI for predictive maintenance	Predicts failures and reduces downtime	Minimizes waste and repairs
Automated train control	Increases train efficiency and reduces fuel use	Decreases emissions and energy use

but also saves the rail system from unnecessary expenditures of maintenance and repair of the tracks or any other worn out parts. This table highlights how technological innovations can address both environmental and social sustainability in rail transport.

6.11.2 Autonomous and Semi-Autonomous Trains

Another area that improves the sustainability of the railway industry is the emergence of automated and partly automated locomotives. These trains do not require much human interaction in their operations thus human error is characteristic of such trains. The trains have onboard sensors and cameras that combine with artificial intelligence to take charge of all essential operations during manoeuvring thereby enhancing safety, ensuring smooth operations and reducing fuel consumption.

6.11.3 Hybrid and Alternative Energy Trains

Trains that utilize renewable sources of energy other than fossil fuels such as hydrogen cells and biofuels are under development to reduce if not eradicate dependence on fossil fuels. For instance, hydrogen-powered trains have as byproduct water vapour as their byproduct and this makes them a great sustainable energy. Another alternative, biofuels also have a cleaner production key to ensuring a more sustainable alternative to fossil fuels and can reduce the carbon emissions associated with rail transport. The power sources employed in railways such as hydrogen fuel cells, biofuels, renewable electricity used in electric rail systems, and hybrid diesel-electric systems have been examined by comparing their environmental opportunities and problems in Table 6.8. Whereas hydrogen fuel cells and biofuels achieve considerable decrease in carbon emissions as compared to diesel, electric rail systems which rely on renewable energy sources are nearly carbon emission-free. The table, however, also targets the problems associated with the increase in the size of these energy sources such as – excessive supply costs, the development of infrastructural facilities, limited availability, etc. The comparison makes it clear that alternative energy sources are vital for the reduction of carbon emissions in the operation of

TABLE 6.8
Comparison of Energy Sources for Sustainable Rail Transport

Energy Source	Environmental Benefit	Challenges
Hydrogen fuel cells	Zero emissions (water vapour only)	High production cost, infrastructure development
Biofuels	Renewable, lower carbon emissions compared to diesel	Limited supply and competition with food production
Electric rail (renewable)	Low emissions when powered by renewable sources (solar, wind, hydro)	Requires electrification of the rail network, high initial infrastructure cost
Diesel-electric hybrid	Reduces fuel consumption and emissions	Still reliant on fossil fuels, though in smaller quantities

rail transport. However, effective and robust infrastructure planning and investment seem to be a prerequisite for overcoming the challenges of mass deployment of the alternatives.

6.11.4 Energy Efficiency in Railways

Railways are already among the most energy-efficient modes of transportation, especially for heavy goods and long-distance travel. Nevertheless, there are still some innovations in the works that seek to improve this efficiency. For instance, technologies like regenerative braking systems which capture and recapture kinetic energy, and aerodynamic train configurations that minimize the resistance caused by the airflow during train movement are putting even more cuts on energy use. Such systems enable the trains to regain and conserve energy that would have been wasted, hence cutting down drastically both fuel and operational costs.

Regenerative brakes, in general, represent a quantum leap in low energy use in train operations. Heat is the major loss during the braking of the train using standard brakes. In contrast, the regenerative braking systems use this energy to charge the batteries of the train or send it back to the grid thereby reducing load requirements on power generation. Table 6.9 assesses more energy efficiency technologies in railway systems and their effect on energy use and environmental performance. It includes regenerative braking systems which save up to 30% of energy consumption, aerodynamic forms that reduce drag forces, and energy-efficient lighting of trains and train stations. The table also proves that aside from energy savings, these technologies assist in the reduction of greenhouse gas emissions and decrease the cost of operations. For instance, the use of regenerative braking aids in capturing and re-using energy which further reduces power generation requirements. This table explains that it helps to improve the energy efficiency of technology in use in order to further reduce the carbon emissions levels of the rail systems and improve operational efficiency.

The environmental sustainability of rail systems is a critical consideration in today's transportation landscape, particularly in the context of climate change and growing urban populations. This addresses the environmental challenge as well through the use of lightweight materials, sustainable construction methods, and the increasing application of modern technologies. These advancements will allow the

TABLE 6.9
Impact of Energy Efficiency Technologies on Railways

Technology	Impact on Energy Consumption	Environmental Benefit
Regenerative braking	Reduces energy consumption by up to 30%	Decreases fuel use and emissions
Aerodynamic train designs	Reduces air resistance, lowering fuel consumption	Decreases CO_2 emissions and energy use
Energy-efficient lighting	Reduces power consumption for station and train lighting	Reduces overall energy demand

railway industry to deliver efficient and cost-effective transportation services and at the same reduce the effects on the current environment. Furthermore, through the application of lifecycle management concepts, the impact that railway systems have on the environment during their usage can be minimal and through planning, pollution, operation, and even disposal can be accounted for. Additionally, the implementation of smart technologies, such as autonomous trains, hybrid propulsion systems, and regenerative braking systems can further improve the sustainability of rail operations.

Transportation systems, especially rail can undertake several changes such as the introduction of greener designs and improvements of implemented drainage systems, as well as the use of recyclable and environmentally friendly materials which all will help in reducing the carbon footprints. As well engaging the target communities around rail systems is very important, to ensure that models adapted do not harm the existing ecosystems, which in turn will promote the establishment of sustainable transport models. In promoting renewable energy no other systems will prove vital over the coming years than the railways. There is great potential and hope for the future of rail transport systems that integrate new technological changes as these will promote sustainable development of the rail sector.

6.12 SUMMARY

The field of railway engineering is being revolutionized by FRPs which have the unique combination of being lightweight, strong, and durable. Their usage as modern train carriages and bogies or as critical structural components is an indication that there is a revolution in the building and maintenance of railway systems. Not only do these materials enhance energy efficiency and reduce maintenance cost in the long run, but they also help to prolong the life of rail components and enable the safety and comfort of passengers. Railways are in the process of transitioning into a new age with strengthened and tougher, FRPs, which are essential for energy efficiency and stability. As well as being able to take onboard aggressive FRPs, the transition is going to shape the future of the railway industry in a new way.

REFERENCES

1. Machado, C. G., Winroth, M. P., & Ribeiro da Silva, E. H. D. (2020). Sustainable manufacturing in Industry 4.0: An emerging research agenda. *International Journal of Production Research*, *58*(5), 1462–1484.
2. Ma, R., & Kishawy, H. A. (2012). Sustainable manufacturing and design concepts, practices and needs. *Sustainability*, *4*(1), 154–174.
3. Sharma, R. C., Palli, S., & Koona, R. (2017). Stress and vibrational analysis of an Indian railway RCF bogie. *International Journal of Vehicle Structures and Systems*, *9*(5), 296–302.
4. Palli, S., Koona, R., Sharma, R. C., & Muddada, V. (2015). Dynamic analysis of Indian railway integral coach factory bogie. *International Journal of Vehicle Structures and Systems*, *7*(1), 16–20.
5. Giunta, M. (2023). Trends and challenges in railway sustainability: The state of the art regarding measures, strategies, and assessment tools. *Sustainability*, *15*(24), 16632.

6. Baker, A. C., Bernstorf, R. A., del Bello, E., Hill, R. J., King, B., Phillips, A. J., Powell, D. G., Shaffner, D., Stewart, G. A., & Grisham, T. (2008). IEEE guide for braced insulator assemblies for overhead transmission lines 60 kV and greater. *IEEE Transactions on Power Delivery*, *23*(2), 785–791.
7. Ceraolo, M., Lutzemberger, G., Meli, E., Pugi, L., Rindi, A., & Pancari, G. (2018). Energy storage systems to exploit regenerative braking in DC railway systems: Different approaches to improve efficiency of modern high-speed trains. *Journal of Energy Storage*, *16*, 269–279.
8. Adeoye, A. O., Quadri, R. O., & Lawal, O. S. (2022). Assessment of biofuel potential of tenera palm kernel shell via fixed bed pyrolysis and thermal characterization. *Results in Surfaces and Interfaces*, *9*, 100091.
9. Adeoye, A. O., Quadri, R. O., & Lawal, O. S. (2023). Biomass waste pyrolysis in the context of Nigeria's oil and gas industry: An overview. *International Journal of Ambient Energy*, *44*, 2525–2541.
10. Adeoye, A. O., Quadri, R. O., Lawal, O. S., & Emojevu, E. O. (2024). Physicochemical characterization, valorization of lignocellulosic waste (Kola nut seed shell) via pyrolysis, and ultrasonication of its crude bio-oil for biofuel production. *Cleaner Waste Systems*, *7*, 100138.
11. Adeoye, A. O., Quadri, R. O., Lawal, O. S., Malomo, D., Emojevu, E. O., Omonije, O. O., Odeniyi, O. K., Fadahunsi, M. O., Yelwa, M. J., Aasa, S. A., Onakpa, A. E., Omoniyi, B. P., Mark, I. N., Usman, J., Ismaila, A. M., & Saidu, A. U. (2024). Advanced techniques in upgrading crude bio-oil to biofuel. In *Energy, Environment, and Sustainability*, edited by Sharma, S. K., Upadhyay, R. K., & Kumar, V., pp. 321–353. Springer, Singapore.
12. Adeoye, A. O., Yelwa, J. M., Imam, N., Quadri, R. O., Lawal, O. S., Malomo, D., Aasa, S. A., Onakpa, A. E., Dorgbaa, R. K., Omoniyi, B. P., Ahmad, N., & Garba, H. A. (2024). Pyrolysis of plastic wastes towards achieving a circular economy: An advanced chemistry and technical approach. In *Energy, Environment, and Sustainability*, edited by Sharma, S. K., Upadhyay, R. K., & Kumar, V., pp. 215–259. Springer, Singapore.
13. Adeoye, A. O., Quadri, R. O., Lawal, O. S., Malomo, D., Aliyu, M. T., Dang, G. E., Emojevu, E. O., Maikato, M. J., Yahaya, M. G., Omonije, O. O., Edidem, V. G., Abubakar, Y. K., Offor, O. F., Sochima, E. H., Peter, B. E., & Hikon, B. N. (2023). Sustainable energy via thermochemical and biochemical conversion of biomass wastes for biofuel production. In *Energy, Environment, and Sustainability*, edited by Sharma, S. K., Upadhyay, R. K., & Kumar, V., pp. 245–306. Springer, Singapore.
14. Adeoye, A. O., Quadri, R. O., & Lawal, O. S. (2023). Wet synthesis, characterization of goethite nanoparticles and its application in catalytic pyrolysis of palm kernel shell in TGA. *Results in Surfaces and Interfaces*, *11*, 100118.
15. Avesh, M., Imam, N., & Adeoye, A. O. (2024). Assessing environmental impacts and mitigation strategies for the DFCCIL rail projects. In *Energy, Environment, and Sustainability*, edited by Sharma, S. K., Upadhyay, R. K., & Kumar, V., pp. 333–353. Springer, Singapore, 2024.
16. Momodu, D. U., Adetuyi, A. O., Adeoye, A. O., Akinlolu, O. S., & Apara, E. O. (2023). Surface functionalization of linen fiber and its dye-fastness assessment using (4-[(E)-(4-amino-2-hydroxyphenyl) diazenyl] benzene-1-sulphonic acid and 4-[(E)-(4-{(Z)-[4-(dimethylamino)phenyl] diazenyl}-2-hydroxyphenyl) diazenyl]benzenesulfonic acid). *Suranaree Journal of Science and Technology*, *30*, 030118.
17. Quadri, R. O., Adeoye, A. O., Ayanda, O. S., & Lawal, O. S. (2024). Conversion of palm kernel shell to sustainable energy and the effect of wet synthesized nanoparticles of iron on its thermal degradation kinetics. *Bioresource Technology Reports*, *27*, 101933.

18. Ribeiro, F. B., Nascimento, F. A. C. D., & Silva, M. A. V. D. (2022). Environmental performance analysis of railway infrastructure using life cycle assessment: Selecting pavement projects based on global warming potential impacts. *Journal of Cleaner Production*, *365*, 132558.
19. Bhardawaj, S., Sharma, R. C., & Sharma, S. K. (2019) A survey of railway track modelling. *International Journal of Vehicle Structures and Systems, 11* (5), 508–518.
20. Goswami, B., Rathi, A., Sayeed, S., Das, P., Sharma, R.C., & Sharma, S.K. (2019). Optimization design for aerodynamic elements of indian locomotive of passenger Train. In *Advances in Engineering Design. Lecture Notes in Mechanical Engineering*, edited by Prasad, A., Gupta, S., & Tyagi, R., Springer, Singapore. https://doi.org/10.1007/978-981-13-6469-3_61
21. Sharma, R. C., Palli, S., & Sharma, S. K. (2023). Ride analysis of railway vehicle considering rigidity and flexibility of the carbody. *Journal of the Chinese Institute of Engineers*, *46*(4), 355–366. https://doi.org/10.1080/02533839.2023.2194918
22. Raju Rallabandi, S., Pilla, D. P., Dowluru, S., Palli, S., Sharma, N., Sharma, S. K., & Sharma, R. C. (2024). Critical evaluation of epoxy-hemp-pineapple-palm fiber composites using hybrid AHM-TOPSIS technique for sustainable structural applications. *Journal of the Chinese Institute of Engineers*, *47*(3), 325–336. https://doi.org/10.1080/02533839.2024.2308250
23. Sharma, S. K., Sharma, R. C., Upadhyay, R. K., & Lee, J. (2024). Enhanced ship engine vibration reduction using magnetorheological dampers and Adaptive Neuro-Fuzzy Control System. *Ships and Offshore Structures*, *20*(2), 143–159. https://doi.org/10.1080/17445302.2024.2335449
24. Choi, S., Lee, J., Sharma, S. K. (2021). A study on the performance evaluation of hydraulic tank injectors. In *Advances in Engineering Design. Lecture Notes in Mechanical Engineering*, edited by Joshi, P., Gupta, S. S., Shukla, A. K., & Gautam, S. S., Springer, Singapore. https://doi.org/10.1007/978-981-33-4684-0_19
25. Sharma, S. K., Lee, J., & Jang, H.-L. (2022). Mathematical modeling and simulation of suspended equipment impact on car body modes. *Machines*, *10*(3), 192. https://doi.org/10.3390/machines10030192
26. Vishwakarma, P. N., Mishra, P., Sharma, S. K., Shukla, A. K. (2023). Formulating and analysing the effect of suspension parameters on critical speeds for various DOFs in rail vehicle. In Advances in Engineering Design. FLAME 2022. Lecture Notes in Mechanical Engineering, Sharma, R., Kannojiya, R., Garg, N., & Gautam, S. S., Springer, Singapore. https://doi.org/10.1007/978-981-99-3033-3_66
27. Benmokrane, B., Xu, H., & Nishizaki, I. (1997). Aramid and carbon fibre-reinforced plastic prestressed ground anchors and their field applications. *Canadian Journal of Civil Engineering*, *24*(6), 968–985.
28. Shakir Abbood, I., Odaa, S. A., Hasan, K. F., & Jasim, M. A. (2021). Properties evaluation of fiber reinforced polymers and their constituent materials used in structures – A review. *Materials Today: Proceedings*, *43*, 1003–1008.
29. Raju, S., Srinivas, L., Gunji, S. R., Srinag, T., Sekhar, M. C., & Pandi, T. (2022). Estimation of effect of cold forging deformational behavior on Al-2024 alloy reinforced with fly-ash particulates. *Advances in Science and Technology*, *120*, 3–14.
30. Rallabandi, S. R., Adapa, S. K., Jagadish, C. J. R., & Yanda, S. (2022). Evaluation of thermo-mechanical behavior of hemp fiber polymer composites. *Composites: Mechanics, Computations, Applications: An International Journal*, *13*(3), 113–132.
31. Sankara Raju, R. S., Panigrahi, M. K., Ganguly, R. I., & Srinivasa Rao, G. (2019). Tribological behaviour of Al-1100-coconut shell ash (CSA) composite at elevated temperature. *Tribology International*, *129*, 55–66.

32. Rallabandi, S. R., Devi Prasad, P., Dowluru, S., Palli, S., Sharma, N., Sharma, S. K., Sharma, R. C., & Sharma, R. C. (2024). Critical evaluation of epoxy-hemp-pineapple-palm fiber composites using hybrid AHM-TOPSIS technique for sustainable structural applications. *Journal of the Chinese Institute of Engineers, 47*(3), 325–336.
33. Rallabandi, S. R., Srinivas, L., Palli, S., Sharma, R. C., Sharma, N., Sharma, A., Sharma, S. K. (2024). A contrastive characterization of pure Mg and AZ91D alloy based on the testing of mechanical, corrosion, wear, and erosion properties. *Engineering Research Express, 6*(1), 015017.
34. Askar, M. K., Hassan, A. F., & Al-Kamaki, Y. S. S. (2022). Flexural and shear strengthening of reinforced concrete beams using FRP composites: A state of the art. *Case Studies in Construction Materials, 17*, e01189.
35. Balamurugan, K., Uthayakumar, M., Ramakrishna, M., & Pillai, U. T. S. (2020). Air jet erosion studies on Mg/SiC composite. *Silicon, 12*(2), 413–423.
36. Ray, P. K., Ganguly, R. I., & Panda, A. K. (2002). Influence of heat treatment parameters on structure and mechanical properties of an HSLA-100 steel. *Steel Research, 73*(8), 347–355.
37. Chauhan, S. R., & Dass, K. (2013). Dry sliding wear behaviour of titanium (Grade 5) alloy by using response surface methodology. *Advances in Tribology, 2013*, 1–9.
38. Chen, J., Yu, Z., & Jin, H. (2022). Nondestructive testing and evaluation techniques of defects in fiber-reinforced polymer composites: A review. *Frontiers in Materials, 9*, 986645.
39. Yangzhou Ze Jia International Trading Co., Ltd. *Cross arm end fitting for composite railway insulator. Zejiapower.* Retrieved July 22, 2025, from https://www.zejiapower.com/showroom/end-fitting-for-railway-insulator-cross-arm.html

7 Adaptive Sliding Mode Controller Development to Improve Rail Network Efficiency through Virtual Coupling

A Case Study

Saikat Dutta

7.1 INTRODUCTION

The concept of virtual coupling has been attracting research interest lately as a next-generation train control operation. In this system, the train control is done entirely by cab-based signalling elements and it relies on the vehicle-to-vehicle communications. Various case study scenarios and roadmaps on how virtual coupling can be introduced into the network are studied [1, 2]. The current focus of the Railways is on enhancing capacity through the construction of high-speed routes, light rail networks, and the integration of additional trains into the network, aiming to accommodate a projected 15% rise in passenger numbers by March 2024 in the UK [3]. One proposed solution involves reevaluating train control systems and virtually coupling trains to increase the number of trains operating within the existing network. This new control system promises to enhance the rail network by enabling virtual coupling and decoupling of trains during operation. Similarly, it aims to optimize road networks by facilitating train platooning in the same direction, thereby improving overall network efficiency. However, this alternative approach brings forth new challenges related to controller development and train operation systems.

7.2 STATE OF ART: AUTOMATIC TRAIN CONTROL

To ensure system safety, rail networks employ various standards of Automatic Train Control systems, predominantly utilizing fixed-block signalling across most operational areas worldwide. A newer development, the moving block system, is also deployed in limited operational environments. Fixed-block signalling divides tracks into fixed-length blocks, establishing safe gaps between trains. Only one train is

DOI: 10.1201/9781003659983-7

permitted to occupy a block at any given time. The fixed length of these blocks restricts flexibility within the network, aiming primarily to enhance reliability and safety while improving network speed.

In the European rail network, the European Train Control System (ETCS) is employed, utilizing fixed-block signalling in both Level 1 and Level 2 systems. Level 2 ETCS introduces cab-based signalling, where train position is communicated via cab-based radio, while movement authority is generated by interlocking systems. Despite these advancements, train positioning continues to rely on trackside balises and adhere to the fixed-block system.

The limitation of not allowing a second train into a section of track defined by the fixed blocks installed, until that is free, restricts the capacity improvement of the network. In the moving block signalling system (shown in Figure 7.1), the track is considered as a continuous line, and it allows trains to run at a safe distance to each other rather than restricting them by the predefined blocks. The safe distance between two trains depends on the speed of the trains, which allows to reduce the gap in low speed. Moving block is in operation in a few urban railways including the Victoria Line, Jubilee line of London Underground, Hong Kong MTR, New York City Subway's BMT Canarsie Line.

ETCS level 3 system, which is in the development stage, operates with fully radio-based communication where the trains transmit information with radio block centre (RBC) without any trackside equipment. Thus, ETCS Level 3 operates in the moving block principle by considering the track as a continuous element rather than dividing it into smaller blocks. The balises are still used to report the position of the train to maintain train integrity. This shift towards a moving block principle in ETCS Level 3, combined with continuous track monitoring, opens the door for advanced

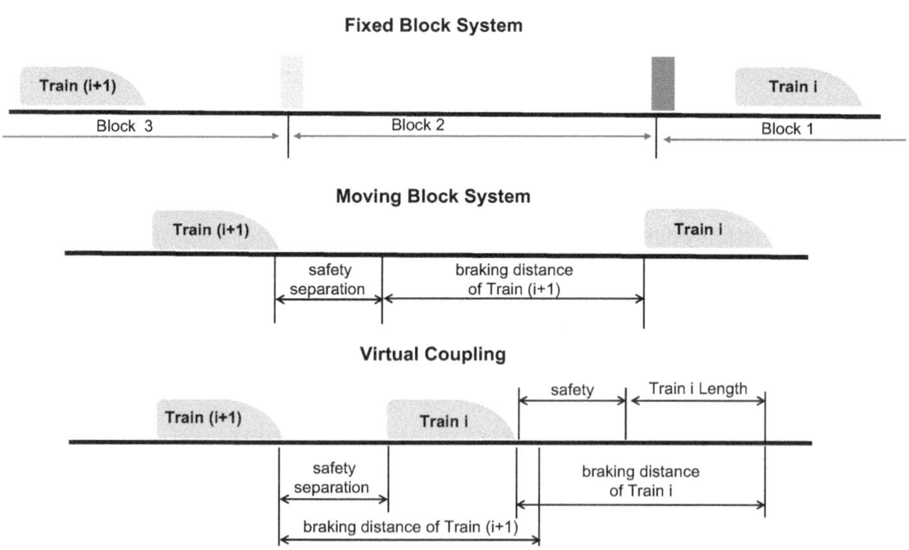

FIGURE 7.1 Moving block signalling and virtual coupling system.

technologies like virtual coupling, which can significantly enhance track capacity, improve operational efficiency, and ensure better safety by allowing trains to operate more closely and dynamically.

7.3 THE CONCEPT: VIRTUAL COUPLING

Virtual coupling was inspired by the car-following model initially proposed for road transport, where multiple vehicles coordinate in a platoon relying on vehicle-to-vehicle (V2V) communications. Following this concept, autonomous rail vehicles can be run to operate in close proximity within a platoon. A critical aspect of such platoons, which are connected (or coupled) virtually is the need for a controller to coordinate the movements of individual vehicles or trains to prevent collisions.

The concept of coupling a set of trains while running and decoupling near the switches/ or stations as required is a potential way to increase the capacity of the network. The virtually coupled trains run as a single coupled train with multiple equally spaced smaller trains. The much smaller gap between two smaller trains ensures more trains on a length of track compared to fixed-block or moving block systems (see Figure 7.1).

In the moving block signalling system, the gap between trains is determined based on the distance it takes for the trains to come to a stop in an emergency. However, this gap could be reduced further if it is based on real-time information about what the train in front is doing and where it will stop if it hits the brakes. In the concept of virtual coupling, two trains communicate information about their changing speed and brake activity so that they can decrease or increase the gap accordingly between them to the minimum necessary.

From a control perspective, studies [3, 4] have introduced cooperative controls into moving block models to enhance the safety performance of multiple train operations. These studies incorporate similar controller components to virtual coupling but still rely on absolute braking distance. Thus, studying different controllers and developing an adaptive controller is one of the needs for virtually coupled trains.

7.4 ADVANTAGES AND LIMITATIONS OF VIRTUAL COUPLING

Virtual coupling, as previously mentioned, is a control strategy that mimics a mechanical link between systems that are physically separated. In train control, this approach creates the effect of trains being coupled together, even though they are not physically connected. This is achieved through control algorithms that synchronize the movements of the trains, ensuring their safe and efficient coordination during operation. To better understand the difference, advantages and limitations of virtual coupling over moving block signalling has been discussed here.

The key difference between virtual coupling and moving block signalling is in their approach to train control. Moving block signalling depends on predefined track segments and strict block occupancy rules. In contrast, virtual coupling employs advanced algorithms to enable a more flexible and efficient operation, allowing trains to coordinate more dynamically.

Moving block signalling offers a concept that is relatively easy to implement while providing a designated level of safety. Virtual coupling, on the other hand, significantly enhances operational efficiency by allowing for higher train densities and shorter headways, thereby increasing capacity. Virtual coupling offers greater flexibility, making it adaptable to various operational scenarios such as dynamic speed limits or priority assignments. This system can also reduce energy consumption by optimizing train movements, leading to more efficient operations.

Virtual coupling, though highly efficient, presents certain challenges. It requires complex control algorithms and advanced infrastructure, making implementation more difficult. The initial and ongoing costs of deploying virtual coupling systems are generally higher than those of traditional systems. Additionally, reliance on modern technology and communication systems introduces potential reliability concerns, as failures could disrupt operations. However, with ongoing advancements in mechatronic switches, mechatronically guided vehicles over switches, and enhanced 5G communication in railways, these challenges are expected to be overcome in the near future.

7.5 RECENT TRENDS IN VIRTUAL COUPLING

In recent years, considerable research has focused on advancing virtual coupling control systems for trains. A proof of concept for Virtual Coupling is presented within the European Rail Traffic Management System (ERTMS/ETCS) framework, along with a coupling control algorithm that accounts for time-varying communication delays, supported by numerical analysis from a case study simulation [5]. Early research demonstrated this concept forms the basis for modern virtual coupling control systems. In the context of Japan's high-speed Shinkansen network, solutions have been proposed to facilitate coupling and uncoupling of trains during operation, improving the line capacity of the network [6]. The effectiveness of this approach was verified using DFSimu, a simulation tool developed by the German Aerospace Center (DLR) [7].

Further feasibility analyses have also identified longitudinal control as a critical element of virtual coupling. Flammini et al. [8] proposed decentralized control strategies based on simplified third-order models of train dynamics in state space. These strategies rely on local actions determined from on-board measurements and information from adjacent trains and infrastructure, while also considering time-varying communication delays and errors.

Model predictive control (MPC) has been widely used in virtual coupling simulations [9–11]. In particular, [9] explored the use of MPC to manage virtually coupled trains, achieving a significant reduction in tracking distances. However, determining the optimal tracking distance to maximize line capacity while ensuring safety remains an open challenge. Other approaches including constraint-force-driven controllers based on the Udwadia-Kalaba method [12], PID controllers [13–16] have also been explored. All these methods have been proposed and shown to reduce communication delays, further enhancing system performance.

However, very few studies have focused on developing sliding mode controllers specifically for virtually coupled trains. Park et al. [17] introduced a gap controller based on sliding mode control for a nonlinear train model with uncertainties, alongside a gap reference generation scheme to manage train merging and separation while adhering to acceleration and jerk constraints. This study addresses position and velocity measurement errors due to wheel diameter inaccuracies by proposing an error correction scheme, both of which are validated through simulations. However, it did not address longer couplings or the impact of parameter uncertainties, which may necessitate the use of adaptive controllers. This study demonstrates the potential of using a sliding mode controller to manage the different modes of virtual coupling: coupling, running, and decoupling.

Adaptive sliding mode control (ASMC) has shown significant potential in truck platooning by effectively handling uncertainties, disturbances, and varying system parameters that are common in such dynamic environments [18, 19]. ASMC can adaptively adjust the control law in response to changes in truck dynamics (e.g., varying loads or road conditions), ensuring robust performance and stability across the platoon. Its ability to mitigate chattering, a common issue in traditional sliding mode control, while maintaining strong disturbance rejection and trajectory tracking, makes it particularly suitable for ensuring smooth and coordinated movement in truck platoons. Kwon and Chwa [20] highlights the importance of the bidirectional strategy in platoon control by demonstrating that, despite its previous drawbacks compared to the leader-predecessor strategy, it offers a more feasible implementation in actual systems, with their proposed adaptive bidirectional control method showing improved performance and guaranteed string stability. Similar work has been demonstrated by [21], where a platoon operates using a bidirectional controller that incorporates information from both preceding and following vehicles.

Considering the robustness to uncertainty and disturbances, and fast response, adaptability, ASMC is chosen as a more preferred solution than MPC as train controller. An ASMC (SMC) is designed to be robust against uncertainties in the system dynamics and external disturbances; whereas MPC is reliant on accurate model parameters. MPC, while effective for systems with accurate models, can be less robust to uncertainties. In addition, In Adaptive Sliding Model Control systems the system's trajectories to converge to a sliding surface, ensuring rapid and stable responses. This is essential for virtual coupling applications where quick and accurate tracking of desired coupling behaviour is required. Whereas MPC's optimization-based approach is slower in response due to the process involved. Moreover, MPC requires more complex computations, involving optimization and prediction horizons, which makes them computationally heavy. Thus, adaptive SMC offers a good combination of robustness, fast response, adaptability, and simplicity, making it a well-suited choice for virtual coupling application.

Given the challenges associated with very short following distances, high speeds, and heavy train masses, controller accuracy and robustness are critical. Simplifications in controller development mean that railway operators demand rigorous validation before real-world implementation. Thus, the need for adaptive bidirectional control arises to explore improved stability and performance in platoon systems.

This chapter focuses on the development of an adaptive controller for virtual coupling systems, with the aim of maintaining string stability and ensuring safe train spacing. By minimizing both track usage and spacing errors, this controller optimizes operational efficiency. One of the key features of this adaptive controller is its ability to dynamically adjust to changing conditions, such as fluctuations in train speeds and network constraints, which enhances its suitability for real-world implementation. The primary innovation of this study lies in the development of an ASMC. Unlike existing robust controllers, which tend to be computationally demanding, this controller offers a more efficient solution that adapts to real-time changes in the railway environment. By dynamically adjusting to varying operational conditions such as train speed variations and network constraints, it significantly enhances the flexibility and practicality of virtual coupling in high-density rail systems.

This chapter contributes to the current research landscape by advancing the practical implementation of virtual coupling technology. By enabling safe and efficient high-density train operations, this study offers a valuable solution for optimizing rail networks and adds to the ongoing research efforts in mechatronic systems and adaptive control within railway engineering.

7.6 MATHEMATICAL MODELLING

7.6.1 Train Longitudinal Model

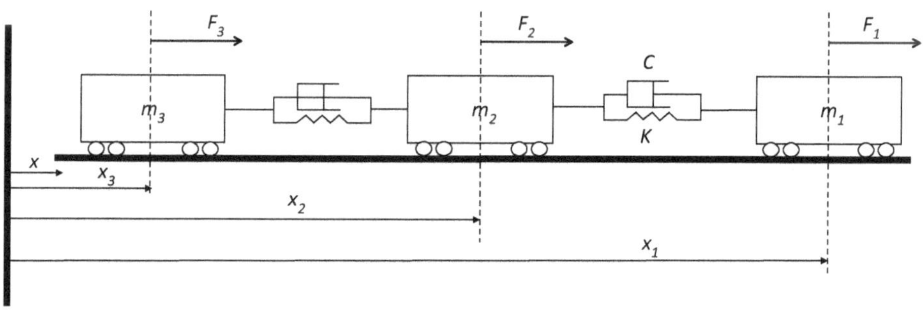

FIGURE 7.2 Schematic model of a three-coach train.

7.6.2 Individual Train Model

For the selected case study, we have considered a Class 195 Northern service which comprises three coaches (DMSL-MS- DMS) (see Figure 7.2). Thus, the first individual train governing equations have been developed to study the dynamic behaviours of the train itself. In this model, the two coaches are considered to be joined using a physical coupling, which is modelled as a combination of stiff spring and damper, and the mathematical equation has been derived as

$$m_1\ddot{x}_1 + K(x_1 - x_2) + C(\dot{x}_1 - \dot{x}_2) = \frac{F}{2} - F_{e1}, \tag{7.1}$$

$$m_2\ddot{x}_2 - K(x_1 - x_2) - C(\dot{x}_1 - \dot{x}_2) + K(x_2 - x_3) + C(\dot{x}_2 - \dot{x}_3) = \frac{F}{2} - F_{e2}, \tag{7.2}$$

$$m_3\ddot{x}_3 - K(x_2 - x_3) - C(\dot{x}_2 - \dot{x}_3) = \frac{F}{2} - F_{e3}, \tag{7.3}$$

$$\dot{F} = \frac{-1}{\tau}(F - u), \tag{7.4}$$

where F is the traction/braking force by the engine; it is considered that the force is distributed equally among the two coaches, thus using $(F/2)$ in the individual equations; u is the controlled force of traction/braking available considering the inertial effect of the system, $F_{e,i}$ are the external forces acting on coaches ($i = 1,2,3$).

The parameters are listed in Table 7.1. The following assumptions were made while constructing the equations:

- the traction/braking force is distributed equally between the two coaches.
- is the inertial lag due to internal dynamics of the system.
- The aerodynamic force is acting on the first coach only.

The external force includes the aerodynamic force, force due to gravity in case of slopes, and force due to curving resistance as follows:

$$F_e = F_{\text{aero}} + F_g + F_R, \tag{7.5}$$

where

$$F_{\text{aero},1} = m_i\left(B_1 + B_2 v_1 + B_3 v_1^{\,2}\right).$$

TABLE 7.1
Parameters of Class 195 Coaches of Northern Trains, UK Running on the Case Study Network in Manchester Victoria – Chester/Leeds Routes [22]

Parameters	Description	Value
m_1, m_3	Mass of the DMSL, and DMS car	41×10^3kg
m_2	Mass of the DM car	39.4×10^3kg
L_1, L_2, L_3	Lengths of the DMS, DMSL cars	24.026 m
τ	Physical delay	0.1 s
L_C	Coupling length	0.85 m
B_{max}	Allowable braking	0.83 m/s^2
A_{max}	Allowable acceleration	0.83 m/s^2
K	Coupling Bounce stiffness	80×10^3
C	Coupling Bounce damping coefficient	100×10^3
B_1	Davis equation coefficient	0.01176 N/kg
B_2	Davis equation coefficient	776.16×10^{-6} N-s/m-kg
B_3	Davis equation coefficient	16×10^{-6} N-s^2/m^2-kg
Slope	Track grade	mm/m
R	Radius of the curved track	m

$$F_g = -m_i g \text{ slope},$$

$$F_R = -\frac{6\ m_i}{R}.$$

The parameters used are explained in Table 7.1.

7.6.3 Equivalent Model

In practice, the two coaches almost run as a single unit, considering the stiff physical coupling, and in this chapter, we are not considering any decoupling of the two coaches within the train. Thus, a simpler mathematical equivalent model can be developed considering the equivalent mass of the two coaches as a single mass, and the following governing equations are constructed.

$$\ddot{x}_e = u - F_{Le}/m_e, \tag{7.6}$$

where m_e is the equivalent mass of the vehicle, the equivalent length of the vehicle is considered as:

$$L_e = L_1 + L_2 + L_3 + 2L_C. \tag{7.7}$$

To compare the performance of the equivalent vehicle against the muti-carriage vehicle model, both were tested against two different traction force conditions, keeping the gravity force, and curvature force as zero.

Figure 7.3 shows that the performance of the equivalent model mimics the multibody model well, with maximum position error of 4 mm for a total displacement of 1.36 km, which is almost negligible. The error in the velocity performance is

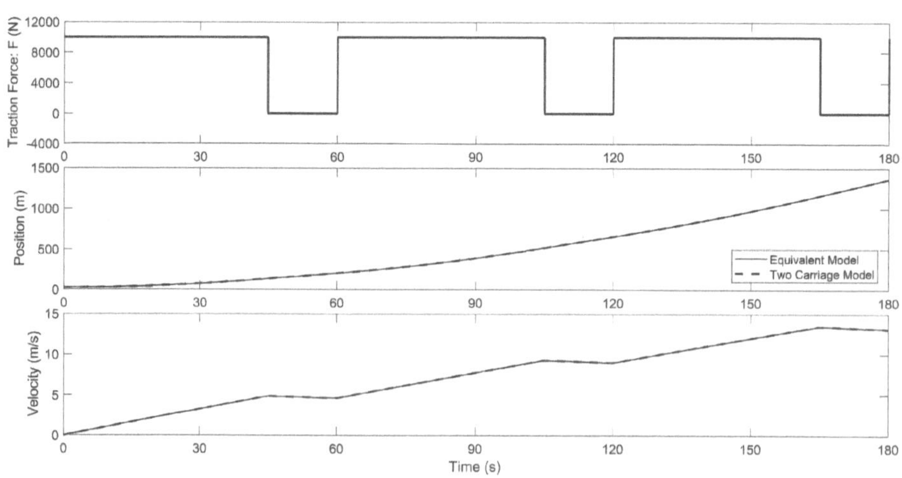

FIGURE 7.3 Performance of equivalent model when subjected to pulse traction input.

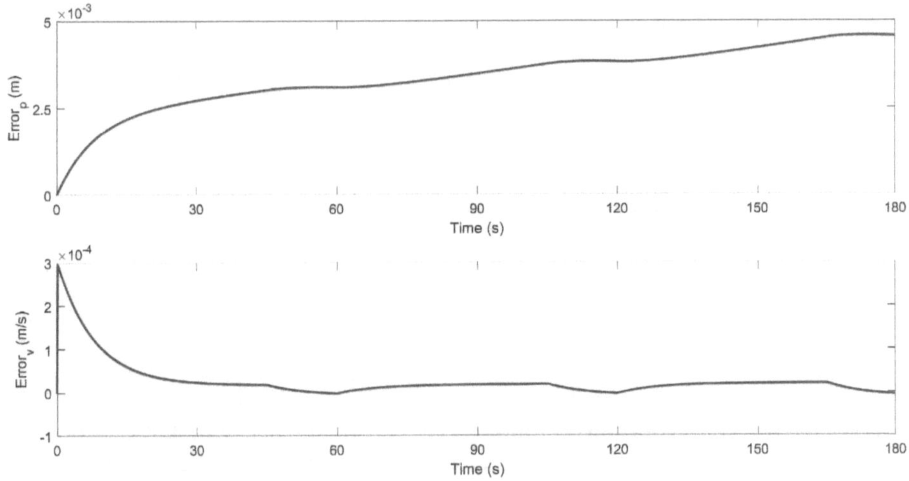

FIGURE 7.4 Error in the performance of two models when subjected to random traction input.

also negligible. However, when a random traction force (Figure 7.4) is applied the performance between the two model deteriorates, with velocity error that results in position error. However, for a longer time, the velocity error converges to zero which reduces the position error to a very small percentage (below 0.01 m) of the total displacement.

The main objective of reducing the model complexity is the increased flexibility of the system to design the controller which can consider the individual trains as a single mass system. From now on, in this chapter, the equivalent train model will be used for study.

7.7 CONTROLLER DEVELOPMENT

To design a longitudinal controller for railway virtual coupling, we need to consider two layers of controllers. Firstly, the upper-level controller, which ensures safe operation by determining the required distance and speed, as well as calculating the control input for individual vehicles. The second layer of control focuses on generating the determined control input within the train. In this chapter, we will primarily focus on the top-level controller, assuming that the second-level controller functions as discussed in Chapter 2, accounting only for the inertia effects of the engine. We will explore various control strategies that can be applied to ensure effective virtual coupling. The performance of these controllers will be evaluated and compared against a random control input to assess their effectiveness.

7.7.1 PID Controller with Preceding Vehicle Connection Only

As the first control strategy, we consider a unidirectional PID controller with a spacing control policy. In this approach, the spacing between vehicles is dynamically determined by the velocity of each vehicle relative to its preceding vehicle, as well as the relative distance between them.

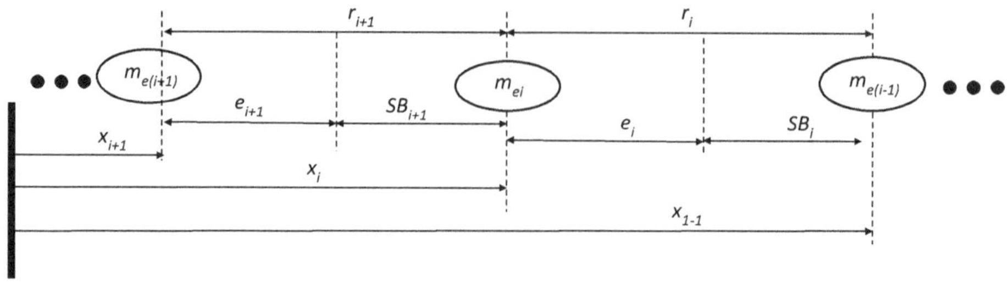

FIGURE 7.5 Schematic of three trains [($i-1$)th, i-th and ($i+1$)-th] in the virtually coupled system of n-trains.

A schematic representation of this system is provided in Figure 7.5. In this section, we consider a system of n trains running in virtual coupling. The first train follows a predefined speed profile of the network, controlled by a PID controller to maintain its speed. The i-th vehicle in the coupling receives information only about the (i−1)-th vehicle (i.e., the preceding vehicle).

The relative spacing between two vehicles is defined as:

$$r_i = x_{i-1} - x_i. \tag{7.8}$$

The spacing error is defined as:

$$e_i = r_i - \text{SB} = x_{i-1} - x_i - \text{SB}. \tag{7.9}$$

The safety buffer (SB) is calculated based on the braking distance of the two vehicles, as well as the length of the preceding train (L). $\text{SB}_i = L + \left(v_{i-1}{}^2 - v_i{}^2\right)/2a_{b,\max}$. Since all the trains have the same length, for simplicity, we omit the suffix i and use a constant L. This assumption allows us to simplify the formulation of the system without introducing additional complexity related to varying train lengths. The SB can also be termed as relative braking distance or movement authority [23].

The purpose of the unidirectional PID controller is to manage the virtually coupled vehicle series, ensuring that each spacing error (e_i) is minimized to zero. This helps maintain the individual stability of each vehicle while also ensuring string stability across the entire system, allowing the virtual coupling to function effectively and safely.

The control law for the i-th vehicle is proposed as:

$$u_i = K_{Pi}\ e_i + K_{Di}\ \dot{e}_i + K_{Ii} \int e_i. \tag{7.10}$$

Now, substituting the control input to the equation of motion, we can write,

$$\ddot{x}_{ei} = u_i - F_{Le}/m_{ei}, \tag{7.11}$$

$$\ddot{e}_i = \ddot{x}_{i-1} - \ddot{x}_i = a_{i-1} - \left(K_{Pi}\ e_i + K_{Di}\ \dot{e}_i + K_{Ii} \int e_i - F_{Le}/m_{ei}\right). \tag{7.12}$$

Rearranging differentiating it

$$\dddot{e}_i = \dot{a}_{i-1} - K_{Pi}\ \dot{e}_i + K_{Di}\ \ddot{e}_i + K_{Ii}\ e_i. \tag{7.13}$$

Now, to ensure the designed controller is stable, we can use Laplace transform, to determine the transfer function as

$$C(s)_i = \frac{E(s)_i}{A(s)_i} = \frac{s}{s^3 + K_{Di}\ s^2 + K_{Pi}\ s + K_{Ii}}. \tag{7.14}$$

For the system to be stable, all poles of the characteristic equation of the transfer function of $C(s)$ must have negative real parts, meaning they should lie in the left half of the complex plane. Using the Routh-Hurwitz Stability Criterion, we chose the control parameters K_{Pi}, K_{Di}, and K_{Ii} such that:

$$\begin{matrix} K_{Pi},\ K_{Di},\ K_{Ii} > 0 \\ K_{Pi},\ K_{Di} > K_{Ii} \end{matrix}. \tag{7.15}$$

These conditions guarantee that all the poles of the system have negative real parts, ensuring stability.

To ensure the designed controller is string stable, we need to consider the amplification of error as it propagates through a virtually coupled system. It is crucial for safety, performance, and scalability to design a formation control architecture that ensures disturbances propagate in a uniformly bounded manner. For any general interconnection of nonlinear systems, string stability is defined as a requirement that the states of the entire system remain bounded over time and spatial index whenever the initial conditions are bounded. The entire system is considered string stable if

$$|G_i(s)|_\infty = \frac{|E_i(s)|_\infty}{|E_{i-1}(s)|_\infty} \le 1. \tag{7.16}$$

To derive the expression for $G_i(s)$, we analyse the control architecture of the given system, as illustrated in Figure 7.6. The system is structured as follows:

$$E_i(s) = \frac{\text{CH}}{1+\text{CH}} E_{i-1}(s) \text{ and } E_2(s) = \frac{1}{1+\text{CH}} E_1(s). \tag{7.17}$$

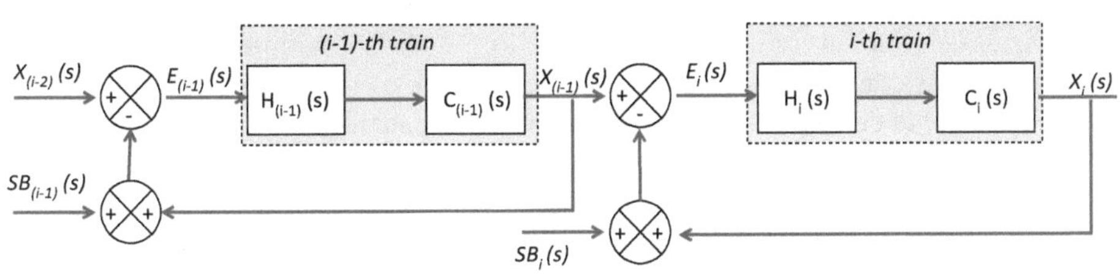

FIGURE 7.6 Control architecture of uni-directionally coupled trains.

It can now be noted that the system to become stable, $\frac{1}{1+\text{CH}}$ needs to be small. In addition, $\left|G_i(s)\right|_\infty \leq 1$ condition needs to be satisfied. These two conditions cannot be guaranteed at all the given frequencies [24]. Thus, a unidirectional PID controller with the information from the preceding vehicle only, is unable to provide the string stability of the system.

7.7.2 PD Controller with Preceding and Following Vehicle Connection

Considering the same system depicted in Figure 7.6, a new control architecture has been developed to account for communication between both preceding and following trains. The control law for i-th train is now designed.

$$u_i = \begin{cases} K_{\text{PF}i}\ e_i + K_{\text{DF}i}\ \dot{e}_i - K_{\text{PF}(i+1)}\ e_{i+1} - K_{\text{DF}(i+1)}\ \dot{e}_{i+1} & \text{for } i = 1,2,\ldots(n-1) \\ K_{\text{PF}n}\ e_n + K_{\text{DF}n}\ \dot{e}_n & \text{for } i = n \end{cases}. \tag{7.18}$$

The last train, i.e, n-th train requires a modified control law in the absence of the preceding vehicle, and only one reference train communication is used to design this.

Considering the time constant for the system dynamics, τ is can be shown that the system is stable if $(K_{\text{DB}} - \tau K_{\text{PB}}) > 0$.

The string stability condition can also be achieved by satisfying the conditions [25]:

$$\begin{aligned} K_{\text{PB}} - \frac{D_P}{\sqrt{2}} &> 0 \\ K_{\text{DB}}\ - \frac{\tau}{4} + \frac{D_D}{2} &< 0 \\ (K_{\text{DB}}\ )^2 > 2\ K_{\text{PB}}\ + D_P\ &+ \frac{D_D{}^2}{2} \end{aligned}, \tag{7.19}$$

where $D_P = K_{\text{PF}} - K_{\text{PB}}$ and $D_D = K_{\text{DF}} - K_{\text{DB}}$.

7.8 ADAPTIVE SLIDING MODE CONTROLLER

As it has been established that the bidirectional control law is best suited to achieve stability, an ASMC has now been designed for virtual coupling. We consider the same virtually coupled train system depicted in Figure 7.5, with n trains, where the first train follows the reference velocity of the network using a PID controller (as per the unidirectional case). We now designed the sliding surface as,

$$\sigma_i = s_i e_i + \dot{e}_i = s_i e_i + (\dot{x}_{i-1} - \dot{x}_i), \tag{7.20}$$

where $s_i > 0$ for each train and considered constant in this series of trains.

Since the information from the preceding and following vehicles can be filtered to obtain their time derivatives, it is reasonable to assume that the time derivatives of position and velocity are available without any loss of generality. Furthermore, because the convergence of the sliding surface for the i-th train doesn't ensure string stability, we need to design a composite sliding surface for the i-th vehicle with communication from preceding and following vehicles. This can be achieved as follows:

$$\sum_i = \begin{cases} \lambda_i \sigma_i - \sigma_{i+1} & i = 1,2\ldots,n-1 \\ \lambda_n \sigma_n & i = n \end{cases}, \tag{7.21}$$

where λ_i is a positive constant. For this problem, where the train parameters are considered similar, we have chosen λ_i as constant for all the trains, thus ignoring the index i from now on.

Lemma 1: $\sum_i$ becomes zero for all $i = 1,2,\ldots n$; if and only if σ_i becomes zero for all $i = 1,2,\ldots n$ at the same time. This holds because $\sum_i = \lambda\sigma_i - \sigma_{i+1}$ with $\lambda > 0$, and $\sum_{n+1}$ doesn't exist and considered as zero to construct the combined sliding surface.

Proof: We can now generate the relationship between σ and $\sum$ as:

$$\sum = \Lambda\sigma, \tag{7.22}$$

where

$$\Delta = \begin{bmatrix} \lambda & -1 & \cdots & 0 & 0 \\ 0 & \lambda & \cdots & 0 & 0 \\ \vdots & \vdots & \ddots & \vdots & \vdots \\ 0 & 0 & \cdots & \lambda & -1 \\ 0 & 0 & \cdots & 0 & \lambda \end{bmatrix},$$

$$\sum = \begin{bmatrix} \sum_1 & \sum_1 & \cdots & \sum_n \end{bmatrix}^T,$$

$$\sigma = \begin{bmatrix} \sigma_1 & \sigma_1 & \ldots & \sigma_n \end{bmatrix}^T.$$

Now, it can be noted that λ is positive constant, which makes the matrix Λ an invertible matrix. Thus, $\sum$ becomes zero for all σ_i become zero.

According to Lemma, the control law is designed to guarantee that the composite sliding surface (CSS) converges to zero. The time derivative of $\sum_i$ becomes:

$$\sum_i = \lambda\dot{\sigma}_i - \dot{\sigma}_{i+1} = \lambda\left[s_i\dot{e}_i + (\ddot{x}_{i-1} - \ddot{x}_i)\right] - \left[s_i\dot{e}_{i+1} + (\ddot{x}_i - \ddot{x}_{i+1})\right]. \tag{7.23}$$

This can be rewritten as

$$\dot{\sum}_i = -\frac{\lambda+1}{m_e}(u_i - F_{Lei}) + \Theta_i, \tag{7.24}$$

where

$$\Theta_i = \lambda \ddot{x}_{i-1} + \ddot{x}_{i+1} + s_i\left[\dot{e}_i - \dot{e}_{i+1}\right].$$

Now, based on this, we design the bidirectional adaptive control law for i-th train as

$$u_i = \begin{cases} \hat{F}_i + \frac{\hat{m}_{ei}}{\lambda+1}\Theta_i + \frac{\mu}{\lambda+1}\sum_i + \frac{\bar{\mu}}{\lambda+1}\operatorname{sgn}\left(\sum_i\right) & \text{for } i = 1,2,\ldots.(n-1) \\ \hat{F}_n + \frac{\hat{m}_{en}}{\lambda}\Theta_n + \frac{1}{\lambda}\left[\mu\sum_n + \bar{\mu}\operatorname{sgn}\left(\sum_n\right)\right] & \text{for } i = n \end{cases}, \tag{7.25}$$

here, $\hat{F}_i$ and $\hat{m}_{ei}$ are estimates of F_{Le} and m_e; the constants μ and $\bar{\mu}$ are selected to ensure the stability.

The adaptation laws for $\hat{F}_i$ and $\hat{m}_{ei}$ are obtained as

$$\dot{\hat{F}}_i = \begin{cases} \zeta_F(\lambda+1)\sum_i & \text{for } i = 1,2,\ldots(n-1) \\ \zeta_F\,\lambda\sum_n & \text{for } i = n \end{cases}. \tag{7.26}$$

$$\dot{\hat{m}}_{ei} = \zeta_m\,\Theta_i\sum_{\mathrm{i}} \quad \text{for } i = 1,2,\ldots n.$$

The two constants, ζ_m and ζ_F are positive constants that regulate the rate of adaptation.

The adaptive mechanisms adjust the parameters $\hat{F}_i$ and $\hat{m}_{ei}$ in real time based on the current state of the system. These estimates are continuously updated to account for variations in the system dynamics and external disturbances. By doing so, the controller maintains high performance and stability under varying operational conditions. The positive constants, ζ_m and ζ_F determine the responsiveness of the adaptation, ensuring that the estimates converge to their true values, thereby improving the overall system's robustness and adaptability.

The convergence of the combined sliding surface can be guaranteed by satisfying the convergence requirements, which are shown in Appendix 7.1.

7.9 COMPARISON OF CONTROLLER PERFORMANCE

To compare the performance of the controllers, a random velocity profile has been generated and applied to train 1.

The results, shown in Figure 7.7, illustrate the performance of five trains following the random velocity profile when using a unidirectional PID controller. It can be observed that the trains are able to follow the preceding trains under normal

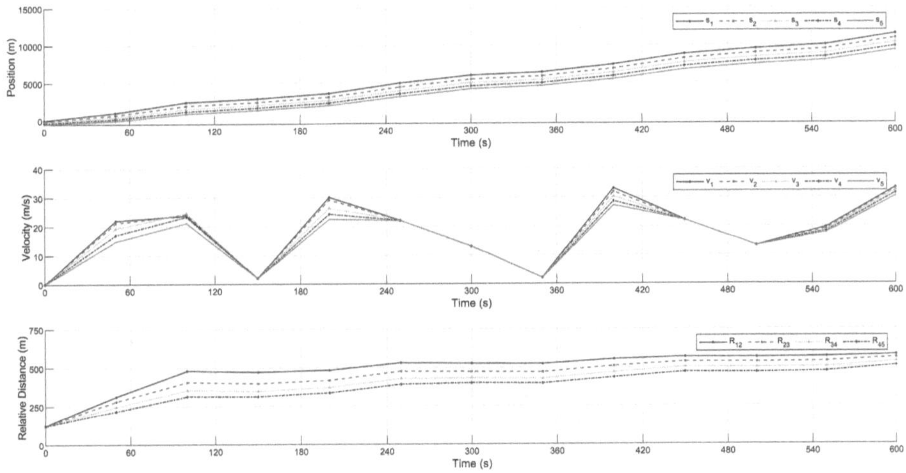

FIGURE 7.7 Control performance of Virtual coupling with PID controller with preceding train information only.

conditions. However, in the presence of disturbances, such as changes in the velocity profile, the overall stability of the system deteriorates. Consequently, the velocity tracking is not as expected, and the relative distance between the trains continues to increase due to unstable behaviour.

Figure 7.8 shows the performance of the virtually coupled system using a bidirectional PD controller. The velocity plot in Figure 7.8 indicates that trains 2 through 5 run together with reduced relative distances between them. Since the control force is calculated using information from two neighbouring trains, the initial acceleration of train 2 from a stationary state is slower than expected, leading to an initial spacing error. The spacing error is lower compared to that in Figure 7.6. The zoomed-in view of the braking distance shows that the spacing error consistently exceeds the braking distance, ensuring the system remains safe. However, the system remains stable, as evidenced by the fact that the relative distances do not increase over time, even in the presence of a changing velocity profile, as shown in Figure 7.8.

For the designed adaptive sliding mode controlled virtually coupled train systems, performance demonstrates a clear improvement (Figure 7.9). The velocity plot reveals that trains 2 through 5 effectively track the velocity profile, and the coupling remains stable even with changing speed profiles. Additionally, the spacing error is significantly lower compared to that observed with other controllers.

Table 7.2 illustrates the performance comparison of the three virtually coupled system. For the given reference velocity of the track. For repeatability, another new set of tracks is used with different velocity profile, slope, and curvatures. The performances are listed as Case 2. The performance indices are calculated as:

- Overall Root Mean Square (RMS) spacing error

$$Er_{\text{RMS}} = \sqrt{\frac{1}{4}\sum_{t}\sum_{i=2}^{5} e_i(t)^2}.$$

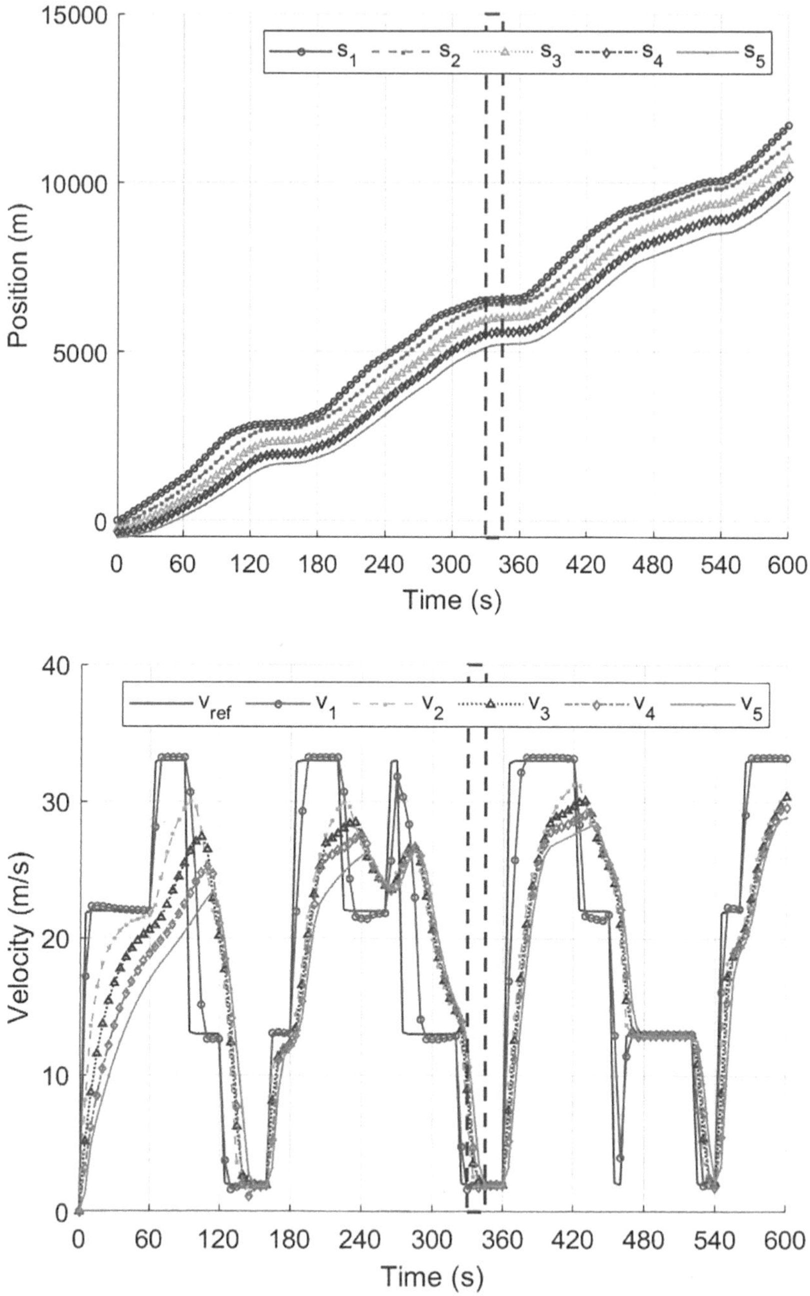

FIGURE 7.8 Control performance of virtual coupling with bidirectional PD controller with preceding and following train information.

(Continued)

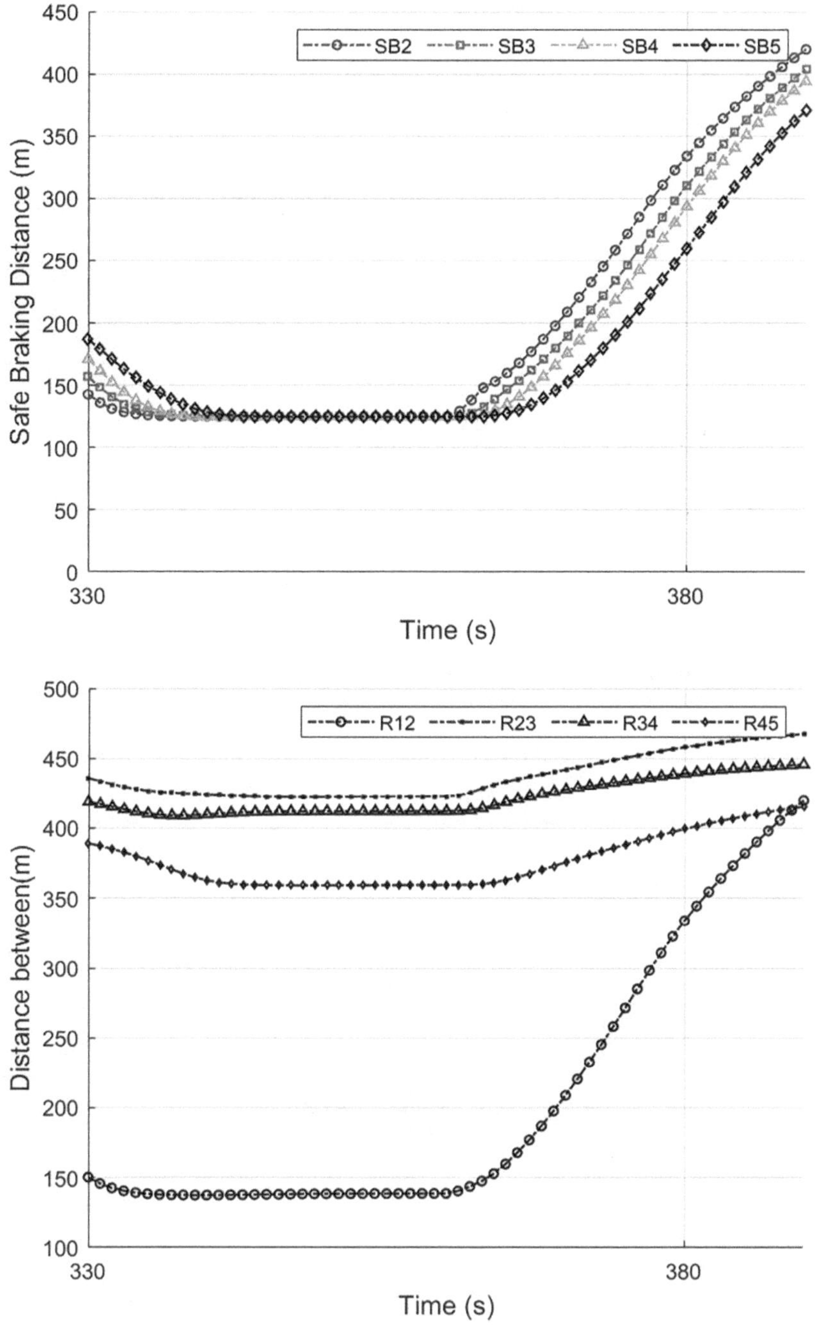

FIGURE 7.8 (*Continued*) Control performance of virtual coupling with bidirectional PD controller with preceding and following train information.

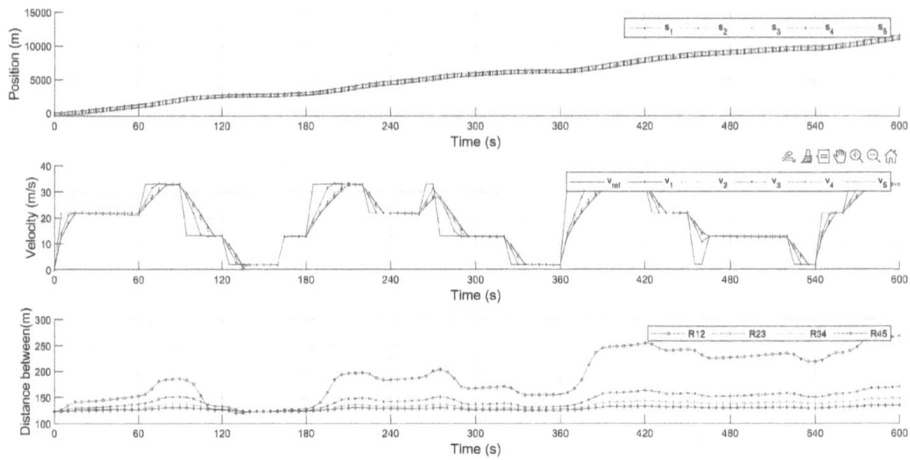

FIGURE 7.9 Control performance of Virtual coupling with an adaptive sliding mode controller with preceding and following train information.

TABLE 7.2
Performance Comparison of the Three Controllers

	Case 1			Case 2		
Index	**Uni-PID**	**Bi-PD**	**ASMC**	**Uni-PID**	**Bi-PD**	**ASMC**
Er_{RMS} (m)	442.1	392.1	152.7	535.3	402.5	161.3
TS_O (km)	10,360.7	90,945.4	35,754.0	15,360.5	84,530.6	41,600.1
Travel distance of train 1 (x_1) (km)	11.41	11.72	11.52	13.5	13.81	13.75
Travel distance of train 5 (x_5) (km)	9.23	9.73	10.79	11.12	11.62	12.01
V_Er_{RMS} (m/s)	2.86	6.28	1.59	5.90	4.99	1.42

- Overall RMS velocity error:

$$V\_Er_{RMS} = \sqrt{\frac{1}{4}\sum_{t}\sum_{i=2}^{5}\left(v_{i-1}(t)-v_i(t)\right)^2}.$$

- Total Space occupied:

$$TS_O = \sum_{t}\left(x_1(t)-x_5(t)\right).$$

In Figure 7.10, it is evident that the ASMC optimizes track usage by allowing trains to run closely together while maintaining string stability throughout the entire virtual coupling system. This approach enhances efficiency and stability, demonstrating effective management of the train series in the virtual coupling scenario.

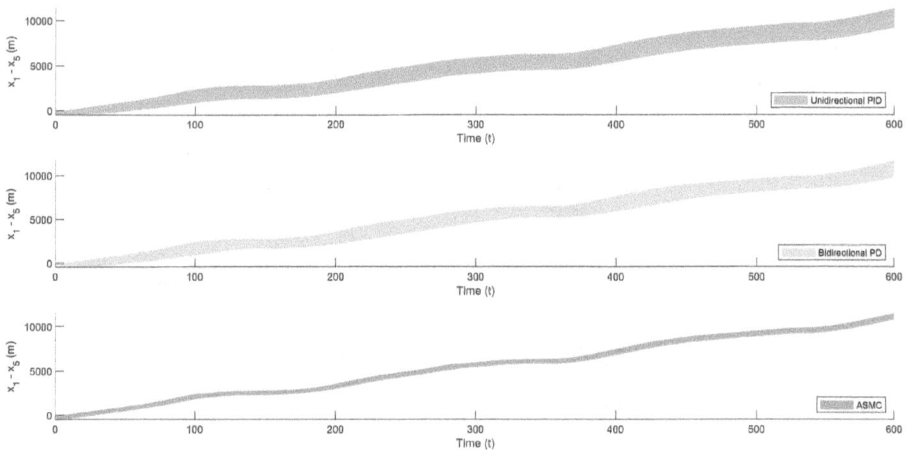

FIGURE 7.10 The use of track for the three virtually coupled systems.

7.10 FRAMEWORK FOR COUPLING AND DECOUPLING

The virtual coupling model is developed with careful consideration of three critical phases: coupling, coupled run, and decoupling. Ensuring system stability throughout these phases is a prime requirement for safe and efficient operation.

In this model, the trains, including the preceding train (i–1) and the following train (i), operate under the moving block signalling system. It is assumed that the states of the trains, such as position and velocity, are known to the operator, along with other essential parameters like emergency braking capabilities and allowable acceleration/deceleration. Below, we define the key terminologies used in the system checks.

- Safety Buffer for i-th train, SB_i
- End of Authority for i-th train, EoA_i
- End of Authority for coupled run, EoA_C
- Absolute Braking Distance, $AB_i = \dfrac{v_i^2}{2\,a_{max}}$
- Relative Braking Distance, $RB_i = AB_i - AB_{i-1}$

Before activating the virtual coupling mode, it is essential that the movement authority verifies that the trains involved are travelling on the same route, ensuring the feasibility and safety of the coupling process.

7.10.1 Coupling State

During the "Coupling" state, the first check ensures that train i's position at the end of the coupling does not exceed the position of the preceding train (i–1). The time required to achieve coupling depends on the velocity difference between the two trains. The time taken for train i to match the speed of train (i–1) is denoted as t_c. The condition for this check (Check 1 in Figure 7.11) can be expressed as follows:

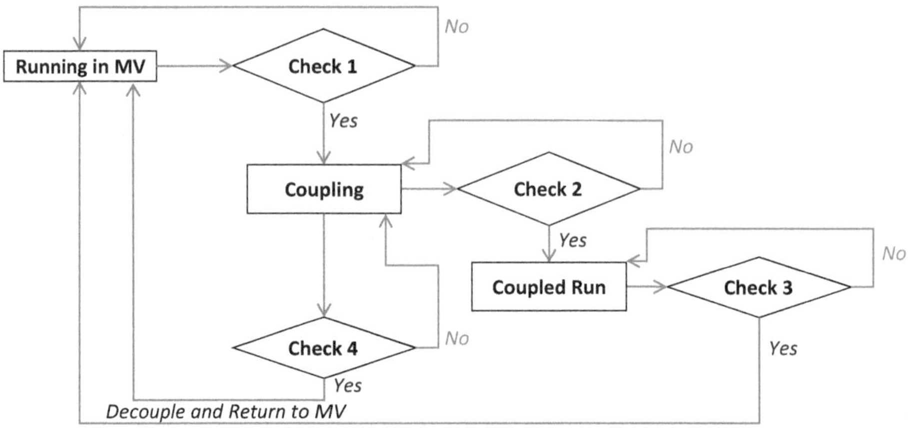

FIGURE 7.11 Flowchart of the virtual coupling: including coupling, coupled run and decoupling phases.

$$x_i(t_c) - \mathrm{SB}_i < x_{i-1}(t_c), \tag{7.27}$$

where

- $x_i(t_c)$ is the position of train i at time t_c,
- $x_{i-1}(t_c)$ is the position of train $(i-1)$ at time t_c.

This equation ensures that train i does not overshoot the preceding train $(i-1)$ during the coupling process, thereby maintaining safe spacing. Once this check is satisfied the train i enters into the coupling state to reach specified speed.

7.10.2 Coupled Run

Train i must now check two critical conditions: speed and position. First, the velocity difference between train i and train $(i-1)$ must fall within the threshold for relative velocity to allow train i to safely move into virtual coupling. Additionally, the spacing error of train i must be greater than the absolute braking distance of train $(i-1)$. This condition ensures that if train $(i-1)$ initiates emergency braking at this moment, train i will not reach train $(i-1)$, thereby maintaining a safe separation that meets the safety threshold.

The two checks for virtual coupling are represented by the following equations:

Check 2.1 – Relative Velocity Check: The velocity difference should be within the allowable threshold for virtual coupling:

$$v_{i-1} - v_i < v_{\mathrm{th}}. \tag{7.28}$$

Check 2.2 – Spacing Error Check: The spacing error between train i and train $(i-1)$ must be greater than the absolute braking distance of train $(i-1)$:

$$e_i < AB_{i-1}.$$

These checks ensure both safe relative speed and sufficient distance to prevent collisions during emergency situations. Once these two checks are satisfied, the two trains start running as coupled using the ASMC designed before.

7.10.3 Decoupling

As train i enters a zone near a junction, the decoupling state will be triggered. The train I initiates the decoupling process, and both trains revert to their original moving block signalling mode. When the distance to the junction falls below the spacing error, train i will decouple and exit the virtual coupling mode. The check equation for this decoupling condition is as follows:

Check 3:

$$d_{ji} < e_i.$$

where d_{ji} is the distance of the junction zone from train i. This condition ensures that decoupling occurs safely before the trains reach the junction, preventing any risk of collision or interference at the junction.

Similarly, while in coupling mode, train i continuously checks this condition (Check 4 in Figure 7.11) to ensure a safe return to moving block signalling as it approaches a junction, particularly in cases where virtual coupling has not been successfully achieved. This ongoing check ensures that the system transitions smoothly and safely out of virtual coupling near critical points like junctions.

To test the developed framework two test cases were developed, and the performance is tested.

7.10.3.1 Test Case 1: Variable Speed

To test this framework, the first test case is designed as a variable speed line with two stations. The simulation was conducted on a track with two stations spanning 15 km as shown by v_{ref} in the top figure at Figure 7.12. The two stations are denoted by two starts at distances 3,500 and 7,500 m. The speed limit near the station zones (shown by shaded areas) is 10 m/s. The test scenario was designed to include a 60-second waiting time for the trains at the platform.

The results of the virtual coupling performance are illustrated in Figures 7.12 and 7.13. The velocity of each train is shown to adhere to the speed limits of the simulated track, with trains successfully forming virtual couplings during the run. In the top figure of Figure 7.12, it is evident that within the junction zones (indicated by the two shaded areas), the trains decouple from the virtual coupling and revert to moving block signalling to stop at the stations.

It is observed that the spacing error is greatest for the second train, as the first train does not use information from the second train to adjust its speed. Specifically, the RMS spacing error between the first and second trains over the two stations is 414.3 m, while the RMS spacing error between the fourth and fifth trains is notably lower at 222.1 m. The RMS speeds of the five trains are as follows: 25.07, 25.00,

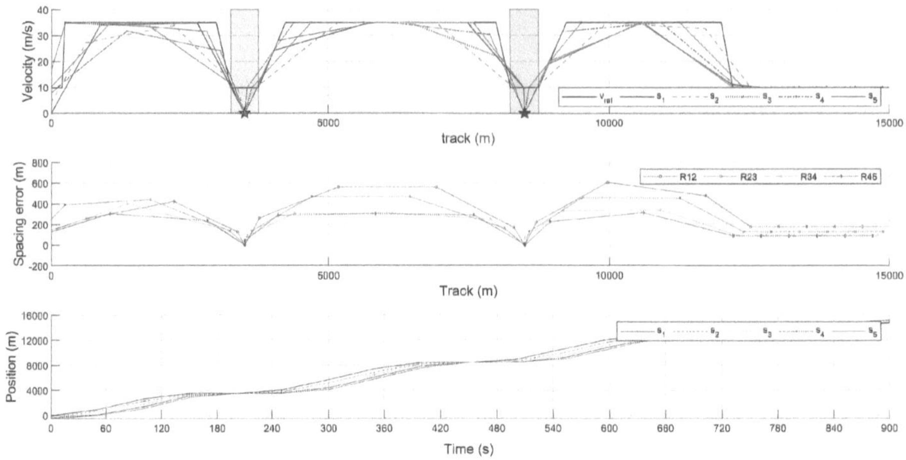

FIGURE 7.12 The performance of the virtually coupled trains: top: velocity of each trains along the track, middle: spacing error between the trains, bottom: position of the trains along the simulated track.

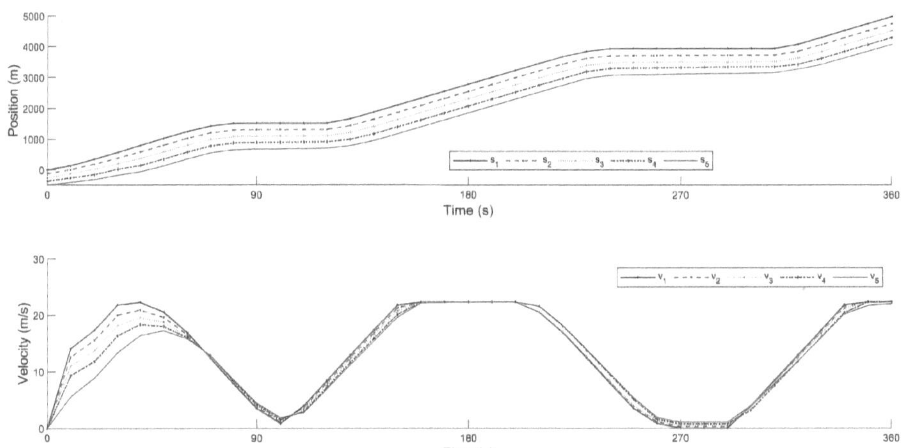

FIGURE 7.13 The performance of the virtually coupled trains for test case 2.

25.06, 25.06, and 25.07 m/s, respectively. These speeds indicate that coupling is well-maintained in this region, demonstrating effective coordination among the trains. Nonetheless, in all cases, the spacing error remains within safe limits, ensuring the secure operation of the system.

7.10.3.2 Test Case 2: Emergency Braking

To evaluate stability and string stability, this test case simulates a route with a speed limit of 50 mph (22.352 m/s). Train 1 performs an emergency stop at 60 seconds while in coupled mode, resuming its journey at 120 seconds and adhering to the speed limit. Another scenario occurs at 200 seconds, when Train 3 stops with an emergency brake, restarting after 90 seconds.

The results in Figure 7.13 indicate the system remains stable. When Train 1 stops, the following trains decelerate gradually and hold a low, constant speed to maintain safe distances. After Train 3 stops, Trains 4 and 5 decelerate as needed without impacting Trains 1 and 2.

However, since this is a simulated track, it is essential to validate these results with a real-world track. This validation will be addressed in the next section.

7.11 CASE STUDY

The case study focuses on the Calder Valley line, which runs between Leeds and Sowerby Bridge within the UK Rail Network's North and East route. The primary objectives of this case study are to evaluate the feasibility of implementing virtual coupling on the Calder Valley line, assess its impact on network flexibility, and optimize train scheduling to enhance overall operational efficiency. Key characteristics include varied track gradients and multiple junctions, which impact train operations and signalling.

This selection is driven by the high volume of daily commuters travelling through three major urban hubs—Leeds, Bradford, and Halifax—in West Yorkshire. The Calder Valley line spans approximately 30 km, featuring a mix of urban and rural settings. The study extends to Sowerby Bridge to explore the potential for adding flexibility to the network.

The Calder Valley line accommodates three different train services:

1. Type 1 – Leeds to Sowerby: Stops at each station along the route.
2. Type 2 – Leeds to Sowerby: Skips stations Bramley (Station 1) and Low Moor (Station 4).
3. Type 3 – Leeds to Halifax: Stops at all station till Halifax, then continues on divergent route.

For this study, five trains have been considered as:

- First Train: Type 2
- Second Train: Type 1
- Third Train: Type 1
- Fourth Train: Type 1
- Fifth Train: Type 3.

The speed profile along the route is shown in Figure 7.14, which provides an overview of the operational speeds for the trains on this line. The route details have been collated from train running timings on this line. Data for five trains of each type were collected, including speed limits and location data at various signal points along the route. This comprehensive data set provides insights into train speeds, spacing, and adherence to signalling requirements.

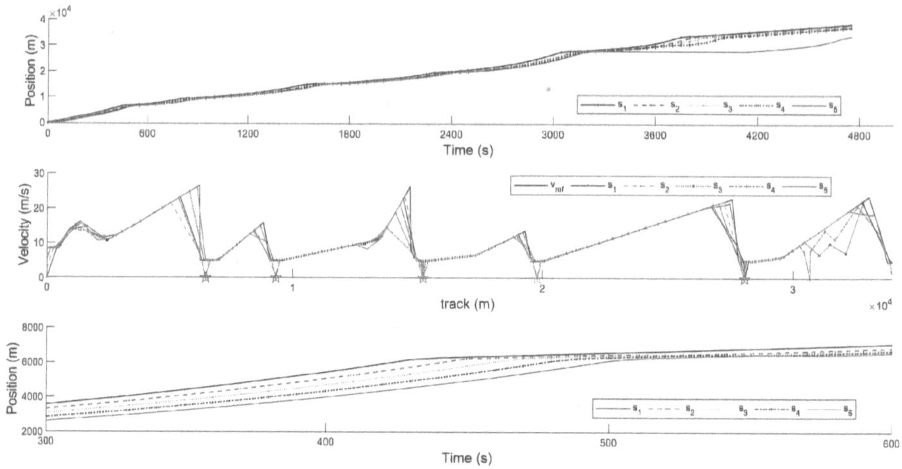

FIGURE 7.14 The performance of the virtually coupled trains: a case study of Leeds-Halifax line, UK.

It is assumed that trains can switch to adjacent lines near junctions. The current switching time within the UK Rail network is approximately 7 seconds. Recent research suggests that this time is expected to decrease in the near future. This assumption is based on ongoing advancements in technology and infrastructure that aim to improve switching efficiency [26, 27]. In the designed case study, the minimum distance between two trains in the virtual coupling mode is always greater than (with minimum of 17 seconds) the switching time. This ensures that there is sufficient time for a train to switch lines safely without interfering with the operation of other trains.

The performance of the virtual coupling framework on the Calder Valley line (in Figure 7.14), shows the behaviour of the trains as they operate under virtual coupling control. The trains follow the virtual coupling pattern on sections of the track outside the station zones. These junction/station zones are highlighted by shaded regions in the Figure 7.14, showing that the coupling is maintained during regular operation but decouples near stations to allow for independent stopping and starting.

The trains are restricted by the velocity profile of the track, ensuring that they adhere to the speed limits imposed by the network. This helps maintain safety and smooth operation across the network. Even after train 1 skips a stop at Bramley (Station 1), the entire system regains its coupled formation. However, there is a noticeable increase in the gap between the first two trains. This is attributed to the low-speed limits on the network, which restricted the motion of train 2 and prevented it from closing the gap immediately after the skipped stop.

Despite the high potential capacity of the network, the current configuration demonstrates low usage of track. The system's performance highlights that while virtual coupling can enhance train coordination, the low-speed limits and operational constraints of the network prevent it from reaching its full capacity potential.

These observations provide valuable insights into the practical application of virtual coupling in a real-world scenario, pointing to areas where improvements in infrastructure or speed limits could further enhance the network's performance.

7.12 SUMMARY AND CONCLUSIONS

Virtual coupling offers a step-change approach to train operations by allowing closely spaced trains to travel together in a coordinated manner, effectively increasing network capacity without the need for significant infrastructure expansion. This is particularly important in high-density rail corridors where conventional fixed or moving block signalling systems limit train frequency and reduce operational flexibility. By enabling dynamic adjustment of train spacing, virtual coupling enhances efficiency, reduces delays, and improves overall network performance, making it a key innovation for the future of rail transportation.

It is shown through this chapter, that the ASMC is able to maintain string stability within the virtual coupling system. It ensures that trains remain safely spaced while minimizing track usage and spacing errors. This controller adapts to changing conditions, such as variations in train speeds and network constraints, to maintain the coupling safety. As demonstrated in the simulations, the ASMC effectively manages the trade-off between operational efficiency and safety, enabling high-density train operations with minimal disruption.

The case study on the Calder Valley line demonstrates the practical application of virtual coupling in a real-world scenario. While the virtual coupling framework shows promise in improving train coordination and capacity, the study highlights several challenges, including the impact of low-speed limits and infrastructure constraints on network performance. Despite these limitations, the system has demonstrated effective coupled formation and spacing management, particularly outside station/junction zones. To unlock the full potential of virtual coupling, further improvements in network infrastructure and operational policies are necessary.

7.13 FUTURE RESEARCH DIRECTION

This case study establishes a foundation for future research and implementation aimed at optimizing rail network performance through advanced control strategies like virtual coupling. As the aim of this chapter is control strategy development, validation is currently confined to simulation environments only. To enhance this, the author is developing a scaled experimental setup to collect representative datasets for validation.

Future research could explore the integration of the proposed control system with emerging technologies, such as self-guided mechatronic bogies, self-adjusting switches, and advanced communication networks. Additionally, the current model does not rigorously account for time delays in state measurement and estimation, incorporating a unit delay in sensor data. A more analytical method should be developed to address these practical considerations in future studies. Overall, these directions aim to ensure that the proposed control systems are safe, practical, and aligned with future advancements in railway technology.

Appendix 7.1

Theorem

When the adaptive control law is applied to the interconnected vehicular system, stability is achieved across the system. This stability is characterized by the following:

Sliding Surfaces (σ_i and $\sum_i$ drives toward zero, and distance error variables (e_i), which represent the deviation from the desired spacing between vehicles, are also driven to zero asymptotically.

Proof:

We first consider a Lyapunov function as:

$$V = \sum_i V_i.$$

The parameter V_i is chosen as:

$$V_i = \frac{m_{ei}}{2} \sum_i^2 + \frac{\tilde{f}_i^2}{2\zeta_F} + \frac{\tilde{m}_{ei}^2}{2\zeta_m}.$$

where $\tilde{f}_i$, $\tilde{m}_{ei}$ are the estimation errrors for i-th train.

Substituting these in the Lyapunov function and differentiating, we get,

$$\dot{V}_i = m_{ei} \sum_i \left[\frac{\lambda+1}{m_{ei}} \tilde{f}_i + \Theta_i - \frac{\tilde{m}_{ei}}{m_{ei}} \Theta_i - \frac{\mu}{m_{ei}} \sum_i - \frac{\bar{\mu}}{m_{ei}} \text{sgn}\left(\sum_i \right) \right] + \frac{\tilde{f}_i \dot{\tilde{f}}_i}{\zeta_F} + \frac{\tilde{m}_{ei} \dot{\tilde{m}}_{ei}}{\zeta_m},$$

$$\le (\lambda+1) \tilde{f}_i \sum_i + \tilde{m}_{ei} \Theta_i \sum_i - \mu \sum_i^2 - \bar{\mu} \left| \sum_i \right| + \frac{\tilde{f}_i \dot{\tilde{f}}_i}{\zeta_F} + \frac{\tilde{m}_{ei} \dot{\tilde{m}}_{ei}}{\zeta_m},$$

$$= -\mu \sum_i^2 - \bar{\mu} \left| \sum_i \right|.$$

So, overall Lyapunov function,

$$\dot{V} = \sum_i \dot{V}_i = \sum_i \left\{ -\mu \sum_i^2 - \bar{\mu} \left| \sum_i \right| \right\}.$$

As the value of μ is positive, the $V(t)$ is bounded for all times. Since $V_i(0)$ is bounded, and $V_i(t)$ bounded, $\sum_i$ will converges to zero.

REFERENCES

1. Goikoetxea, J. 2016. "Roadmap towards the Wireless Virtual Coupling of Trains." In *Communication Technologies for Vehicles: 10th International Workshop*, Nets4Cars/Nets4Trains/Nets4Aircraft 2016, Springer, Cham.
2. Aoun, J., E. Quaglietta, and R. M. Goverde. 2020. "Investigating Market Potentials and Operational Scenarios of Virtual Coupling Railway Signaling." *Transportation Research Record* 2674 (8): 799–812.
3. Network Rail. 2018. *CP6-Strategic-Business-Plan-Comprehensive-Executive-Summary*, Network Rail, London.
4. Zhu, L., D. Huang, X. Li, and Q. Wang. 2023. "Cooperative Operation Control of Virtual Coupling High-Speed Trains with Input Saturation and Full-State Constraints." *IEEE Transactions on Automation Science and Engineering* 21 (3): 3497–3510.
5. Mitchell, I., E. Goddard, F. Montes, P. Stanley, R. Muttram, W. Coenraad, J. Poré, S. Andrews, and L. Lochman. 2016. "ERTMS Level 4, Train Convoys or Virtual Coupling." *IRSE News* 219: 14–15.
6. Schumann, T. 2016. "Increase of Capacity on the Shinkansen High-Speed Line Using Virtual Coupling." *International Journal of Transport Development and Integration* 1 (4): 666–676.
7. Homay, A., M. de Sousa, and L. Almeida. 2017. "Nash Equilibrium for Proactive Anti-Jamming in IEEE 802.15.4e (Emerging Wireless Sensor Actuator Technologies for I4.0)." In *2017 IEEE 15th International Conference on Industrial Informatics (INDIN)*, Emden, Germany.
8. Flammini, F., S. Marrone, R. Nardone, A. Petrillo, S. Santini, and V. Vittorini. 2018. "Towards Railway Virtual Coupling." In *2018 IEEE International Conference on Electrical Systems for Aircraft, Railway, Ship Propulsion and Road Vehicles & International Transportation Electrification Conference (ESARS-ITEC)*, Nottingham.
9. Felez, J., Y. Kim, and F. Borrelli. 2019. "A Model Predictive Control Approach for Virtual Coupling in Railways." *IEEE Transactions on Intelligent Transportation Systems* 20 (7): 2728–2739.
10. Chen, M., J. Xun, and Y. Liu. 2020. "A Coordinated Collision Mitigation Approach for Virtual Coupling Trains by Using Model Predictive Control." In *2020 IEEE 23rd International Conference on Intelligent Transportation Systems (ITSC)*, Rhodes, Greece.
11. Luo, X., T. Tang, H. Liu, L. Zhang, and K. Li. 2021. "An Adaptive Model Predictive Control System for Virtual Coupling in Metros." *Actuators* 10 (8): 178.
12. Wang, B., D. Yang, X. Zhang, and X. Jia. 2022. "Constraint-Force Driven Control Design for Rail Vehicle Virtual Coupling." *Journal of Vibration and Control* 28 (5–6): 551–563.
13. Duan, H., Y. Yang, Y. Duan, and J. Zhang. 2020. "Research on Virtual Coupling Train Operations Based on Moving-Block and Vehicle-to-Vehicle Communication." In *Journal of Physics: Conference Series*, IOP Publishing, Bristol.
14. Rawlings, C., S. Dutta, and R. Dixon. 2022. "Exploring the Virtual Coupling of Railway Vehicles." In *World Congress of Railway Research*, Birmingham, UK.
15. Qiu, S., N. Li, X. Wei, and G. Li. 2023. "Research on the Stability of Virtual Coupling Train Formation Based on PID Controller Under the Influence of Time Delay." *Transportation Safety and Environment* 6(3): tdad037.
16. Wu, Q., X. Ge, Q.-L. Han, B. Wang, H. Wu, C. Cole, and M. Spiryagin. 2023. "Dynamics and Control Simulation of Railway Virtual Coupling." *Vehicle System Dynamics* 61 (9): 2292–2316.
17. Park, J., B.-H. Lee, and Y. Eun. 2020. "Virtual Coupling of Railway Vehicles: Gap Reference for Merge and Separation, Robust Control, and Position Measurement." *IEEE Transactions on Intelligent Transportation Systems* 23 (2): 1085–1096.

18. Yadav, R. D., V. N. Sankaranarayanan, and S. Roy. 2021. "Adaptive Sliding Mode Control for Autonomous Vehicle Platoon under Unknown Friction Forces." In *2021 20th International Conference on Advanced Robotics (ICAR)*, Ljubljana, Slovenia.
19. Zhu, Y., Y. Li, H. Zhu, W. Hua, G. Huang, S. Yu, S. E. Li, and X. Gao. 2022. "Joint Sliding-Mode Controller and Observer for Vehicle Platoon Subject to Disturbance and Acceleration Failure of Neighboring Vehicles." *IEEE Transactions on Intelligent Vehicles* 8 (3): 2345–2357.
20. Kwon, J.-W., and D. Chwa. 2014. "Adaptive Bidirectional Platoon Control Using a Coupled Sliding Mode Control Method." *IEEE Transactions on Intelligent Transportation Systems* 15 (5): 2040–2048.
21. Goli, M., and A. Eskandarian. 2014. "A Systematic Multi-Vehicle Platooning and Platoon Merging: Strategy, Control, and Trajectory Generation." In *Dynamic Systems and Control Conference*, American Society of Mechanical Engineers, San Antonio, TX.
22. Northern Railway. 2024. "Our Train Types." https://www.northernrailway.co.uk/before-you-travel/our-train-types. Accessed February 7, 2024.
23. Pan, J., Q. Peng, S. Zhan, and J. Bai. 2021. "Multiscenario-Based Train Headway Analysis under Virtual Coupling System." *Journal of Advanced Transportation* 2021: 5642709.
24. Shaw, E. E.-W. 2005. *Heterogeneous String Stability for Vehicle Formation Control*, University of California, Berkeley.
25. Ghasemi, A., R. Kazemi, and S. Azadi. 2013. "Stable Decentralized Control of a Platoon of Vehicles with Heterogeneous Information Feedback." *IEEE Transactions on Vehicular Technology* 62 (9): 4299–4308.
26. Dutta, S., T. Harrison, C. P. Ward, R. Dixon, and T. Scott. 2020. "A New Approach to Railway Track Switch Actuation: Dynamic Simulation and Control of a Self-Adjusting Switch." *Proceedings of the Institution of Mechanical Engineers, Part F: Journal of Rail and Rapid Transit* 234 (7): 779–790.
27. Olaby, O., S. Dutta, T. Harrison, C. P. Ward, and R. Dixon. 2021. "Realisation of a Novel Functionally Redundant Actuation System for a Railway Track-Switch." *Applied Sciences* 11 (2): 702.

8 Railway Transition Zones

Avni Jain, Andrei Metrikine, and Karel van Dalen

8.1 INTRODUCTION

Railway tracks are subjected to constant degradation over the operational period. In addition to this, railway transition zones degrade about 4–8 [1] times faster than the open track leading to high maintenance and operation costs. Moreover, in the Netherlands, it is estimated that 40% of the maintenance costs are associated with preserving the geometry of railway tracks [1, 2]. Transition zones in railway tracks can be generally described as structural and geometrical discontinuities along the track which are vulnerable to deterioration causing an increase in the cost of maintenance of these zones. More specifically, transition zones are areas where the railway track crosses a different transportation modality (road, waterway, etc.) or where the rail experiences major changes in the type of track support structure. The most common types of transitions (Figure 8.1) are embankment-bridge transition [3–19], transition with culvert [20], embankment-bridge with approach slabs [11] and transition

FIGURE 8.1 The most common types of railway transition zones: (a) a tunnel transition, (b) a culvert transition, (c) and a railroad crossing transition (d) a bridge-embankment transition.

DOI: 10.1201/9781003659983-8

with culvert and approach slabs [1, 21, 22], embankment-bridge transition with a wedge-shaped backfill [8, 10, 24], railroad crossing [24, 25], and transition with underpass [26–28].

Over the years, several studies [29, 30] have suggested that strong amplifications in stresses and strains are induced by moving axles of the train over the inhomogeneity along the longitudinal direction of the railway track. This inhomogeneity can be attributed to changes in geometry, discontinuity in forces, the configuration of track components, and the constitutive properties of the materials. These variations are inherent characteristics of transition zones and are believed to be major reasons for rapid degradation in these zones. Track degradation has been defined and categorized in several works done in the past [31] and can be mainly described as damage of track components [2, 31–34] (rail surface defects, broken fasteners, cracks in concrete sleepers, breakage of ballast particles, etc.) and deterioration of track geometry [32, 35–37]. The operation-induced dynamic response amplification leads to phenomena like but not limited to rapid compaction/ densification, abrasion and crushing of ballast, and increased settlement of embankment and subgrade layers, eventually leading to the occurrence of hanging sleepers in the vicinity of the transition interface. On one hand, several studies have related these phenomena in transition zones to two reasons [3, 14, 18]: uneven track stiffness and differential settlement. On the other hand, some other detailed studies [11, 32, 33] on railway transition zones (RTZs) classify the factors as listed below:

- Primary Factors: uneven track stiffness, geotechnical issues, subgrade failure, progressive shear failure, excessive plastic settlement, soil water response, wetting and shrinkage cycles.
- Secondary Factors: height of embankment, type of bridge joint, type of abutment in the approaching structure, axle load, and train speed.

The extensive list of primary factors listed above, in essence, can be reduced to uneven track stiffness and differential settlement. A quantitative comparison of these primary factors and their influence on the degradation of trackbed layers has been done in [5]. According to several studies, [3, 5, 6, 18, 25, 38, 39] the uneven track stiffness is one of the most influential factors leading to degradation in RTZs. At the same time, factors like soil water response, wetting and shrinkage cycles lead to changes in elasticity modulus and frictional properties of track components [4, 18] and eventually contribute to uneven track stiffness. Likewise, the remaining primary causes in the abovementioned list namely geotechnical issues, subgrade failure, progressive shear failure, and excessive plastic settlement ultimately result in the differential settlement. In addition to this, the secondary factors listed above can be attributed to the type of transition and vehicle characteristics under consideration. Therefore, these factors can be dealt with a robust design solution for specific site conditions. To conclude the above discussion, the main factors that lead to an increased degradation in RTZs are uneven track stiffness and differential settlement. However, uneven track stiffness and differential settlement are not independent of each other [2] but lead to self-perpetuating mechanisms. In addition to this, the uneven track stiffness is claimed to be the reason for the 'initiation' [3, 5, 6, 18, 25, 38, 39] mechanisms (transition radiation [40–44]) that lead to the rapid degradation of RTZs. Furthermore, the operation-driven differential settlement can be minimized or eliminated by avoiding dynamic amplification in the RTZs. Several attempts have been made in the past to minimize the degradation in transition zones. However, the track components or trackbed layers in the transition zones are designed similarly to open tracks with some interventions addressing the uneven track stiffness or differential settlement. Some of the most studied interventions to mitigate transition effects include auxiliary rails [38, 45, 46], larger and closely spaced sleepers [38, 45, 47–52], optimized rail pads [46, 47, 49, 53, 54], approach slabs [22, 23, 55–57], wedge-shaped backfills [58–63], etc. However, these measures proved to be inefficient as there is a lack of understanding of the primary requirements to design a robust mitigation measure aimed at minimizing the degradation due to the amplified dynamic response of RTZs. These requirements include a robust design/evaluation criterion, identification of the most influential parameters governing design and understanding of the key phenomena behind the amplified degradation of RTZs. In the following sections, firstly, the two main factors (stiffness jump and differential settlement) discussed above will be elaborated based on the literature. Secondly, the existing practices adopted for the design of track components (rail, softening pads, sleepers, ballast, embankment and subgrade) in RTZs are discussed. Finally, a novel design methodology is proposed aimed at minimizing the operation-driven dynamic amplifications in RTZs.

8.2 DIFFERENTIAL SETTLEMENT AND HANGING SLEEPERS

One of the two major factors that contribute to the rapid degradation of RTZs is the phenomenon of differential settlement which leads to uneven track levels in these zones [64, 65]. Although this unevenness in track level is inevitable at transition zones as ballasted railway tracks will settle as a result of permanent deformations (static) in the ballast and the underlying material layers, the settlement of bridge or structure on pile foundation is almost negligible [66]. However, reducing the dynamic amplifications in these zones can delay the process of degradation and settlement (dynamic) resulting in less frequent maintenance and associated costs. The periodic measurements of relative transient and permanent vertical displacements were performed at various depths and locations along the vertical plane of the railway track [65]. The results in [65] show that the majority of permanent vertical displacements occur in the region between the top of the sleepers and the bottom of the ballast layer. The amplified degradation in RTZ is mainly due to the occurrence of hanging sleepers as a consequence of differential settlement [64, 65]. Over the years several studies have been conducted both experimental [67–69] and numerical [39, 66, 67, 70–72] to study the amplification of loads/ stresses due to hanging sleepers or voids between sleeper and ballast layer. Even though the hanging sleepers are found not only in transition zones but throughout the ballasted track, their occurrence is more likely in RTZs. In the study reported in [66, 72], it is found that one single hanging sleeper with a 1 mm gap between sleeper and ballast can cause the sleeper-ballast contact force at the sleeper adjacent to the hanging one to increase by 70% and the displacement by 40%. Consequently, this uneven loading of the ballast bed may induce an irregular settlement of the trackbed worsening the problem even further. The worst case in an open track, considering the local settlement was identified as when one supported sleeper is surrounded by two unsupported sleepers. Sleeper voids also tend to increase the stresses (nearing 40%) experienced by the rail in the vicinity of the void [71]. The effect of sleeper voids was proved to have a far more significant effect on the forces imposed on the track components than on the wheel-rail contact. To summarize this discussion in the context of transition zones, it can be concluded that the occurrence of hanging sleepers cannot be avoided but can be delayed by minimizing the operation-driven dynamic response amplification of the track components and hence delaying the process of differential settlement or unevenness in track level. In addition to this void identification methods can be helpful in determining the need for maintenance [67] and controlling the occurrence of hanging sleepers.

8.3 VARIATION OF VERTICAL STIFFNESS AND MITIGATION MEASURES

Track stiffness is an important factor to consider when designing railway infrastructure since it impacts the loads transmitted to the track. On one hand, low stiffness values could lead to excessive settlement of rail. On the other hand, very high stiffness values can lead to excessive dynamic overloads [73] and ultimately lead to track wear and fatigue. Hence, it is very important to find an optimal value of the vertical track stiffness for the speeds under consideration. In this section, firstly the importance of choosing an optimum track stiffness is discussed and then the influence of an abrupt change in vertical track stiffness on the track performance is highlighted.

It has been found that the subgrade soil condition, ballast/sub-ballast thickness and vertical stiffness of sleepers are the most influential factors that affect the track stiffness whereas the track maintenance procedures like rail surfacing and ballast tamping have negligible effect. The effect of each track component on track modulus is studied in [74]. The correlations between track modulus and vertical deflections and bending moments of rail, sleeper, and subgrade surface have been studied and it can be seen that an increase in track modulus results in a decrease in all deflections and moments. The studies [74] also show that an increase in the track modulus results in an increase of the maximum vertical stress and the maximum deviator stress at the subgrade surface and that the location of these maximum stresses is at the outer end of the sleepers. Although higher track modulus leads to higher stresses in the subgrade, but does not imply poor track performance as it depends not only on the generated stresses but also on the soil strength. A higher track modulus also implies a stiffer subgrade and hence a higher strength to resist the higher stresses. The value of track modulus that is optimum to ensure no excessive stresses and deformation depends on many factors which will be discussed in detail in the following sections.

The above discussion highlights the importance of choosing the appropriate track stiffness for optimal performance of a track and the associated challenges in doing so. In addition to this, railway transition zone design is even more sensitive to fluctuations in track stiffness as an abrupt stiffness jump in these zones is a major cause for the initiation of the mechanisms leading to amplified degradation in these zones. Several attempts have been made in the past to provide a gradual stiffness gradient in these zones to minimize the degradation. Despite these attempts at smoothening the variation [24] in vertical stiffness along the longitudinal direction of the track by either increasing the stiffness on the softer side or by decreasing the stiffness of the stiffer side, the benefits were limited. The approach that involves a decrease in stiffness on the stiffer side was not effective in the mitigation of excessive settlement of subgrade due to heavy axle load, and other factors [11]. At the same time increasing the stiffness on the softer side led to a smoother stiffness transition and decrease in displacements but can lead to an increase in stresses. Moreover, there is a lack of understanding about the required uniformity of the system properties in the longitudinal direction to avoid or substantially reduce degradation. There have been some trials both at the site and in numerical models to test the mitigation measures aiming at gradual change of stiffness along the longitudinal axis of the track. The most popular ones include transition wedges, improvement of foundation by insertion of piles and approach slabs. The transition wedges are aimed at smoothening the track transition [58–63] and were proved to be effective to some extent for embankment bridge type of transition zone [45]. But for an effective implementation of transition wedges, a detailed analysis is required to make recommendations for thickness, slope, material and layer distribution of the wedge [75]. A comparative study of different wedge transitions based on material configuration and geometric shapes can be found in [59]. Some studies [76] have found an exponential increase in degradation after a year of operation for transition zones with wedges. In RTZ, ground improvement techniques include the insertion of piles which not only provide a gradual stiffness transition but also reduce the settlement on the softer side of these zones. The piles can be of different materials like concrete, steel, gravel, timber, sand or stone columns and depending on the length and material of these piles [49] the costs associated with this particular intervention can be very high. The effectiveness of piles at a transition zone must be ensured by arranging them in a proper pattern and by varying their lengths depending on the requirements of the particular transition. This could possibly prove to be an effective mitigation measure provided that it is well-designed and implemented for each specific transition zone. The construction of horizontal or inclined approach slabs [21, 22, 55–57] on both sides of the structure in RTZs is the most common practice in Europe to achieve a smoother transition. However, recent research by [56] shows that the track on an approach slab has about four and eight times higher vertical displacements than on an embankment and culvert, respectively. It is reported [77] in a detailed study that these higher displacements could be due to hanging sleepers on the approach slab and the tracks rocking under train movement due to a pivoting action of the approach slabs around the edges of the stiff culverts, all of which leads to high impact loading. A detailed overview of the abovementioned mitigation measures can be found in [32, 33] and the authors in [32] conclude that these mitigation measures have been utilized without adequate theoretical reasoning, and therefore cannot overcome the need for frequent maintenance. Although these mitigation measures or design practices aim towards reducing the inhomogeneity in RTZ but they have been implemented without the knowledge of required uniformity of the system properties or tolerance of the track components towards amplified stresses. An optimum solution must ensure that all the track components are within the allowable limits of stresses and deformations to insure reduced degradation and maintenance in long term. The following section will discuss the current design practices and mitigation measures concerning each track components in detail.

8.4 DESIGN OF TRACK COMPONENTS

The following sections discuss the standard design practices for each track component, performance evaluation in the context of transition zones and mitigation measures implemented in the past related to each of these track components (see Figure 8.2 for the most commonly used mitigation). The following critical components (shown in Figure 8.2) of railway tracks and more specifically transition zones will be discussed in this section:

Superstructure:

- Rail
- Sleepers or ties

Substructure:

- Ballast
- Embankment
- Subgrade or natural terrain.

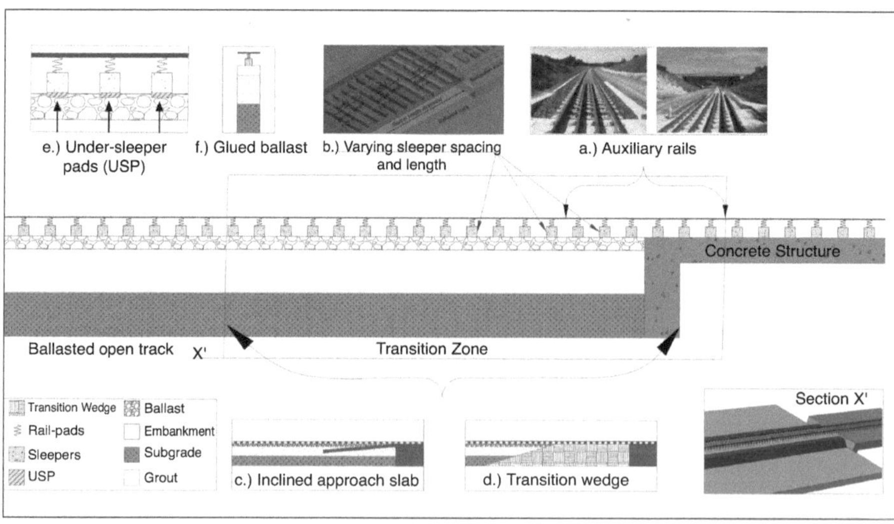

FIGURE 8.2 Cross-section details of an embankment-bridge transition showing some commonly used mitigation measures like (a) auxiliary rail [78], (b) larger and closely spaced sleepers [45], (c) inclined approach slab, (d) transition wedges, (e) glued ballast, (f) under-sleeper pads.

8.5 SUPERSTRUCTURE

8.5.1 Rail

The rail behaves as a beam and can be characterized by the bending parameters, stiffness and mass. The standard rail sections used in Europe are 54E1 and 60E1. Even though the stresses, deformations and wheel-rail interaction forces in transition zones at the rail level are much higher compared to the open tracks, the geometric and physical properties used to design the rail profiles in these zones are the same as open tracks. It is suggested that an increase in the height of the rails can result in a higher bending stiffness which implies a better distribution of loads over a higher number of sleepers, reducing the settlement of sleepers and the support layers (substructure). Even so, it is not a common practice to adopt different rail sections. Among other design modifications concerning rails that have been implemented, include the use of auxiliary rails [38, 45, 46] aimed at improving the bending stiffness of the track and reducing the stress in ballast by guaranteeing a better distribution of loads over the sleepers [45]. The comparison of rail deflections for different numbers of auxiliary rails (0,2,4,6) was made and it was concluded that the use of two auxiliary rails leads to lower rail deflection variations for all analysed speeds [46]. The addition of more than two auxiliary rails further decreases the rail deflection variation along the railway transition but the reduction is negligible. The numerical models suggest that the amplification of rail deflections in transition zones relative to open tracks changed from 35% to 40% (low-high velocity) to 28%–33% by the usage of auxiliary rails. However, the site measurements show a negligible effect of auxiliary rail on the degradation of RTZs. Even though the studies suggest a reduction in vertical rail displacements by adding additional rails, its influence on the permanent deformations of the trackbed layers is minimal.

8.5.1.1 Sleepers

Sleeper dimensions of 2.6 m × 0.24 m × 0.24 m and a spacing of 0.6 m is a standard for railway tracks which is also used in the case of transition zones most often. However, larger in size [38] and closely spaced sleepers were used in the transition area [49, 51, 52] to reduce the abrupt change in track stiffness. The larger sleepers were effective in reducing the ballast settlement and providing a better distribution of contact force between sleeper and ballast [45, 51] over a larger area. However, these design modifications did not affect dynamic load amplification in the transition zones [50]. The analysis of different sleeper configurations at the transition was done to evaluate track support separation in the vicinity of the slab track while maintaining good track performance which is assessed in terms of sleeper vertical displacements and stresses on ballast [48]. In addition to these, efforts have been made to minimize the maximum wheel-rail contact force and the maximum pressure between the sleeper and foundation [47] to obtain the most efficient distributions of sleeper spacing (and rail pad stiffness) adjacent to the transition. These superstructure-level mitigation measures are of great interest as they are cost-effective and installation of such interventions is easy and straightforward. However, in all these studies, the solutions have not been studied or evaluated for long-term performance and costs. Also, the effect of these measures is limited to reduced responses at the rail level or sleeper-ballast interface and it does not ensure or verify reduced degradation in ballast, embankment and subgrade.

8.5.1.2 Softening Pads

The other solutions that are easily implemented at the superstructure level are rails pads and under-sleeper pads [79]. The usage of rails pads improves the damping properties of the track [46, 47, 49, 53, 54] and under-sleeper pads mitigate the transition effects by minimizing the dynamic load impact [5, 6, 34, 50, 52–54, 80–82]. Although both rail pads and under-sleeper pads have been effective to some extent in reducing vertical rail displacements and/ or contact forces, their performance has been unsatisfactory based on a low benefit-cost ratio. Moreover, the influence of these pads on the permanent deformation of trackbed layers in RTZ is limited.

8.5.2 Substructure

8.5.2.1 Ballast

The ballast is the most vulnerable component of the track in transition zones, and unfortunately, the major part of the rail infrastructure consists of the ballasted track. The ballast is widely used in railway tracks due to the fact that it helps to distribute the loads to the underlying soil, dampens the dynamic loads and provides rapid drainage at low cost [83]. The ballast layer was found to be the primary cause of track settlement based on an analysis of the track settlement data (or permanent deformation) at the instrumented sites. Generally, ballast layers accounted for over 50% of the transient deformations experienced by the track under train load, based on a quantitative analysis of each layer's contribution [84].

The required depth of the ballast beneath the sleepers depends on the track use and operation requirements [85]. As a general rule, the minimum acceptable depth of ballast is 150 mm [85] and the recommended depth is 250–300 mm [83, 85] but it can go up to 500 mm for high-speed trains. Moreover, to ensure the lateral stability of the track, shoulder width and slope are two important parameters that play a major role. For speeds lower than 160 km/h, the recommended width is between 450 and 500 mm and larger widths (700 mm) must be adopted for higher speeds [86]. As for the slope of the ballast shoulder, an inclination of 2:3 (33°) is recommended which is lower than the average angle of internal friction of crushed stone ballast, to ensure ballast bed stability. In the end, different depths of ballast result in different vertical track stiffness and improper geometry (width and slope) leads to running away of ballast and these must be checked for their influence on the performance of RTZs. Although these standards have been established for years now, they refer to specifications for normal track conditions and have not been particularly designed for transition zones.

In RTZs, the design modifications that have aimed at reducing the dynamic effects on the ballast include concrete confinement walls and glued ballast. On one hand, some studies [49, 87] have shown that these measures are effective in dealing with problems like reduction in lateral movement and loosening of ballast. On the other hand, some results [33] have proved them to be counterproductive as increased confinement leads to an increased ballast breakage. Also, some studies have claimed that the use of hot mix asphalt under the ballast [49, 61, 87, 88] reduces the stress in ballast [89] and subgrade [90], increases drainage capacity of tracks [61], and hence reduces maintenance cycles. There is no evidence of these measures being effective in mitigating the operation-induced amplified degradation in RTZs.

8.5.2.2 Embankment and Subgrade

The required depth of the embankment or sub-ballast beneath the ballast layer is 150–450 mm [83] and the minimum acceptable depth of sub-ballast is 100 mm. Underlying the ballast bed and embankment, there is natural soil at the site of the railway track or other granular material that can be used to replace natural soil for better track performance. The subgrade must have sufficient bearing strength and stiffness to support the weight of the rail track structure and the vehicle loads and have high compactness to minimize the settlements due to these loads [83, 85].

The design standards and specifications mentioned above are the basis for the design of ballasted railway tracks but for the design requirements of transition zones specifically, a detailed analysis to obtain the optimal material properties of these trackbed layers is lacking in the literature. The authors in [91] have developed mathematical expressions to obtain vertical track stiffness as a function of height of embankment (h_{EM}) and the modulus of elasticity of embankment (E_{EM}) and natural ground (E_{NG}). This has been done by graphical analysis of results from a finite element simulation of the track using elastic rail, rail pads, sleepers and perfectly elastoplastic ballast, embankment and natural ground. In total 120 possible combinations of h_{EM} (3, 5, 7, 10 m), E_{EM} (25, 40, 60, 80, 160 MPa), E_{NG} (25, 40, 60, 80 MPa) with and without formation layer (E_{FL}) of 0.6 m and a fixed depth of 6 m for natural terrain. The study concludes that the vertical stiffness variation is inversely proportional to the height of the embankment and the dependence on this height is independent of the elasticity modulus of the embankment and natural ground. At the same time, this stiffness variation is dependent on the combination of the elasticity modulus of the embankment and natural soil irrespective of the embankment height variation. The results of [73, 91] emphasized the importance of geometric (depths and slopes) and physical properties (elasticity modulus) of ballast, embankment and natural terrain on the vertical track stiffness (static) and established important relationships and design recommendations. At the same time, these studies have been performed with an assumption that for K ranging from 50 to 60 kN/mm, the dynamic value of K is only 2% larger than static. Additionally, the identification of influential geometric and material parameters of embankment and subgrade layer, that significantly affect the dynamic response of the RTZs is lacking and hence the choice of studied parameters has no basis. Moreover, the studies discussed above are all aimed at establishing standard height and elastic modulus of embankment and subgrade for ballasted railway tracks and the standard values of these parameters are unknown for RTZs. As these parameters eventually affect the vertical track stiffness, they are bound to influence the behaviour of RTZs and need to be investigated.

8.6 A COMPREHENSIVE DESIGN METHODOLOGY

To summarise the discussion presented thus far and establish a common ground for a comprehensive design methodology, the following points can be drawn out:

- The abrupt change in vertical stiffness and differential settlement are the main reasons for amplified response in transition zones. On the one hand, the difference in autonomous settlement of the ballasted track and the concrete structure is unavoidable. On the other hand, the operation-driven dynamic amplifications due to abrupt changes in vertical stiffness can be controlled to a great extent with a robust design solution. The consequences of the autonomous differential settlement must be controlled by advanced monitoring systems and timely maintenance.
- No existing design practices aim at a specialized design of all track components to address the mechanisms associated with the amplified degradation of RTZs. The existing design involves mitigating the transition effects either at the substructure or superstructure level.
- Different track components affect the equivalent vertical track stiffness to various degrees. Geometry and material parameters of trackbed layers significantly affect the vertical track stiffness and the track performance. However, the effects of these are unknown for RTZs.
- There is no design methodology that aims towards the reduction of both induced stresses and deformations in "all" track components simultaneously for RTZs.
- There is a lack of clarity regarding the mechanisms leading to an amplified stress-strain state in transition zones.
- The current literature lacks consensus and required information on an appropriate criterion for the evaluation/ design of RTZs.
- A comprehensive evaluation of existing mitigation measures using one single criterion is missing in the literature.
- The behaviour of RTZs is sensitive to critical load states like changes in load velocity.

Based on the points mentioned above and the insights obtained from the existing literature, it is established that there is a need for a comprehensive design methodology to deal with the problem of amplified degradation in RTZs. Figure 8.3 shows the systematic methodology for the design of a transition structure aimed at minimizing the operation-driven degradation of RTZs. This section summarizes the work done in each step of the proposed design methodology eventually leading to a final design of an embankment-bridge transition.

The first step (Section 8.6.1) is the formulation of a criterion to evaluate the existing mitigation measures and to design a novel mitigation measure that aims at minimizing the degradation due to transition effects.

The energy-based criterion proposed in Section 8.6.2 is used to evaluate the performance of existing mitigation measures on both substructure levels (Section 8.6.2) and superstructure (Section 8.6.3) using several numerical models. On one hand, the evaluation of existing mitigation measures on the superstructure level (large or closely spaced sleepers) proved them to be inefficient according to the energy-based criterion. On the other hand, the evaluation of interventions aiming at gradual vertical stiffness change at the substructure level in RTZs like approach slabs and traditional transition wedge proved that smoothening the stiffness gradient only cannot lead to mitigation of dynamic amplifications in RTZs. It was found that the energy was channelled to the different parts of the system (most prone to degradation) due to the geometry of these transition structures (approach slabs and transition wedge). This led to the investigation of key phenomena of energy absorption (by using infinite elements at the transition interface) and uniform energy distribution within the trackbed layers (by using a homogeneous foundation) in Section 8.6.4. The results of these abovementioned investigations (Sections 8.6.2, 8.6.3 and 8.6.4) led to the preliminary design of the transition structure. This preliminary design is based on energy-limiting design principles that aim at guiding the energy away from the transition zones such that the energy is neither concentrated nor obstructed in any part of the system and maintains a smooth stiffness gradient at the same time. A two-dimensional model of an embankment-bridge transition equipped with a safe hull-inspired energy limiting design (SHIELD) is used to evaluate this new mitigation measure against the traditional mitigation measures for ideal (straight track) and non-ideal conditions (with hanging sleepers). SHIELD outperformed all the most commonly used mitigation measures like horizontal and inclined approach slabs, transition wedges and modified sleeper spacing and configurations.

The preliminary design proposed in the previous step is then subjected to rigorous optimization and verification processes. The optimization is mainly carried out in two parts for parameters associated with material characterization (Section 8.6.5) and geometry (Section 8.6.6) using surrogate models obtained by polynomial chaos expansion and various three-dimensional finite element models respectively. Subsequently, the optimized design of SHIELD was subjected to critical loading conditions (Section 8.6.7) to propose the final design of the transition structure for the embankment-bridge transition under study.

In Section 8.6.8, the proposed design methodology (Figure 8.3) was extended to another type of embankment-bridge transition where the ballast layer continues over the bridge demonstrating a wider applicability of the proposed design methodology and the possibility of applying the same design principles and methods to the other types of RTZs. The following subsections discuss briefly the contents of the steps shown in the flowchart (Figure 8.3).

8.6.1 Design/Evaluation Criteria

In [92], a two-step method is outlined for formulating a preliminary design criterion to delay uneven track geometry in RTZs. It begins with a systematic evaluation of all track components in an RTZ using a finite element model in terms of variations in kinematic responses, stresses, and kinetic and strain energies. The study then proposes an energy-based criterion, using a model with linear elastic behaviour, indicating that amplified strain energy near the transition interface suggests greater degradation compared to open tracks. This correlation between amplified strain energy and excessive

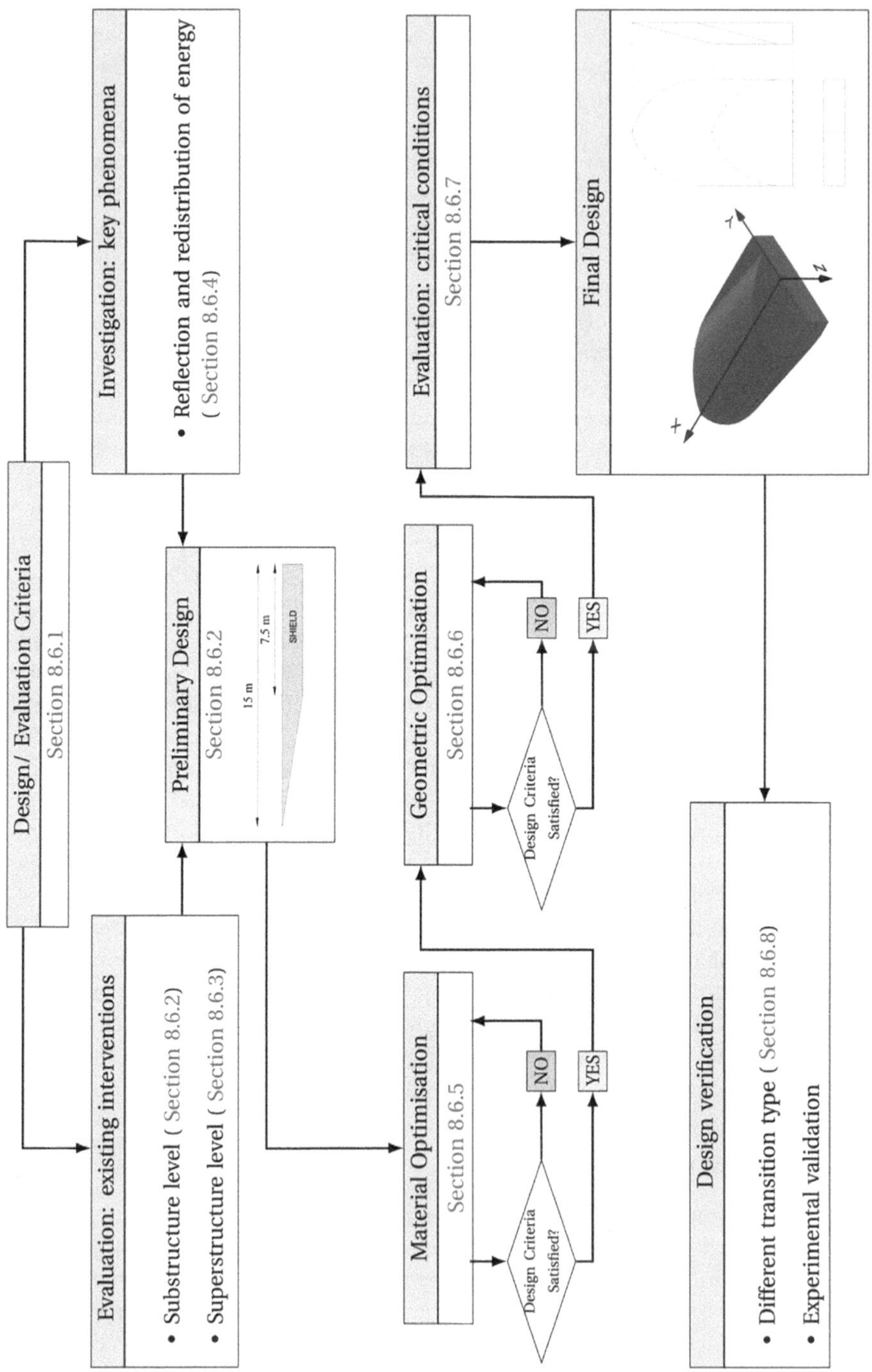

FIGURE 8.3 Methodology for evaluation and design of railway transition zones.

permanent deformations is further confirmed through a non-linear elastoplastic model of the ballast layer. The study concludes that minimizing total strain energy and ensuring its uniform distribution along the longitudinal direction of the track can prevent degradation and maintain a uniform track geometry.

8.6.2 Evaluation of Existing Interventions: Substructure Level

In [93], the authors assess the effectiveness of standard substructure-level mitigation techniques and introduce a novel SHIELD for transition structures. The evaluation begins with traditional transition structures such as horizontal and inclined approach slabs and transition wedges, using established criteria based on minimizing total strain energy. The study then examines the preliminary design of the SHIELD transition structure using the same energy-based criterion. A comprehensive analysis using a 2-dimensional finite element model demonstrates the superiority of SHIELD in managing energy flow at transition zones and explains why other common transition structures fall short. This step confirms the effectiveness and versatility of the new energy-based criterion across different transition zones and underscores SHIELD's potential as a robust design solution.

8.6.3 Evaluation of Existing Interventions: Superstructure Level

In [94], the study explores the influence of various sleeper configurations and reduced sleeper spacing on dynamic amplifications in RTZs, using a criterion based on total strain energy across trackbed layers (ballast, embankment, and subgrade). It also examines the effects of sleepers losing contact with the ballast (hanging sleepers), typically due to differential settlement. The first part of the research provides insights into how design interventions by adjusting sleeper configurations and spacing influence the dynamic behaviour of transition zones, while the second part emphasizes the necessity of addressing sleeper-ballast contact loss. The findings indicate that completely mitigating transition effects with these superstructure-level interventions is not possible.

8.6.4 Investigation of Key Phenomena Governing Design

In [95], several finite element models are employed to investigate the amplification of total strain energy due to the phenomena of reflection and redistribution of energy close to the transition interface. The models used in this work include non-reflective boundary (representing an energy sink) and homogeneous material along the vertical direction of the track, and the responses are studied for individual and combined effects in comparison to a benchmark case. The results show that eliminating the phenomena results in no dynamic amplification in total strain energy in RTZs and a potential design solution must aim at addressing these phenomena.

8.6.5 Material Optimization

In [96], the study evaluates the design challenges posed by variations in material properties across the depth and length of the track, and how these change over time due to external factors during operation. It explores the impact of these variations, particularly stiffness distributions, on the behaviour of RTZs, tests the performance of a new transition structure called SHIELD, and develops a method to optimize material properties to reduce dynamic amplifications in these zones. Findings reveal that stiffness variations significantly affect RTZs' dynamic responses. The study suggests a permissible range of stiffness ratios to manage strain energy amplification in RTZs, both initially and over time. Additionally, the optimization framework demonstrates SHIELD's adaptability to a wide range of material properties.

8.6.6 Geometric Optimization

In [97], a heuristic approach is used to optimize the geometry of the SHIELD transition structure and assess its impact on strain energy redistribution in RTZs. Several three-dimensional finite element models evaluate different SHIELD geometric profiles systematically. Modifications to SHIELD's geometry along longitudinal, transversal, and vertical directions explore how these changes affect strain energy redistribution across trackbed layers (ballast, embankment, and subgrade) in terms of spatial and temporal energy distributions. The study finds that SHIELD's longitudinal profile significantly impacts strain energy distribution in the transition zones, the transversal profile mainly affects the ballast layer, and the vertical profiles enhance local strain energy distribution uniformity near the transition interface. This preliminary optimization suggests how full-scale optimization might achieve transition structure shapes that evenly distribute energy without concentration and/ or obstruction of energy in RTZs.

8.6.7 Evaluation for Critical Conditions

In this phase [98], SHIELD is tested under critical conditions including various velocities (sub-critical, critical, super-critical), moving load directions, combinations of inertial effects with track imperfections (like non-straight rails and hanging sleepers), and multiple axle passages using comprehensive vehicle and finite element models of the RTZs. Results show that SHIELD effectively reduces dynamic amplifications and achieves a uniform strain energy distribution across different velocity regimes and movement directions, suggesting minimized operation-induced degradation longitudinally. Moreover, although designed for ideal track conditions, SHIELD also successfully mitigates issues from track imperfections such as hanging sleepers and non-straight rails, proving its robustness under all tested critical conditions.

8.6.8 Design Verification

Jain et al. [99], a SHIELD of a transition structure was proposed for an embankment-bridge transition (without a ballast layer over the bridge) to deal with operation-induced degradation. In this particular step, the scope of this mitigation measure is extended for an embankment-bridge transition with ballast running over the bridge and its performance is evaluated using a strain-energy criterion. It was concluded in the end, that the SHIELD can effectively mitigate the operation-induced dynamic amplification for more than one type of railway transition zone.

Figures 8.4 and 8.5 show the energy fields moving with the moving load (one axle) in an embankment-bridge transition with and without SHIELD. On one hand (Figure 8.4), it can be clearly seen that the energy is obstructed and amplified (due to reflections and interference) in the proximity of the transition interface in the absence of SHIELD. On the other hand (Figure 8.5), the presence of SHIELD facilitates an uninterrupted energy flow leading to a smoother transition of energy from one zone to the other.

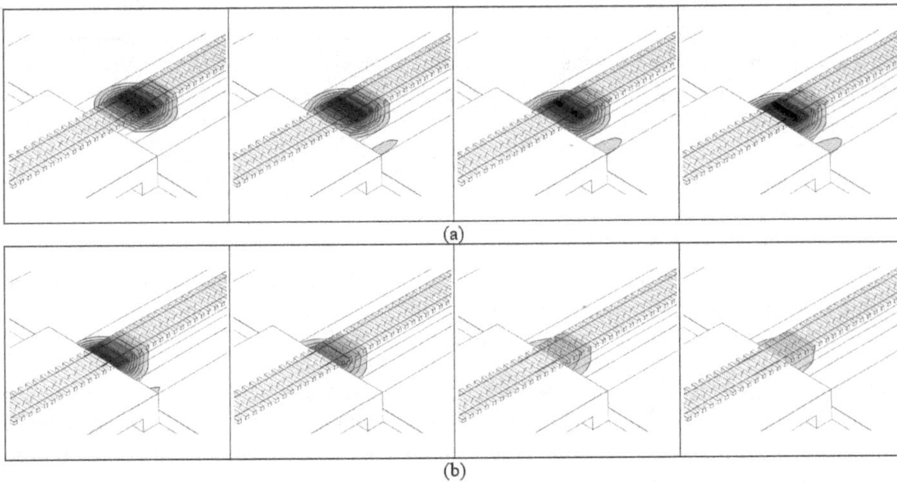

FIGURE 8.4 Energy fields in an embankment-bridge transition for (a) a moving load approaching a transition and (b) after the moving load has moved past the transition interface.

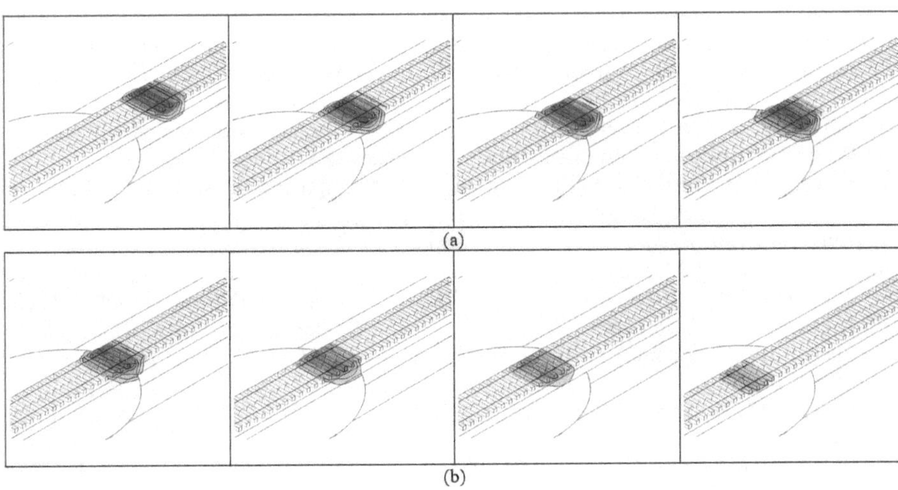

FIGURE 8.5 Energy fields in an embankment-bridge transition equipped with SHIELD for (a) a moving load approaching SHIELD and (b) after the moving load has moved past the interface with SHIELD.

8.7 SUMMARY AND FUTURE SCOPE

The proposed methodology provides a preliminary design and a good starting point to designers and engineers for practical implementation of SHIELD to mitigate transition effects. However, further investigations are needed to develop a detailed material model of SHIELD and to suggest real materials that could be representative of the properties proposed in [96]. Furthermore, the energy-based design/evaluation criterion must be quantified by laboratory experiments for the materials forming trackbed layers and SHIELD. This can be done by identifying the energy limits associated with the failure of the corresponding materials. Future investigations must include evaluation of the long-term performance of SHIELD and the proposed methodology must be adapted according to specific site conditions and the type of transition considered.

REFERENCES

1. Coelho, B., P. Hölscher, J. Priest, W. Powrie, and F. Barends. 2011. "An Assessment of Transition Zone Performance." *Proceedings of the Institution of Mechanical Engineers, Part F: Journal of Rail and Rapid Transit* 225 (2): 129–139.
2. Coelho, B. E. Z. 2011. *Dynamics of Railway Transition Zones in Soft Soils.* PhD diss., Delft University of Technology.
3. Paixão, A., E. Fortunato, and R. Calçada. 2015. "A Numerical Study on the Influence of Backfill Settlements in the Train/Track Interaction at Transition Zones to Railway Bridges." *Proceedings of the Institution of Mechanical Engineers, Part F: Journal of Rail and Rapid Transit* 230 (3): 866–878.
4. Wang, H., M. Silvast, V. Markine, and B. Wiljanen. 2017. "Analysis of the Dynamic Wheel Loads in Railway Transition Zones Considering the Moisture Condition of the Ballast and Subballast." *Applied Sciences* 7 (12): 1208.
5. Jain, A., K. N. van Dalen, A. V. Metrikine, A. B. Faragau, and M. J. M. M. Steenbergen. 2023. "Comparative Analysis of the Dynamic Amplifications Due to Inhomogeneities at Railway Transition Zones." In *Proceedings of The Fifth International Conference on Railway Technology: Research, Development and Maintenance*, edited by J. Pombo, Vol. CCC 1. Civil-Comp Press, Edinburgh, Scotland.
6. Wang, H., and V. Markine. 2018. "Corrective Countermeasure for Track Transition Zones in Railways: Adjustable Fastener." *Engineering Structures* 169: 1–14.
7. Paixão, A., E. Fortunato, and R. Calçada. 2013. "Design and Construction of Backfills for Railway Track Transition Zones." *Proceedings of the Institution of Mechanical Engineers, Part F: Journal of Rail and Rapid Transit* 229 (1): 58–70.
8. Wang, H., and V. Markine. 2019. "Dynamic Behaviour of the Track in Transition Zones Considering the Differential Settlement." *Journal of Sound and Vibration* 459: 114863.
9. Labrado, P. M., F. R. Barceló, F. R. Llario, and J. R. Herráiz. 2018. "Effect of Vehicle Speed on the Dynamics of Track Transitions." *Journal of Vibration and Control* 24 (21): 5118–5128.
10. Gallage, C., B., S. S. S. Dareeju, and M. Dhanasekar. 2013. "State-of-the-Art: Track Degradation at Bridge Transitions." In *Proceedings of the 4th International Conference on Structural Engineering and Construction Management 2013*, edited by K. P. P. Pathirana, pp. 40–52. Nethwin Printers Ltd, Sri Lanka.
11. Wang, H., V. Markine, and X. Liu. 2017. "Experimental Analysis of Railway Track Settlement in Transition Zones." *Proceedings of the Institution of Mechanical Engineers, Part F: Journal of Rail and Rapid Transit* 232 (6): 1774–1789.

12. Wang, H., and V. L. Markine. 2018. "Methodology for the Comprehensive Analysis of Railway Transition Zones." *Computers and Geotechnics* 99: 64–79.
13. Wang, H., and V. Markine. 2018. "Modelling of the Long-Term Behaviour of Transition Zones: Prediction of Track Settlement." *Engineering Structures* 156: 294–304.
14. Li, D., D. Otter, and G. Carr. 2010. "Railway Bridge Approaches under Heavy Axle Load Traffic: Problems, Causes, and Remedies." *Proceedings of the Institution of Mechanical Engineers, Part F: Journal of Rail and Rapid Transit* 224 (5): 383–390.
15. Stark, T. D., and S. T. Wilk. 2015. "Root Cause of Differential Movement at Bridge Transition Zones." *Proceedings of the Institution of Mechanical Engineers, Part F: Journal of Rail and Rapid Transit* 230 (4): 1257–1269.
16. Wang, H., L. Chang, and V. Markine. 2018. "Structural Health Monitoring of Railway Transition Zones Using Satellite Radar Data." *Sensors* 18 (2): 413.
17. Li, D., and D. Davis. 2005. "Transition of Railroad Bridge Approaches." *Journal of Geotechnical and Geoenvironmental Engineering* 131 (11): 1392–1398.
18. Paixão, A., E. Fortunato, and R. Calçada. 2014. "Transition Zones to Railway Bridges: Track Measurements and Numerical Modelling." *Engineering Structures* 80: 435–443.
19. Ognibene, G., W. Powrie, L. L. Pen, and J. Harkness. 2019. "Analysis of a Bridge Approach: Long-Term Behaviour from Short-Term Response." In *15th Railway Engineering Conference*, Edinburgh, Scotland.
20. Ribeiro, C. A., R. Calçada, and R. Delgado. 2017. "Calibration and Experimental Validation of a Dynamic Model of the Train-Track System at a Culvert Transition Zone." *Structure and Infrastructure Engineering* 14 (5): 604–618.
21. Varandas, J. N., P. Hölscher, and M. A. G. Silva. 2011. "Dynamic Behaviour of Railway Tracks on Transition Zones." *Computers & Structures* 89 (13–14): 1468–1479.
22. Varandas, J. N., P. Hölscher, and M. A. G. Silva. 2016. "Three-Dimensional Track-Ballast Interaction Model for the Study of a Culvert Transition." *Soil Dynamics and Earthquake Engineering* 89: 116–127.
23. Real-Herráiz, J., C. Zamorano-Martín, T. Real-Herráiz, and S. Morales-Ivorra. 2016. "New Transition Wedge Design Composed by Prefabricated Reinforced Concrete Slabs." *Latin American Journal of Solids and Structures* 13 (8): 1431–1449.
24. Lei, X., and B. Zhang. 2010. "Influence of Track Stiffness Distribution on Vehicle and Track Interactions in Track Transition." *Proceedings of the Institution of Mechanical Engineers, Part F: Journal of Rail and Rapid Transit* 224 (6): 592–604.
25. Le Pen, L., G. Watson, W. Powrie, G. Yeo, P. Weston, and C. Roberts. 2014. "The Behaviour of Railway Level Crossings: Insights through Field Monitoring." *Transportation Geotechnics* 1 (4): 201–213.
26. Paixão, A., C. A. Ribeiro, N. Pinto, E. Fortunato, and R. Calçada. 2014. "On the Use of Under Sleeper Pads in Transition Zones at Railway Underpasses: Experimental Field Testing." *Structure and Infrastructure Engineering* 11 (2): 112–128.
27. Ribeiro, C. A., A. Paixão, E. Fortunato, and R. Calçada. 2014. "Under Sleeper Pads in Transition Zones at Railway Underpasses: Numerical Modelling and Experimental Validation." *Structure and Infrastructure Engineering* 11 (11): 1432–1449.
28. Ramos, A., A. G. Correia, R. Calçada, and D. P. Connolly. 2022. "Ballastless Railway Track Transition Zones: An Embankment to Tunnel Analysis." *Transportation Geotechnics* 33: 100728.
29. Barbosa, J. M. de O., and K. N. van Dalen. 2019. "Dynamic Response of an Infinite Beam Periodically Supported by Sleepers Resting on a Regular and Infinite Lattice: Semi-Analytical Solution." *Journal of Sound and Vibration* 458: 276–302.
30. Metrikine, A. V., A. R. M. Wolfert, and H. A. Dieterman. 1998. "Transition Radiation in an Elastically Supported String: Abrupt and Smooth Variations of the Support Stiffness." *Wave Motion* 27 (4): 291–305.

31. Wang, H. 2019. *Measurement, Assessment, Analysis and Improvement of Transition Zones in Railway Track*. PhD thesis, Delft University of Technology.
32. Indraratna, B., M. B. Sajjad, T. Ngo, A. G. Correia, and R. Kelly. 2019. "Improved Performance of Ballasted Tracks at Transition Zones: A Review of Experimental and Modelling Approaches." *Transportation Geotechnics* 21: 100260.
33. Sañudo, R., L. dell'Olio, J. A. Casado, I. A. Carrascal, and S. Diego. 2016. "Track Transitions in Railways: A Review." *Construction and Building Materials* 112: 140–157.
34. Ferreira, J. N. V. da S. 2013. *Long-Term Behaviour of Railway Transitions under Dynamic Loading: Application to Soft Soil Sites*. PhD thesis, Delft University of Technology.
35. Coelho, L. 2008. *Structure/Embankment Transitions in Railway Infrastructures: Behaviour and National and International Practices*. PhD thesis, Delft University of Technology.
36. Selig, E. T., and J. M. Waters. 1994. *Track Geotechnology and Substructure Management*. Thomas Telford Publishing, London.
37. Li, D., D. Otter, and G. Carr. 2010. "Railway Bridge Approaches under Heavy Axle Load Traffic: Problems, Causes, and Remedies." *Proceedings of the Institution of Mechanical Engineers, Part F: Journal of Rail and Rapid Transit* 224 (5): 383–390.
38. Shahraki, M., and K. J. Witt. 2015. "3D Modeling of Transition Zone between Ballasted and Ballastless High-Speed Railway Track." *Journal of Traffic and Transportation Engineering* 3 (4): 1–12.
39. Banimahd, M., P. K. Woodward, J. Kennedy, and G. M. Medero. 2012. "Behaviour of Train–Track Interaction in Stiffness Transitions." *Proceedings of the Institution of Civil Engineers - Transport* 165 (3): 205–214.
40. Vesnitskii, A. I., and A. V. Metrikin. 1992. "Transition Radiation in One-Dimensional Elastic Systems." *Journal of Applied Mechanics and Technical Physics* 33 (2): 202–207.
41. Metrikine, A. V., and A. C. W. M. Vrouwenvelder. 2000. "Surface Ground Vibration Due to a Moving Train in a Tunnel: Two-Dimensional Model." *Journal of Sound and Vibration* 234 (1): 43–66.
42. Metrikine, A. V., A. R. M. Wolfert, and H. A. Dieterman. 1998. "Transition Radiation in an Elastically Supported String: Abrupt and Smooth Variations of the Support Stiffness." *Wave Motion* 27 (4): 291–305.
43. Van Dalen, K. N., and A. V. Metrikine. 2008. "Transition Radiation of Elastic Waves at the Interface of Two Elastic Half-Planes." *Journal of Sound and Vibration* 310 (3): 702–717.
44. Van Dalen, K. N., A. Tsouvalas, A. V. Metrikine, and J. S. Hoving. 2015. "Transition Radiation Excited by a Surface Load That Moves over the Interface of Two Elastic Layers." *International Journal of Solids and Structures* 73–74: 99–112.
45. Shahraki, M., C. Warnakulasooriya, and K. J. Witt. 2015. "Numerical Study of Transition Zone between Ballasted and Ballastless Railway Track." *Transportation Geotechnics* 3: 58–67.
46. Heydari-Noghabi, H., J. N. Varandas, M. Esmaeili, and J. Zakeri. 2017. "Investigating the Influence of Auxiliary Rails on Dynamic Behavior of Railway Transition Zone by a 3D Train-Track Interaction Model." *Latin American Journal of Solids and Structures* 14 (11): 2000–2018.
47. Aggestam, E., and J. C.O. Nielsen. 2019. "Multi-Objective Optimisation of Transition Zones Between Slab Track and Ballasted Track Using a Genetic Algorithm." *Journal of Sound and Vibration* 446: 91–112.
48. Sañudo, R., M. Cerrada, B. Alonso, and L. dell'Olio. 2017. "Analysis of the Influence of Support Positions in Transition Zones: A Numerical Analysis." *Construction and Building Materials* 145: 207–217.
49. Read, D., and D. Li. 2006. *Design of Track Transitions*. The National Academies Press, Washington, DC.

50. Nicks, J. E. 2009. *The Bump at the End of the Railway Bridge*. PhD thesis, Texas A&M University.
51. Namura, A., and T. Suzuki. 2007. "Evaluation of Countermeasures against Differential Settlement at Track Transitions." *Quarterly Report of RTRI* 48 (3): 176–182.
52. Namura, A., Y. Kohata, and S. Miura. 2004. "Effect of Sleeper Size on Ballasted Track Settlement." *Quarterly Report of RTRI* 45 (3): 156–161.
53. Dahlberg, T. 2004. *Railway Track Settlements: A Literature Review*. https://api.semanticscholar.org/CorpusID:40237712
54. Sol-Sánchez, M., F. Moreno-Navarro, and C. Rubio-Gámez. 2015. "The Use of Elastic Elements in Railway Tracks: A State-of-the-Art Review." *Construction and Building Materials* 75: 293–305.
55. Cai, C. S., X. M. Shi, G. Z. Voyiadjis, and Z. J. Zhang. 2005. "Structural Performance of Bridge Approach Slabs Under Given Embankment Settlement." *Journal of Bridge Engineering* 10 (4): 482–489.
56. Coelho, B. Z., J. Priest, and P. Hölscher. 2018. "Dynamic Behaviour of Transition Zones in Soft Soils during Regular Train Traffic." *Proceedings of the Institution of Mechanical Engineers, Part F: Journal of Rail and Rapid Transit* 232 (3): 645–662.
57. Sasaoka, D., and D. Davis. 2005. *Implementing track transition solutions for heavy axle load service*. https://www.semanticscholar.org/paper/Implementing-Track-Transition-Solutions-for-Heavy-Sasaoka-Davis/a26bc934a173cdb910211c65fae6fc1545b4251f
58. Coelho, B., P. Hölscher, J. Priest, W. Powrie, and F. Barends. 2011. "An Assessment of Transition Zone Performance." *Proceedings of the Institution of Mechanical Engineers, Part F: Journal of Rail and Rapid Transit* 225 (2): 129–139.
59. Fortunato, E., A. Paixão, and R. Calçada. 2013. "Railway Track Transition Zones: Design, Construction, Monitoring and Numerical Modelling." *International Journal of Railway Technology* 2 (4): 33–58.
60. Hu, P., C. Zhang, S. Wen, and Y. Wang. 2019. "Dynamic Responses of High-Speed Railway Transition Zone with Various Subgrade Fillings." *Computers and Geotechnics* 108: 17–26.
61. Li, D., and D. Davis. 2005. "Transition of Railroad Bridge Approaches." *Journal of Geotechnical and Geoenvironmental Engineering* 131 (11): 1392–1398.
62. Paixão, A., E. Fortunato, and R. Calçada. 2015. "A Numerical Study on the Influence of Backfill Settlements in the Train/Track Interaction at Transition Zones to Railway Bridges." *Proceedings of the Institution of Mechanical Engineers, Part F: Journal of Rail and Rapid Transit* 230 (3): 866–878.
63. Paixão, A., J. N. Varandas, E. Fortunato, and R. Calçada. 2018. "Numerical Simulations to Improve the Use of Under Sleeper Pads at Transition Zones to Railway Bridges." *Engineering Structures* 164: 169–182.
64. Hunt, H. E. M. 1997. "Settlement of Railway Track near Bridge Abutments." *Proceedings of the Institution of Civil Engineers - Transport* 123 (1): 68–73.
65. Stark, T. D., and S. T. Wilk. 2015. "Root Cause of Differential Movement at Bridge Transition Zones." *Proceedings of the Institution of Mechanical Engineers, Part F: Journal of Rail and Rapid Transit* 230 (4): 1257–1269.
66. Lundqvist, A., and T. Dahlberg. 2005. "Load Impact on Railway Track Due to Unsupported Sleepers." *Proceedings of the Institution of Mechanical Engineers, Part F: Journal of Rail and Rapid Transit* 219 (2): 67–77.
67. Sysyn, M., M. Przybylowicz, O. Nabochenko, and L. Kou. 2021. "Identification of Sleeper Support Conditions Using Mechanical Model Supported Data-Driven Approach." *Sensors* 21 (11): 3609.

68. Zhu, J. Y., D. J. Thompson, and C. J. C. Jones. 2011. "On the Effect of Unsupported Sleepers on the Dynamic Behaviour of a Railway Track." *Vehicle System Dynamics* 49 (9): 1389–1408.
69. Takahashi, R., K. Hayano, T. Nakamura, and Y. Momoya. 2019. "Integrated Risk of Rail Buckling in Ballasted Tracks at Transition Zones and Its Countermeasures." *Soils and Foundations* 59 (2): 517–531.
70. Jing, G., Q. Luo, Z. Wang, and Y. Shen. 2015. "Micro-Analysis of Hanging Sleeper Dynamic Interactions with Ballast Bed." Journal of Vibroengineering 17 (1): 444.
71. Bezin, Y., S. D. Iwnicki, M. Cavalletti, E. de Vries, F. Shahzad, and G. Evans. 2009. "An Investigation of Sleeper Voids Using a Flexible Track Model Integrated with Railway Multi-Body Dynamics." *Proceedings of the Institution of Mechanical Engineers, Part F: Journal of Rail and Rapid Transit* 223 (6): 597–607.
72. Sysyn, M., O. Nabochenko, and V. Kovalchuk. 2020. "Experimental Investigation of the Dynamic Behavior of Railway Track with Sleeper Voids." *Railway Engineering Science* 28 (3): 290–304.
73. Gallego Giner, I., and A. L. Pita. 2009. "Numerical Simulation of Embankment–Structure Transition Design." *Proceedings of the Institution of Mechanical Engineers, Part F: Journal of Rail and Rapid Transit* 223 (4): 331–343.
74. Selig, E., and D. Li. 1994. "Track Modulus: Its Meaning and Factors Influencing It." *Transportation Research Record* 1470 (4): 47–54.
75. Gallego Giner, I., A. L. Pita, E. W. V. Chaves, and A. M. R. Álvarez. 2012. "Design of Embankment Structure Transitions for Railway Infrastructure." *Proceedings of the Institution of Civil Engineers - Transport* 165 (1): 27–37.
76. Fara, A. 2014. *Transition Zones for Railway Bridges: A Study of the Sikan Bridge.* Master's thesis, Luleå University of Technology.
77. Froio, D., E. Rizzi, F. M. F. Simões, and A. P. Da Costa. 2018. "Universal Analytical Solution of the Steady-State Response of an Infinite Beam on a Pasternak Elastic Foundation Under Moving Load." *International Journal of Solids and Structures* 132–133: 245–263.
78. Sañudo, R., I. Jardí, J.-C. Martínez, F.-J. Sánchez, M. Miranda, B. Alonso, L. dell'Olio, and J.-L. Moura. 2022. "Monitoring Track Transition Zones in Railways." *Sensors* 22 (1): 76.
79. Insa, R., P. Salvador, J. Inarejos, and A. Roda. 2011. "Analysis of the Influence of Under Sleeper Pads on the Railway Vehicle/Track Dynamic Interaction in Transition Zones." *Proceedings of the Institution of Mechanical Engineers, Part F: Journal of Rail and Rapid Transit* 226 (4): 409–420
80. Abadi, T., L. L. Pen, A. Zervos, and W. Powrie. 2019. "Effect of Sleeper Interventions on Railway Track Performance." *Journal of Geotechnical and Geoenvironmental Engineering* 145 (4): 04019009.
81. Insa, R., P. Salvador, J. Inarejos, and A. Roda. 2012. "Analysis of the Influence of Under Sleeper Pads on the Railway Vehicle/Track Dynamic Interaction in Transition Zones." *Proceedings of the Institution of Mechanical Engineers, Part F: Journal of Rail and Rapid Transit* 226 (4): 409–420.
82. Navaratnarajah, S. K., B. Indraratna, and N. T. Ngo. 2018. "Influence of Under Sleeper Pads on Ballast Behavior under Cyclic Loading: Experimental and Numerical Studies." *Journal of Geotechnical and Geoenvironmental Engineering* 144 (9): 04018068.
83. Esveld, C. 2001. *Modern Railway Track.* 2nd ed. MRT Productions, Brahampura, Muzaffarpur.
84. Mishra, D., E. Tutumluer, H. Boler, J. P. Hyslip, and T. R. Sussmann. 2014. "Railroad Track Transitions with Multidepth Deflectometers and Strain Gauges." *Transportation Research Record* 2448 (1): 105–114.

85. Bonnett, C. 2005. *Practical Railway Engineering.* Imperial College Press, London. https://doi.org/10.1142/p373.
86. Rodrigues, A. F. da S. 2017. *Viability and Applicability of Simplified Models for the Dynamic Analysis of Ballasted Railway Tracks.* PhD thesis, https://www.scipedia.com/public/Rodrigues_2017a.
87. Wilk, S. T., T. D. Stark, and J. G. Rose. 2016. "Evaluating Tie Support at Railway Bridge Transitions." *Proceedings of the Institution of Mechanical Engineers, Part F: Journal of Rail and Rapid Transit* 230 (4): 1336–1350.
88. Labrado, P. M., F. R. Barceló, F. R. Llario, and J. R. Herráiz. 2018. "Effect of Vehicle Speed on the Dynamics of Track Transitions." *Journal of Vibration and Control* 24 (21): 5118–5128.
89. Rose, J. G., L. A. Walker, and D. Li. 2001. *Heavy-Haul Asphalt (HMA) Underlayment Trackbeds: Pressures/Deflections/Materials Properties Measurements.* CSX Transportation, Lexington, KY.
90. Sayeed, Md. A., and M. A. Shahin. 2018. "Design of Ballasted Railway Track Foundations Using Numerical Modelling. Part I: Development." *Canadian Geotechnical Journal* 55 (3): 353–368.
91. Gallego Giner, I., J. Muñoz, A. Rivas, and S. Sánchez-Cambronero. 2011. "Vertical Track Stiffness as a New Parameter Involved in Designing High-Speed Railway Infrastructure." *Journal of Transportation Engineering* 137 (12): 971–979.
92. Jain, A., A. V. Metrikine, M. J. M. M. Steenbergen, and K. N. van Dalen. 2024. "Design of Railway Transition Zones: A Novel Energy-Based Criterion." *Transportation Geotechnics* 46: 101223.
93. Jain, A., A. V. Metrikine, M. J. M. M. Steenbergen, and K. N. van Dalen. 2023. "Railway Transition Zones: Evaluation of Existing Transition Structures and a Newly Proposed Transition Structure." *International Journal of Rail Transportation* 12 (6): 1–21.
94. Jain, A., A. V. Metrikine, M. J. M. M. Steenbergen, and K. N. van Dalen. 2024. "Dynamic Amplifications in Railway Transition Zones: Performance Evaluation of Sleeper Configurations Using Energy Criterion." *Frontiers in Built Environment* 10: 1285131.
95. Jain, A., K. N. van Dalen, M. J. M. M. Steenbergen, and A. V. Metrikine. 2024. "Dynamic Amplifications in Railway Transition Zones: Investigation of Key Phenomena." *Journal of Physics: Conference Series* 2647 (15): 152002.
96. Jain, A., Y. Marykovskiy, A. V. Metrikine, and K. N. van Dalen. 2024. "Quantifying the Impact of Stiffness Distributions on the Dynamic Behaviour of Railway Transition Zones." *Transportation Geotechnics* 45: 101211.
97. Jain, A., A. V. Metrikine, and K. N. van Dalen. 2024. "Energy Redistribution in Railway Transition Zones by Geometric Optimisation of a Novel Transition Structure." *Transportation Geotechnics* 49: 101383.
98. Jain, A., A. V. Metrikine, M. Steenbergen, and K. N. van Dalen. 2024. "Railway Transition Zones: Energy Evaluation of a Novel Transition Structure for Critical Loading Conditions." *Research Square* 13: 15.
99. Jain, A., A. Metrikine, M. Steenbergen, and K. van Dalen. 2024. "Evaluation of a Novel Mitigation Measure for Two Different Types of Railway Transition Zones." In *Proceedings of the Sixth International Conference on Railway Technology: Research, Development and Maintenance*, Vol. 7, pp. 1–11. Civil-Comp Press, Edinburgh, UK.

9 Navigating the Cybersecurity Landscape in Railway Infrastructure
Measures, Challenges, and Innovation

Ajay Kumar Verma and Shilpi Lavania

9.1 INTRODUCTION

Railway is one of the major commodities to carry passengers and freight around the world. There are more than 1.3 million route-kilometres in the global railway network. Some projections estimate that the size of the worldwide rail market will increase from US$780.6 million in 2022 to US$1112.2 million in 2029, with a compound annual growth rate (CAGR) of 1,112.2 between 2023 and 2029 [1].

Railway operations and maintenance are undergoing substantial changes due to digitalization, affecting several aspects such as sustainability, availability, dependability, maintainability, capacity, safety, and security, including cybersecurity. Railway stakeholders, however, view the changes brought about by digitalization as both an opportunity and a challenge. Data collection, transformation, modelling, processing, visualization, safety, security, quality, and information assurance are some of the problems associated with digitalization. A new mindset among railroad employees, digital skills, and the creation of a plan to thwart cyberattacks and safeguard railroad assets, which calls for specialized knowledge of digital technology are among the other difficulties. In asset management, an asset is often defined as any object, thing, or entity that has the potential or actual to provide value to an organization. Examples of such items include servers, data, laptops, databases, apps, people, buildings, and physical systems [2, 3].

Today, the digital transformation of railways business is a real challenge, especially for data security. The increasing complexity of IT systems and the development of multi-dimensional digital environments have led to the emergence of an increasing number of connected devices and communication channels between employees, customers, equipment and multi-level interconnected systems. Therefore, for an organization, the protection of its own information is an essential element for running a successful business [4]. As organizations use more and more data and integrate reporting mechanisms and automated work processes into their day-to-day operations, the evaluation and protection of stored data is becoming increasingly

 DOI: 10.1201/9781003659983-9

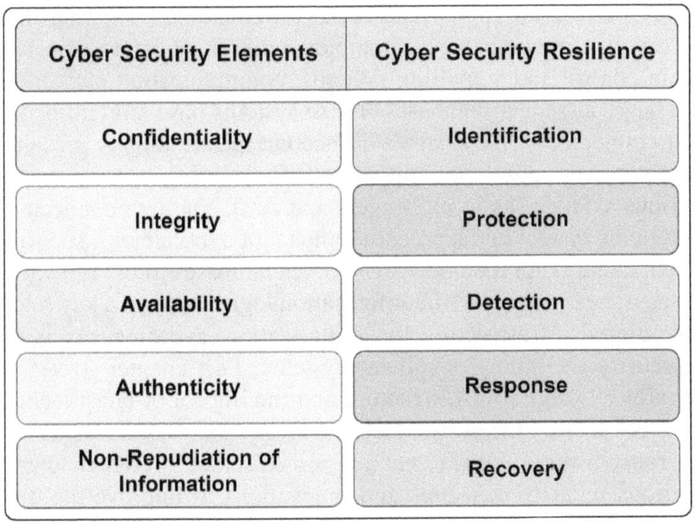

FIGURE 9.1 Cybersecurity and cyber resilience key elements.

important. There is a significant difference between the volume of data generated daily that requires protection and the amount of data that is actually protected [5]. Figure 9.1 shows the key elements of cybersecurity and cybersecurity resilience.

Infrastructure maintenance managers organize the planning and decision-making processes in their respective organizations in radically diverse ways, with minimal synergy. The convergence of Information and Communication Technology (ICT) and operating technologies is necessary to support maintenance planning effectively and efficiently. These technologies and solutions typically result in inadequate synchronization with service demand because they are improperly integrated into an entire asset management process. This makes it more difficult to fully use the immense promise of digitalization. Organizations responsible for maintaining railway infrastructure must regularly check and monitor the state of the tracks. In many situations, manual measurements may be replaced by sensor technology [3]. Numerous issues, such as data collection, transformation, modelling, processing, visualization, safety, security, quality, and information assurance, confront a digitalized railway. The railway sector must develop policies that allow for a seamless transition from the current setup to a digitalized system to address these issues. The digitalized railway impacts liveware (i.e., people) in addition to the technical components of the system, such as hardware and software [6].

Internal and external cybersecurity threats are growing, which could pose a serious security risk. Railway transportation is going through a major development in terms of higher capacity requirements and a greater amount of communication, computer, and control technologies that are applied in train control systems. In traditional railway systems security is rarely considered, despite safety being the primary consideration in railway design. Cybersecurity poses a danger to key infrastructure and industrial control systems' regular operations hence, cybersecurity issues should receive more attention. Additionally, railway is a complex technical system of systems (SoS) and a multi-stakeholder environment. The existing digital infrastructure in railways encompasses a complex network of interconnected systems, including train control and signalling systems, communication networks, ticketing systems, and huge passenger databases. These systems have streamlined operations and essential connectivity, they have also become prime targets for cyber threats. These threats not only jeopardize the smooth functioning of railway services but also pose serious safety risks to passengers and staff. The interconnected nature of the railway systems increases the potential impact of cyberattacks, security and data breaches which can disrupt the operations, financial losses to the railway sector and compromise passenger safety. While information logistics plays a key role in providing a suitable digital infrastructure for digitalization, cybersecurity is essential to the overall security and safety of railway systems. This chapter strives to navigate through the railway's digital infrastructure and the impact of latest technologies on the digital infrastructure of railways. Key threats such as DDoS (Distributed Denial of Services), ransomware, malware etc. are also discussed. This chapter also highlights the various security measures and innovations to improve the digital infrastructure security to minimize these risks.

9.2 RAILWAY AS WEB OF INTERCONNECTED SYSTEM

Railways, in a broader perspective, can be seen as an integration of several systems used to transport people and freight using wheeled vehicles that operate on tracks. These subsystems can be grouped as,

- Railway Infrastructure: It includes all the tracks, buildings, facilities and the systems that guarantee a steady flow of traffic. The facility and premises include stations, depots, and other like establishments as well as wayside systems that run alongside the lines. These include signalling systems, electrification facilities which are intricately linked to the overall security of the electrical grid and level crossings.
- Rolling Stocks: These comprise engineering vehicles, trailer vehicles, and powered vehicles such as locomotives, single rail cars, shunters, etc.
- Railway Operations: The railway operations comprise the technical tasks carried out to guarantee train movement as well as the commercial activities carried out by railway corporations to guarantee profits [7].

9.2.1 Current Railway Infrastructure

The railway industry is continuously evolving and as mentioned earlier, with the dense interconnected systems, it is getting technologically more and more complex. Railway infrastructure is made of several (sub) systems. The most critical ones are the Interlocking system (I×L) and the Control Centre or Automatic Train Supervision (ATS). The operation, maintenance and performance evaluation of these interconnected systems are essential for an efficient rail transportation. Figure 9.2 depicts the existing interconnected systems in railway for effective operation. In a railways transportation system, most tasks require all three of the subsystems at the

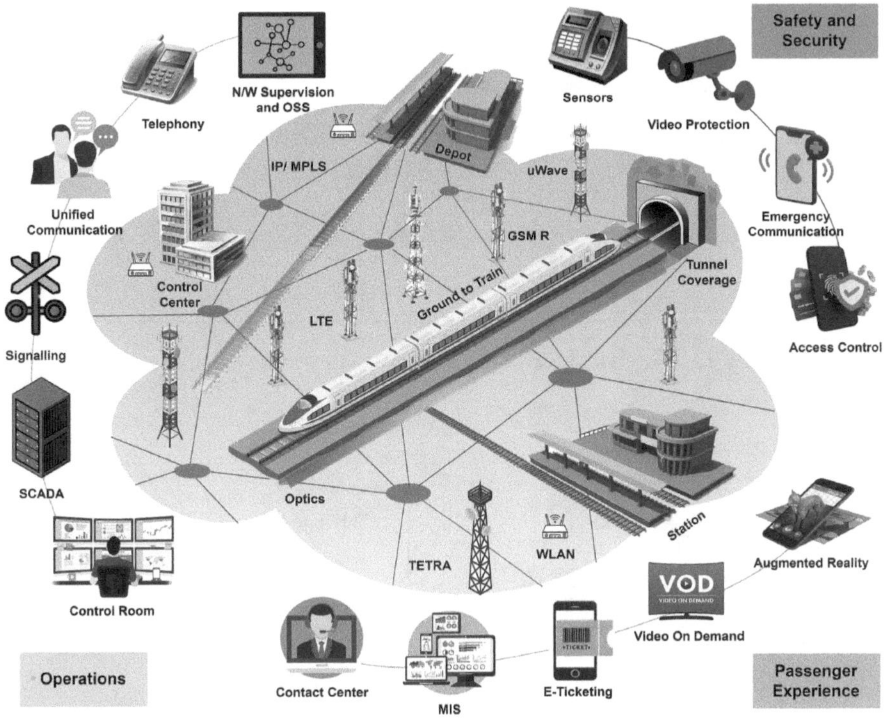

FIGURE 9.2 Railways interconnected system.

same time. This means that "holistic" approaches, which prioritize system security generally, are better than those that concentrate on protecting a particular system component at the expense of the system's total capacity to carry out its numerous functions [7].

9.2.1.1 Railway Signalling and Control System (RSCS)

Railway is an interconnection of subsystems, and these subsystems must communicate to each other in multiple layers for a specific task to be done. Hence the signalling system is the backbone of the entire communication and wherever the signal is processing through interconnected system the risk of cyber threat has high probability. To understand how these signals are vulnerable towards the threats, it is required to understand the proper working of signalling system. Figure 9.3 provides an insight of the communication system in railways [8].

A variety of control systems, such as SCADA systems, distributed control systems (DCSs), and other control system topologies such Programmable Logic Controllers (PLCs), are included under the one unit called Industry Control System (ICS). Most ICS deployments are hybrids, combining features from both DCSs and SCADA systems to obfuscate the distinction between them. These systems can be highly distributed systems that are used to manage geographically dispersed assets, which are frequently dispersed across thousands of square miles. Centralized data gathering

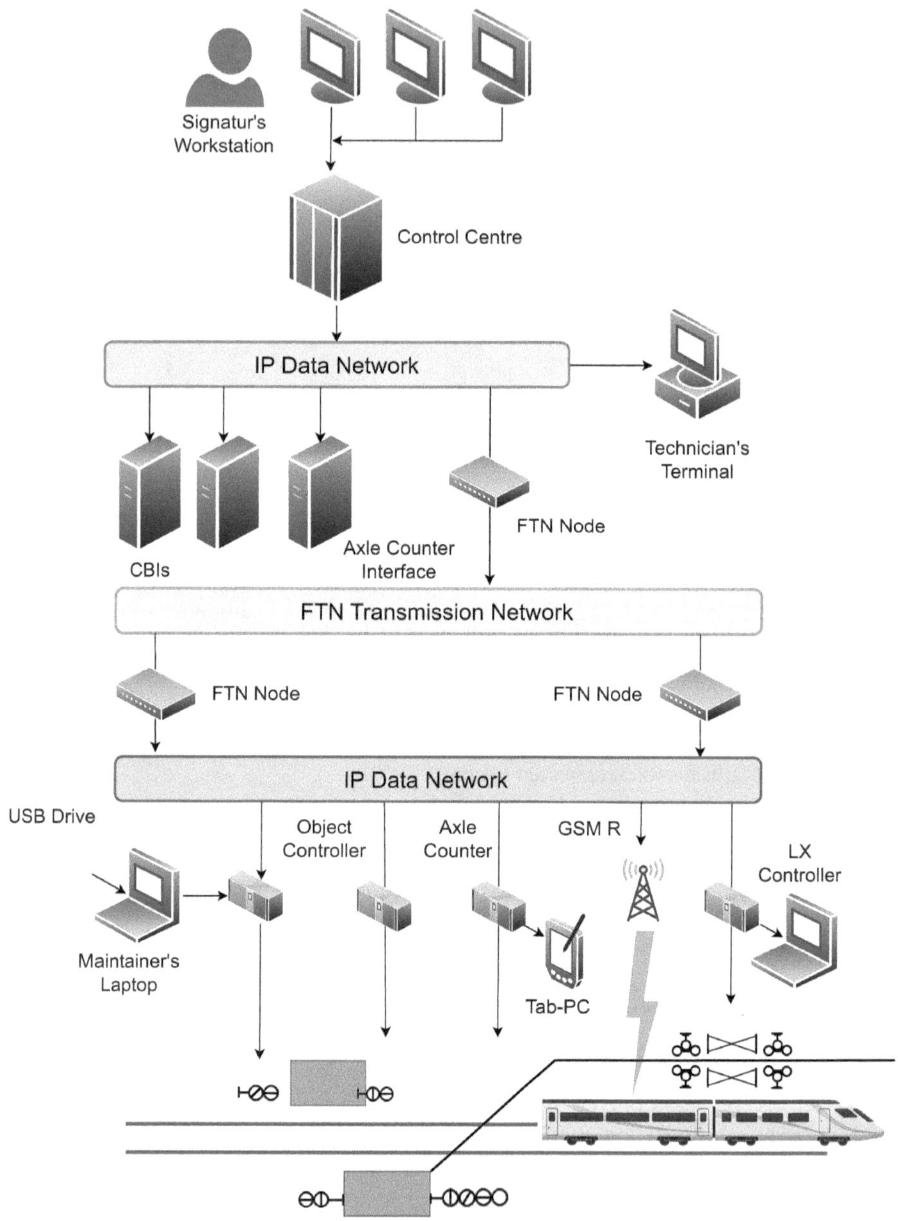

FIGURE 9.3 Railway signalling system.

and control are essential to the proper functioning of these systems. Through long-distance communication networks, a SCADA control centre manages and monitors field stations centrally [8].

Figure 9.3 clearly illustrates how the contemporary RSCS has every attribute of an ICS. Based on data from the axle counter system, trackside object, and level crossing

(LX) controllers, the control centre system processes controls and indications for the signaller. Computer-Based Interlockings (CBIs) verify all safety-critical tasks. The LX controllers, CBIs, and object controllers may be PLC devices or specialized hardware that has been set up for their respective purposes. IP data networks at the control centre and the trackside, which are also connected, enable communication between the geographically scattered assets [9].

There is a significant role of wireless communication, which is called Communication-Based Train Control (CBTC). Figure 9.4 depicts the CBTC system that is required for controlling the train control system via wireless communication [10, 11]. Track circuits are the primary means of communication between trains and the control centre. In a classical train control system without wireless communication is typically regarded as closed. This is because track circuits are physical private networks. However, the use of wireless communication in CBTC systems leaves them susceptible to potential security vulnerabilities [12].

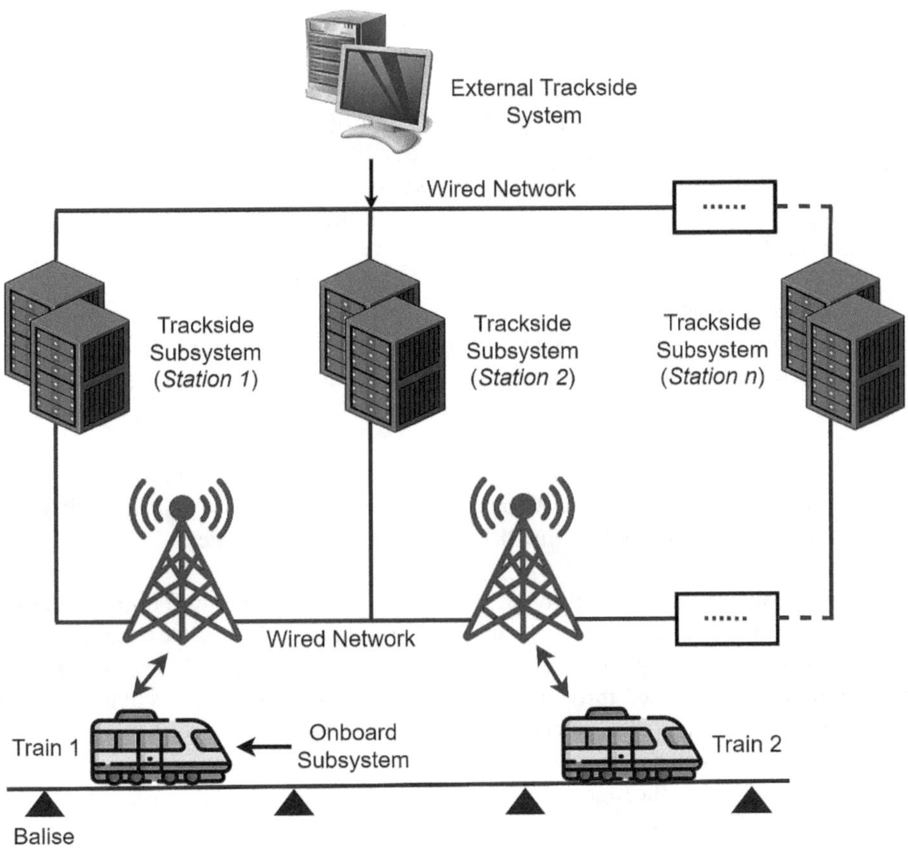

FIGURE 9.4 CBTC system and signalling mechanism.

The security situation for CBTC can be described as follows [13, 14]:

- The utilization of wireless communications results in a more effective CBTC system but malicious attackers could be able to get unauthorized access to the control network. This could bring unpredictable consequences to the control system.
- Since the railway networks are widely deployed throughout the nation, there are many options for attackers, and there is a significant likelihood of security attacks on CBTC systems, particularly when open wireless connections are used.
- The transmission protocols are not verified through security assessment, which means some potential vulnerabilities could be used by malicious attackers, and it can result in catastrophic accidents.

9.2.1.2 Passenger Data Handling and Security

Information about one's identity must now be disclosed in many online exchanges. For both individuals who reveal the knowledge and those who receive it, this presents significant issues. Previously, a rudimentary model that involves an identity service provider or trusted third party disclosing identification information to a recipient-referred to as the relying party-on behalf of a user has been used to solve this issue. By verifying certain parts of the trade, the identity service provider lowers risk and increases efficiency. Three categories may be used to broadly categorize the wide range of cybercrimes that have emerged with the development and adoption of the Internet [15],

- Consumer Risks: It includes identity theft, financial fraud, endangering children, and hacked computers being used for unauthorized activities
- Enterprise Threats: financial data theft, economic information espionage, loss of personally identifiable information, denial-of-service attacks-based extortion, and
- Cyberwar and Information Warfare: It poses a danger to the government. Four characteristics such as Global connection, Anonymity, Untraceability, and Lucrative targets are what draw criminals to the Internet.

Features that may be accessed through technology are referred to as a digital identity [16]. Computer-based systems are capable of storing and retrieving these characteristics. A digital identity might be something like a username or email address. A subject's identity may be established through a collection of distinct digital traits that make up their digital identity [16]. The effect of fraud and identity theft due to inadequacies in Internet security systems is enormous. The primary issue that most of the organizations are encountered with is the high implementation and management costs associated with authentication. With nearly all user data being available online, protecting personal privacy and personally identifiable information becomes increasingly important. Identity management has been a significant issue ever since the Internet was created [17]. The original purpose of Identity Management Systems (IdMS) was to manage resource access and match individuals with appropriate roles in predefined scenarios [16]. The features of IdMS confront several technological and societal problems with the introduction of inter-organizational systems, social networks, e-commerce, m-commerce, service-oriented computing, and automated agents [15, 18].

The majority of cybercrimes committed nowadays are primarily motivated by online identity theft [18]. Therefore, to combat online risks, an effective online identification system that offers distinctive identities is required. These systems need to offer appropriate answers for problems with accuracy, volume, security, and speed [19]. Since the railway network is very wide in the world, hence passenger reservation data security from cyberattacks is also a major concern and various frameworks and protocols are implemented to secure it. In the next section, the attacks on the railway network to steal the passenger data are also discussed.

9.3 CYBER THREATS IN RAILWAY

Railways have experienced a major transition by adopting technology developments for comfort and efficiency in an increasingly digitalized environment. To maintain effective operations, critical infrastructure systems like railroads are depending more and more on digital technology and networked systems. Although there are many advantages to these developments, they also open new doors of vulnerabilities in the form of cyberattacks. Railway systems cybersecurity has become a major problem, requiring preventative steps to safeguard these essential transportation networks. It is essential to defend vital infrastructure and data against cyberattacks to maintain public confidence, operational integrity, and passenger safety.

9.3.1 Common Cybersecurity Threats

Malware: Malicious software that grants illegal access or damages a computer is referred to as "malware." They frequently spread through downloads that appear authentic or unsolicited email attachments. Followings are the different types of malwares [20],

- Virus: A virus is a self-replicating program that enters a computer system, attaches itself to a clean file, and spreads to other files, infecting them with harmful code.
- Trojan: Trojans are a kind of malware that poses as authentic software. Cybercriminals deceive people into downloading Trojans, which then harm or gather data from their machine.
- Spyware: Spyware is a type of software that surreptitiously logs user actions in order to profit from hackers. Spyware has the ability to get credit card information in a few instances.
- Ransomware: Ransomware is a type of malware that encrypts and locks down a user's files and data, threatening to remove them unless a ransom is paid.

- Adware: Software for advertisements that has the potential to disseminate malware.
- Botnets: Botnets are collections of malware-infected machines that hackers employ to carry out Internet operations without the user's consent.

Phishing: Phishing is the practice of cybercriminals posing as trustworthy sources in emails to victims, requesting sensitive information (such as a password or credit card number) from the recipient. It is the most typical kind of online assault [20].

Vishing: Vishing refers to the attempt made by fraudsters to get personal information over the phone by calling victims' ID, CVV, ATM PIN, Net Banking password, OTP, and card expiration date, among other details.

Man-in-the-Middle Attack: A type of cyberthreat, in which a hacker intercepts and then relays messages between two parties to steal data. An attacker may, for instance, intercept data being sent between the victim's device and the network via an insecure Wi-Fi network.

Denial-of-Service Attack: A denial-of-service attack aims to overload a network, server, or website with so much traffic that it crashes. This makes the system useless, making it impossible for an organization to do essential tasks.

Social Engineering: Cybercriminals employ social engineering as a method to fool people into disclosing personal information. It is the simplest hack and requires no knowledge of computers.

Advanced Persistent Threats (APTs): An APT occurs when an intruder or group of intruders enters a system and stays hidden for a long time. In order to eavesdrop on corporate activities and collect confidential data without setting off defensive countermeasures, the hacker leaves networks and systems unbroken [20]. The complex nature of cybersecurity strategies is increasing together with the number of digital gadgets that are connected. Additionally, the many, complex, and heterogeneous systems that make up railway transportation make cybersecurity management more difficult. Almost all information systems are vulnerable to attack because of the numerous vulnerabilities and weak protocols included in these kinds of devices and communication networks [21]. Malicious actors take advantage of these flaws and are always looking for ways to compromise railway systems' security. For example, enemies may intercept, manipulate, corrupt, or erase data connected to railways, including operational, signalling, and passenger data, by focusing on equipment with limited resources or unsecured wireless communications. Table 9.1 provides a more thorough explanation of the potential for various cyberattacks connected to the attacker's source, his goal, and the compromised security aspect.

9.3.2 Case Studies of Various Cyberattacks around the World

The railways system around the globe faces new risks because of the generalization of electronic components and information and communication technologies, specifically in the areas of rolling stock and infrastructure (particularly in the

TABLE 9.1
Impact of Cyberattacks and the Compromised Security Elements [21]

SL. No.	Impact of Cyberattack	Source	Security Element
1.	Data network and power supply malfunctions.	Internal	Availability
2.	Tapping, snooping, scavenging, shoulder surfing and traffic analysis, and traffic operational data.	Internal/ External	Confidentiality
3.	Modification, Replay, and Repudiation of acquired data.	Internal/ External	Integrity
4.	DoS and Procedural violation, Labour unrest.	Internal/ External	Availability
5.	Using wireless networks carelessly, sharing private information on blogs and discussion forums, sending confidential emails and instant messages, sensitive material not being disposed of properly and not logging out before leaving the workstation.	Internal	Confidentiality
6.	Data input mistakes and omissions related to failure and maintenance.	Internal	Integrity
7.	Programming faults, such as issues with syntax and logic.	Internal	Availability
8.	Breaching emanations, listening in, and stealing control of a legitimate session.	Internal/ External	Confidentiality
9.	Jamming in communications.	Internal/ External	Availability

control-command portion). These risks make it harder to draw a clear line between cybersecurity and railway operating safety. Precision scheduling is made possible by the connection of previously isolated operational technology (OT), such as railway signal control systems, with advanced technology, such as the cloud and the Internet of Things (IoT) [22]. Due to this convergence, OT systems are vulnerable to the same cyberthreats that affect IT systems. Through OT systems, nation-state assaults and denial-of-service breaches can impact train operations, severely interrupting and even damaging vital rail infrastructure, and perhaps resulting in fatalities. For this reason, OT security is essential in the digital era [23, 24]. Following are the various cases of cyber threats in different countries,

Case 1: Cyberattack on Rail Network in Poland in August 2023

Overnight, transportation in the country's northwest was disrupted by hackers who gained access to railway frequencies. Recordings were inserted between the signals. A significant conduit for the transportation of western armaments to Ukraine is Poland. The incident happened when hackers sent out a signal that caused trains close to the city of Szczecin to halt urgently [25].

Case 2: Ransomware Attack on UK's Northern Ticket Machines in July 2021

A cyberattack using ransomware was believed to target the new self-service ticket machines of Northern Rail. For a week, the system was not accessible. It happens barely 2 months after 420 stations in the north of England have 621 of the touch-screen systems installed. Together with its provider, Flowbird, the government-run operator claimed to have taken "swift action", and that no payment or customer information had been stolen. The only people impacted were the servers that run the ticket machines. Reports unequivocally show that a ransomware cyberattack affected the ticket machine service [26].

Case 3: Cyberattack on US Railroad Company in December 2022

The incident took place on December 25, 2022. According to the reports, threat actors obtained the personal data of the business's suppliers, present and past workers, and their dependents. Names, dates of birth, social security numbers, driver's license numbers or other government-issued identifying numbers, employer tax identification numbers, and banking information were reportedly among the stolen data. Highly sensitive data, including birth or marriage documents, drug test results, employment evaluations, and medical and health insurance information, was also compromised. 7,413 persons, according to the Office of the Maine Attorney General, were impacted by the attack. ARRC has provided free credit monitoring and identity theft prevention services to those who are impacted [26].

Case 4: Cyberattack on German Railways in October 2022

Deutsche Bahn, a German train operator, reported that a worldwide cyberattack had compromised its systems, causing computer disruption across around 100 nations. In a statement, it claimed that while train services were unaffected, several electronic displays at stations that display arrivals and departures encountered issues. Government computer systems were unaffected, according to German Interior Minister Thomas de Maiziere. Ransomware was employed by cyber-extortionists to encrypt the machines' data, and they demanded $300–600 in exchange for access to be restored. Malicious virus attachments to spam emails that purported to be bills, job offers, security alerts, or other official files were used to deceive victims into opening them [27].

Case 5: Cyberattack on Ukrainian Railway in December 2016

Ukrzaliznytsia, the Ukrainian railway corporation, declared on December 13, 2016, that it had been the victim of a hack that disrupted its online ticketing system, delaying train schedules and ticket sales. The attack had wider effects on the economy of the nation in addition to the railway system since the rail network is essential to the movement of people and products across Ukraine. The attack's perpetrators, dubbed "Sandworm," were subsequently found to be a state-sponsored hacker collective with ties to the Russian government. Renowned for its intricate cyber-espionage operations, Sandworm has been linked to other high-profile assaults, such as the malware that targeted Ukraine's power infrastructure in 2015. Virus was used in the hack against Ukraine, more especially the "KillDisk" wiper virus. KillDisk is a data-wiping program that overwrites files and prevents them from being recovered, rendering compromised computers unusable [28].

Case 6: Cyberattack on Denmark Railway in November 2022

All trains run by DSB, the biggest rail operator in Denmark, reportedly came to a stop and were unable to move again for many hours, according to Danish broadcaster (DR). The operational technology (OT) system was affected, according to the claims. A ransomware assault was launched against Supeo, a Danish firm that offers enterprise asset management solutions to railroads, operators of transportation infrastructure, and public passenger authority. Supeo made the decision to terminate its servers due to the hacking assault. As a result, a component of software that train drivers were using stopped functioning. Although the DSB computer network was not the target of the cyberattack, they still suffered from the infection in their supply chain management [29].

Case 7: Cyberattack on Iran's National Railway System in July 2021

The official IRNA news agency said that a "cyber-disruption" in computer systems caused the websites of Iran's Ministry of Transportation and Urbanization to go offline. Messages regarding alleged train cancellations or delays emerged on notice boards at Iranian railway stations, suggesting that the nation's rail system had again been the target of a cyberattack. There have been reports of electronic train tracking failures in Iran. Although it's unclear exactly what caused the disruption, both passenger and freight transit services were impacted [30].

Case 8: Cyberattack on Indian Railway in December 2022

Indian railways was reported to be hit by a cyberattack and the data of around 30 million customers was up for sale on the dark web. According to several reports, the hacker has claimed to have "one of the biggest railways databases in India" but hasn't revealed the source of the data. The information allegedly included "Username, Email, Verified Mobile Number, Unverified Mobile Numbers, Gender, City Id, City Name, State ID, Language Preferences. "The hacker was quoted as adding, "There are a lot of government emails and important people in the 30 million data." The hacker stated in a different sample that there is an additional endpoint that "discloses all user history of travel information, include a lot of data like PNR number, invoice pdf-which includes all the information like username, mobile number, train number, gender, nationality, email and all passenger's information." There were also some of the December 31, 2022, invoices in the compromised data collection [31].

Case 9: Cyberattack on Sweden Railway in 2017

On two different days, DDoS assaults bring down multiple IT systems used by Sweden's transportation organizations, resulting in delays for trains. The train order management IT system was shut down by the initial attack. During the attack, the agency had to halt or postpone trains. The problem was exacerbated as travellers were unable to make bookings or get updates on the delays due to the failure of Trafikverket's website and email system. Facebook was utilized by the organization to notify and handle the crisis for tourists. While some may have initially believed this to be an isolated incident, a similar DDoS attack targeted the websites of two other government agencies the following day. The Sweden Transport Agency (Transportstyrelsen) and public transport operator Västtrafik serve parts of Western Sweden with bus, train, ferry, and tram services [32].

Case 10: WannaCry Ransomware Attack on Deutsche Bahn in 2017

German rail operator Deutsche Bahn was hit by the WannaCry virus. It was a massive attack that compromised computer systems in around 100 nations. Though the train services were not interrupted, according to reports and railway officials, however, the attack had an impact on some electronic announcing boards at stations that displayed arrivals and departures. In addition to the disruption of display of train timings, the hackers also posted a message demanding ransom money on the passenger information screens at stations to restore access. Interlocking and track safety remained unaffected. The cybersecurity experts explained that the WannaCry virus was spread by the systems running Windows XP that had not been updated in a decade. The computers were protected and employees were restricted from having access. Soon after the systems were hit by the WannaCry virus, the organization promptly updated the systems.

This incident came out as a "wake-up-call". According to cybersecurity experts around the globe, while defining and determining the policy, it is important to involve all stakeholders i.e. internal as well as third party. All stakeholders must be aware of the processes and activities. Also, they must be aware of how these processes are related to cybersecurity. These attacks have urged a solution for proactive measures so that these cyberattacks can be avoided.

9.4 SECURITY MEASURES AND EVOLUTION IN DIGITAL INFRASTRUCTURE OF RAILWAYS

9.4.1 Security Measures for Railways

For railway systems to be secure, dependable, and private for both operations and passenger data, cybersecurity is essential. Here are some measures that must be considered for cybersecurity of railways:

Network Segmentation: To separate important systems from less secure ones, divide the railway network into pieces. This stops hackers from moving laterally and lessens the effect of a cyberattack.

Intrusion Detection Systems (IDS) and Firewalls: Utilization of IDS to keep an eye out for any unauthorized access attempts or suspicious activity and deploy firewalls to regulate traffic between network segments.

Encryption: To stop bad actors from intercepting and interfering with communication channels, especially those carrying sensitive data and control signals, encrypt them.

Access Control: To limit access to important systems and data, apply the least privilege principle and make use of robust authentication methods (such as multi-factor authentication).

Patch Management: To fix known vulnerabilities, update firmware and software on a regular basis. This is true for both IT infrastructure and operational technology (OT) systems.

Physical Security: To avoid unwanted tampering or access, secure physical access is necessary to vital infrastructure components including data centres, control rooms, and communication equipment.

Monitoring and Logging: To quickly identify abnormalities or possible security breaches, continuously monitor network traffic, system logs, and user activity.

Incident Response Plan: Create and keep up an incident response plan that describes how to find, address, and resolve cybersecurity problems. To make sure you are prepared, practice drills often.

Training and Awareness: Employees and contractors should get cybersecurity best practices training as well as awareness campaigns to help them spot social engineering scams, phishing efforts, and other possible dangers.

9.4.2 Innovation in Security Measures and Protocols for Improving the Cybersecurity

Increased dangers from cyberattacks and the increasing digitalization of railway systems, railway cybersecurity has seen a major evolution in security measures. From early digitalization to early cybersecurity measures like firewall, network segregation to protect critical circuits, integration of IT and operational technology (OT) to protect signalling and control system, and strong incident response plans, backup systems, and redundancy are becoming common components of cybersecurity strategies to reduce downtime and guarantee security, the cybersecurity measures and protocols are enhanced to secure railway infrastructure. Table 9.2 represents various protocols/frameworks and affected assets of railways.

TABLE 9.2
Various Protocols and Frameworks for Railway Security

SL. No.	Protocols/ Framework	Assets	Result	References
1.	WSN, NS2	Railway Communication	Responsible for end-to-end communication for wireless communication.	[33]
2.	WSN, GRNET virtual machine	Freight and Tracks	Scalability and security.	[34]
3.	Safety and Security Impact Reference Model (SSIRM) for CBTC	Railway Signalling	Functional integrity levels.	[35]
4.	Knowledge Acquisition for Objective Satisfaction Framework	Automatic operation of Train	Safety and Security.	[36]
5.	SHILED framework	Surveillance and monitoring	Security, dependability and privacy.	[37]
6.	MILS and Pike OS	Data communication, Signalling	Separation between reliable and unreliable components.	[38]
7.	Data communication protocol	Signalling system	It enables the control of separate system components via a two-wire differential bus that may be extended across the system's central component.	[39]
8.	Cloud technology	Cloud service for IoT enabled devices in railway network	Encryption scheme for cloud network.	[40]
9.	WebGME	Resilience of railway operations	Supports executing large simulations in the cloud.	[41]
10.	RaSTA	Signalling system	Testbed for simulating EULYNX-defined standards for railway signalling network connection.	[42]

9.5 VARIOUS CHALLENGES IN RAILWAY CYBERSECURITY

Identifying a suitable cybersecurity strategy for the railway sector necessitates having a thorough understanding of related issues.

Lack of Initiatives: The application of more proactive and predictive technologies and methodologies, such as security-by-design, PSA, and threat intelligence, is limited in railway systems. In order to address the current cybersecurity circumstances, where attackers are more dynamic in initiating assaults, the railway must apply modern security analytics and be proactive and dynamic [43].

Ignorance of Third-Party Security: Railways use a range of digital devices, including third-party provided IoT, surveillance equipment, and WSNs, to deliver continuous, day-to-day services. Thus, third-party security is also a big challenge in railway systems [40].

Failure in Assessing LTE-R Security: High-speed wireless voice and data communications for railroad infrastructure and rolling stock are made possible by the Long-Term Evolution-R digital wireless communications network, which is specifically designed for railway applications. For the future safety of railroads, it is therefore increasingly necessary to evaluate LTE-R and 5G networks for railroads in the context of cybersecurity [44].

Resilience and Incident Response:

- To reduce the impact of cyber events on railway operations, it is imperative to develop and maintain strong incident response and recovery capabilities.
- Maintaining service continuity and passenger safety requires quick discovery, containment, and mitigation of cyber threats.

Budget Constraints:

- The increase in global attacks and interruptions on rail networks is the main driver of development in investment in rail cybersecurity solutions. Based on the many ideas, eleven factors influence organizations' implementing cybersecurity technology can be outlined (i) Compatibility, (ii) perceived utility, (iii) trialability, (iv) observability (v) IT modularity, (vi) organizational flexibility, (vii) top management support, (viii) collaborative board supervision, (ix) competitive pressure, (x) trialability and (xi) vendor support are the first eleven factors. The deployment of dedicated cybersecurity solutions in the current railway infrastructure is very high and thus it imposes a serious budget constraint on the railway organizations.
- Also, with the integration of IoT in various domains of railways and in deploying cybersecurity solutions comes at a high initial cost, requiring large initial investments to set up field-level devices, replace ageing infrastructure, introduce cutting-edge technologies like artificial intelligence (AI), big data, cloud computing, and predictive maintenance in urban rail, set up transmission networks between end users, and manage both new and old systems within buildings. Thus, on the one side, cybersecurity solutions are a need of the hour in the railway industry, and on the other side, it demands great investment.

Lack of Skilled Personnel:

- In any organization, Information security is significantly affected by various human factors. According to Gartner research, "a lack of talent or human failure" would be the source of 50% of all cybersecurity breaches by 2025. As the railway industry embraces new digital technology for maintenance and operations, the current workforce's competence must be improved. Employees need to have a decent level of training, education, and experience in cybersecurity. To develop a cybersecurity culture, railway companies need to set up and maintain policies, strategies, and controls. To achieve the necessary threshold of knowledge among the task force, cybersecurity awareness and training should be a regular practice. Organization can fight the assault of fraudsters and other threat actors looking to access, disrupt, and steal from their IT networks, by hiring or growing cybersecurity expertise and offering security hygiene training to all non-IT staff.

  The following are the other challenges that appear in implementation of cybersecurity [45]:
  - Lack of cybersecurity awareness in railway sector
  - The challenge of balancing the domains of cybersecurity and safety
  - The digital modernization of the main railway operations
  - Reliance on the supply chain for security measures
  - The distribution of railway infrastructure geographically and the presence of older systems
  - Finding a balance between competitiveness, security, and operational efficiency
  - The complexity of cybersecurity regulations.

9.6 APPLICATIONS OF AI IN RAILWAYS

Maintenance and Inspection: There are various applications of AI tackling diverse problems including (i) defect detection, (ii) fault detection, (iii) defect prediction, (iv) maintenance planning, and (v) autonomous maintenance. AI has assisted the railways in all of the areas mentioned above [46].

Risk Management: Risk management describes the frequency, severity, and possibility of repercussions for each hazardous occurrence using a five-phase model based on the fuzzy reasoning method and the fuzzy analytical hierarchy decision-making process [47].

Estimating Disruption and Abnormality Estimation: A track circuit is a train detection tool that determines whether or not a block or segment of railroad tracks is in use. AI has been applied to failure prediction and maintenance assessment [48].

Energy-Efficient Driving: Approaches based on genetic algorithms (GAs), approximate dynamic programming (ADP), reinforcement learning (RL), etc. are being developed to address energy-efficient driving optimization [49].

Verification and Validation and Data Source: Artificial intelligence (AI) has the potential to learn from expert knowledge in railway control verification and validation, even in the absence of an expert. For this, fuzzy-based models have been used. Additionally, Machine Learning models may be modified to produce more optimal outcomes with the use of data obtained from the past expertise of experts [50].

Anomaly Detection: Islam et al. [51] has suggested a security solution extended neural network (ENN) for Internet of Railways (IoR) network. The IoR consists of various sensors, transducers, network layers, actuators and communication system. The key to the IoR is effective communication. A network of railways exchanges and transmits information with one another via various protocols. The entire system is vulnerable to attacks because of the extensive use of wireless technologies on trains. These intrusions in communication can lead to adverse behaviour of IoR. The proposed ENN model outperformed the existing convolutional neural network (CNN), long short-term memory (LSTM) and deep neural network (DNN) [51]. Another study suggested an AI-based model that uses data gathered by in-service trains to identify anomalies along the infrastructure. Track condition information is gathered from drive-by data from accelerometers mounted on the train's car-body and bogie. This framework makes use of a one-class support vector machine (OCSVM). According to the results, the OCSVM performs better in terms of recall, precision, and F1-score than the other algorithms based on Isolation Forest (IF), Local Outlier Factor (LOF), and Robust Mahalanobis Distance (RMD) [52].

There are numerous other areas where AI is assisting the railways cybersecurity and increasing its efficiency. Following are some other areas of applications:

- Strategical Planning [53]
- Tactical Planning [54]
- Traffic Analysis [55]
- Rescheduling and Disruptions [56]
- Passenger Flow prediction [57]
- Risk Assessment
- Real-time train dispatching

9.7 FUTURE RESEARCH DIRECTIONS AND INTERNATIONAL STANDARD TO SUPPORT CYBERSECURITY

Railway is a very complex network of interconnected systems. In this pre-existing network, lack of proactiveness, lack of cybersecurity quantification, lack of a comprehensive viewpoint, an inability to distinguish between safety-critical and non-safety-critical systems, a failure to evaluate Long-Term Evolution-R (LTE-R) security for wireless communication, and third-party security ignorance are the few gaps in the effective implementation of cybersecurity solutions. Therefore, future research directions are outlined and must be of primary concern in order to provide a secure rail environment:

- Adoption of AI to detect threats and secure the existing rail network
- Focused research on railway cyber-physical systems and evaluation of the impacts of cyberattacks on safety and operation of rail.
- Cyber risk identification and priority assessment
- Evaluation of cybersecurity knowledge among railway employees at various levels
- An identification of methods to promote cybersecurity culture within railway

Additionally, various international standards are also empowering cybersecurity to the railway infrastructure. Cybersecurity standards are important for (i) Threat identification and assessment, (ii) Security control including intrusion information, network segmentation and access control (iii) Procedure to overcome the cyberattack, and (iv) keeping the system updated for the advanced security fixes. New standards in railway network offer effective and reliable tools for creating multi-layered protection against cyberattacks. The record for mixed dispersed systems, which is an essential component of railway systems and was missing in IE 62443, is covered in CENELEC technical standard TS 50701. Furthermore, TS 50701, is laying the foundation of the international standard (IEC 63452) which would be the first international standard dedicated to the railway network cybersecurity and the future of cybersecurity.

9.8 SUMMARY

Worldwide, digitalization is changing the railway industry. The growing vulnerability of any company, including railways, to cyberattacks is one of the key factors to be considered while transforming digitally. The amount of digital goods and connections between digital and physical components in railway systems is constantly growing, making the railway sector susceptible to these attacks. To ensure the railway system's resilience to cybersecurity challenges, such as a lack of proactiveness, a lack of a holistic perspective, and the obsolescence of safety systems exposed to the current and future landscape of cyber threats, an increased number of items and interfaces calls for new frameworks, concepts, and architectures. As a result, future cybersecurity attacks may have a major effect on the railway system. Railway transportation communications networks include safety-critical equipment. As a result, a security concept is required to guarantee the stability and resistance of railroad communications networks against cyberattacks. This chapter has provided insight into cybersecurity in railway systems and their evolution over time. It has also shown how cyberattacks around the world have happened, which shows the vulnerabilities even when the technology is so advanced. Involvement of AI, ML, and IoT has provided a more secure medium in railways. With the advancement and introduction of new protocols and frameworks cybersecurity has improved in recent years. In the future, AI and IoT will make more ways to secure the railways' cybersecurity infrastructure.

REFERENCES

1. Railway Research Market. 2023. *Global Rail Market Growth 2023–2029*. Rail Market Research, Germany.
2. Jägare, V., R. Karim, P. Söderholm, P.-O. Larsson-Kråik, and U. Juntti. 2019. "Change Management in Digitalised Operation and Maintenance of Railway." *Proceedings: International Heavy Haul Association Conference* 236 (7): 904–911.
3. Kour, R., A. Patwardhan, A. Thaduri, and R. Karim. 2023. "A Review on Cybersecurity in Railways." *Proceedings of the Institution of Mechanical Engineers, Part F: Journal of Rail and Rapid Transit* 237 (1): 3–20.
4. Kumar, S., U. Espling, and U. Kumar. 2008. "Holistic Procedure for Rail Maintenance in Sweden." *Proceedings of the Institution of Mechanical Engineers, Part F: Journal of Rail and Rapid Transit* 222 (4): 331–44.
5. Asplund, M., S. M. Famurewa, and M. Rantatalo. 2016. "Data Quality Assessment of Automatic Wheel Profile Measurement Systems." *Proceedings of the Third International Conference on System Safety and Cyber Security* 53: 717–738.
6. Predescu, S. G., D. Savu, and V. E. Badea. 2022. "Cybersecurity in the Railway Sector." *Romanian Cyber Security Journal* 4 (2): 95–104.
7. Predescu, S. G., D. Savu, and V. E. Badea. 2023. "Cybersecurity in the Railway Sector." *Romanian Cyber Security Journal* 5 (1): 95–104. https://doi.org/10.54851/v4i2y2022010
8. Stouffer, K., V. Pillitteri, S. Lightman, M. Abrams, and A. Hahn. 2015. *Guide to Industrial Control Systems (ICS) Security*. National Institute of Standards and Technology, Gaithersburg, MD.
9. Bastow, M. D. 2014. "Cyber Security of the Railway Signalling & Control System." In *9th IET International Conference on System Safety and Cyber Security*, pp. 5.2.1–5.2.1. Institution of Engineering and Technology, Uttar Pradesh, India.
10. Pascoe, R., and T. Eichorn. 2009. "What Is Communication-Based Train Control?" *IEEE Vehicular Technology Magazine* 4 (4): 16–21.
11. Zhao, H., L. Zhu, H. Jiang, and T. Tang. 2014. "Design and Performance Tests in an Integrated TD-LTE Based Train Ground Communication System." In *Proceedings of the 17th International IEEE Conference on Intelligent Transportation Systems (ITSC)*, pp. 747–750. IEEE, New York.
12. Craven, P. V. "A Brief Look at Railroad Communication Vulnerabilities." In *Proceedings of the 7th International IEEE Conference on Intelligent Transportation Systems (IEEE Cat. No.04TH8749)*, pp. 245–249. IEEE, New York.
13. Gr, J., T. K. Madsen, and H. P. Schwefel. 2008. "Safe Wireless Communication Solution for Driver Machine Interface for Train Control Systems." In *Proceedings of the Third International Conference on Systems (ICONS 2008)*, pp. 208–213. IEEE, New York.
14. Yu, Z., H. Wang, and F. Chen. 2023. "Security of Railway Control Systems: A Survey, Research Issues and Challenges." *High-Speed Railway* 1 (1): 6–17.
15. Thangavel, S., S. Gayathri, and T. Anusha. 2020. "We Bring Your Identity: A Secure Online Passenger Identity Protocol (SOPIP) for Indian Railways Using Aadhaar Number." In *Proceedings of the Conference*, 745–758.
16. Hovav, A., and R. Berger. 2019. "Tutorial: Identity Management Systems and Secured Access Control." *Communications of the Association for Information Systems* 25 (1): 531–70.
17. Romney, G. W., and D. W. Parry. 2006. "A Digital Signature Signing Engine to Protect the Integrity of Digital Assets." In *Proceedings of the 2006 7th International Conference on Information Technology Based Higher Education and Training*, pp. 800–805. IEEE, New York.
18. SSA Publications. 2016. *Identity Theft and Your Social Security Number*. Social Security Administration, Woodlawn, MD.

19. Sadique, K. M., R. Rahmani, and P. Johannesson. 2020. "IMSC-EIoTD: Identity Management and Secure Communication for Edge IoT Devices." *Sensors* 20 (22): 6546.
20. Telang, A. 2022. *Booklet on IT/ Cyber Security – Best Practices* [Publication no. CAMTECH/S/PROJ/2022-23/SP2A]. Government of India, Ministry of Railways, CAMTECH, Gwalior. https://rdso.indianrailways.gov.in/uploads/files/Booklet%20on%20IT%20Cyber%20Security%20-%20Best%20Practices(1).pdf
21. Thaduri, A., M. Aljumaili, R. Kour, and R. Karim. 2019. "Cybersecurity for E-Maintenance in Railway Infrastructure: Risks and Consequences." *International Journal of System Assurance Engineering and Management* 10 (2): 149–159.
22. Voronko, I. 2024. "The Security of IoT Systems in Railway Transport." *Transport Systems and Technologies*, 43 (June): 90–99.
23. DXC Technologies. 2024. *Protecting Critical Rail Infrastructure from Cyber Attacks.* DXC Technologies, Ashburn, Virginia.
24. Tzanakakis, K. 2021. *Managing Risks in the Railway System.* Vol. 18, Springer Tracts on Transportation and Traffic. Springer Cl Publishing, Cham.
25. Reuters. 2023. *Poland Investigates Hacking Attack on State Railway Network.* Reuters, London.
26. Robinson, C. 2021. *Northern's Ticket Machines Hit by Ransomware Cyber Attack.* BBC, London.
27. Reuters. 2017. *German Rail Operator Affected by Global Cyber Attack.* Reuters, London.
28. Wall, T. 2023. *Throwback Attack: Ukrainian Railway Hit by Cyberattack, Stranding Passengers.* Industrial Cyber Security Pulse, Chicago, IL.
29. Kovacs, E. 2022. *Cyberattack Causes Trains to Stop in Denmark.* SecurityWeek, Denmark.
30. Staff and Agencies. 2021. *Cyber-Attack Hits Iran's Transport Ministry and Railways.* The Guardian, Iran.
31. Times of India. 2022. *Indian Railway Data Leak: 30 Million Railway Customers' Data for Sale on the Dark Web.* Times of India, Film City, Noida.
32. Cimpanu, C. 2017. *DDoS Attacks Cause Train Delays Across Sweden.* BleepingComputer, Sweden.
33. Hatzivasilis, G., I. Papaefstathiou, K. Fysarakis, and I. Askoxylakis. 2017. "SecRoute: End-to-End Secure Communications for Wireless Ad-Hoc Networks." In *Proceedings of the 2017 IEEE Symposium on Computers and Communications (ISCC)*, pp. 558–563. IEEE, New York.
34. Hatzivasilis, G., K. Fysarakis, S. Ioannidis, I. Hatzakis, G. Vardakis, N. Papadakis, and G. Spanoudakis. 2021. "SPD-Safe: Secure Administration of Railway Intelligent Transportation Systems." *Electronics* 10 (1): 92.
35. Pawlik, M. 2017. "Communication Systems' Safety and Security Challenges in Railway Environment." In *Proceedings of the Advances in Intelligent Systems and Computing Conference*, pp. 96–109. Springer, Cham.
36. Ponsard, C., J. Grandclaudon, P. Massonet, and M. Touzani. 2018. "Assessment of Emerging Standards for Safety and Security Co-Design on a Railway Case Study." In *Proceedings of the Advances in Intelligent Systems and Computing Conference*, pp. 130–145. Springer, Cham.
37. Priscoli, F. D., A. Di Giorgio, M. Esposito, A. Fiaschetti, F. Flammini, S. Mignanti, and C. Pragliola. 2017. "Ensuring Cyber-Security in Smart Railway Surveillance with SHIELD." *International Journal of Critical Computer-Based Systems* 7 (2): 138.
38. Procházka, J., P. Novobilsky, and D. Procházkova. 2020. "Cybersecurity of Railway Network Management and Partitioning." *Problemy Kolejnictwa - Railway Reports* 64 (189): 125–131.

39. Xu, W. 2017. "High-Speed Railway Signal System Information Security Analysis and Countermeasures." In *2017 Chinese Automation Congress (CAC)*, pp. 4862–4866. IEEE, New York.
40. Zheng, Q., X. Wang, M. K. Khan, W. Zhang, B. B. Gupta, and W. Guo. 2018. "A Lightweight Authenticated Encryption Scheme Based on Chaotic SCML for Railway Cloud Service." *IEEE Access* 6: 711–22.
41. Potteiger, B., H. Abdel-Aziz, H. Neema, and X. Koutsoukos. 2019. "Simulation-Based Evaluation of Security and Resilience in Railway Infrastructure." In *Proceedings of the 6th Annual Symposium on Hot Topics in the Science of Security*, pp. 1–2. ACM, New York.
42. Heinrich, M., T. Arul, and S. Katzenbeisser. 2020. "Demo: Railway Signalling Security Testbed." In *Proceedings of the 2020 IEEE Vehicular Networking Conference (VNC)*, 1–2. IEEE, New York.
43. Kour, R. 2020. *Cybersecurity in Railway: A Framework for Improvement of Digital Asset Security*. Master's thesis, Luleå University of Technology.
44. Wen, T., Q. Ge, X. Lyu, L. Chen, C. Constantinou, C. Roberts, and B. Cai. 2021. "A Cost-Effective Wireless Network Migration Planning Method Supporting High-Security Enabled Railway Data Communication Systems." *Journal of the Franklin Institute* 358 (1): 131–50.
45. ENISA. 2024. *Railway Cybersecurity from ENISA*. Railway Cybersecurity Training, Lille, France.
46. Tang, R., L. De Donato, N. B. inović, F. Flammini, R. M. P. Goverde, Z. Lin, R. Liu, T. Tang, V. Vittorini, and Z. Wang. 2022. "A Literature Review of Artificial Intelligence Applications in Railway Systems." *Transportation Research Part C: Emerging Technologies* 140 (July): 103679.
47. An, M., Y. Chen, and C. J. Baker. 2011. "A Fuzzy Reasoning and Fuzzy-Analytical Hierarchy Process Based Approach to the Process of Railway Risk Information: A Railway Risk Management System." *Information Sciences* 181 (18): 3946–3966.
48. Zilko, A. A., D. Kurowicka, and R. M. P. Goverde. 2016. "Modeling Railway Disruption Lengths with Copula Bayesian Networks." *Transportation Research Part C: Emerging Technologies* 68 (July): 350–368.
49. Brenna, M., F. Foiadelli, and M. Longo. 2016. "Application of Genetic Algorithms for Driverless Subway Train Energy Optimization." *International Journal of Vehicular Technology* 2016 (February): 1–14.
50. Nowakowski, W., P. Bojarczak, and Z. Łukasik. 2018. "Verification and Validation of Railway Control Systems Using an Expert System." *Proceedings of the Advances in Intelligent Systems and Computing Conference*, vol. 5, pp. 43–50.
51. Islam, U., R. Q. Malik, A. S. Al-Johani, M. R. Khan, Y. I. Daradkeh, I. Ahmad, K. A. Alissa, Z. Abdul-Samad, and E. M. Tag-Eldin. 2022. "A Novel Anomaly Detection System on the Internet of Railways Using Extended Neural Networks." *Electronics* 11 (18): 2813.
52. Ghiasi, R., M. A. Khan, D. Sorrentino, C. Diaine, and A. Malekjafarian. 2024. "An Unsupervised Anomaly Detection Framework for Onboard Monitoring of Railway Track Geometrical Defects Using One-Class Support Vector Machine." *Engineering Applications of Artificial Intelligence* 133 (July): 108167.
53. Pu, H., T. Song, P. Schonfeld, W. Li, H. Zhang, J. Hu, X. Peng, and J. Wang. 2019. "Mountain Railway Alignment Optimization Using Stepwise & Hybrid Particle Swarm Optimization Incorporating Genetic Operators." *Applied Soft Computing* 78 (May): 41–57.
54. Tormos, P., A. Lova, F. Barber, L. Ingolotti, M. Abril, and M. A. Salido. 2008. "A Genetic Algorithm for Railway Scheduling Problems." In F. Xhafa and A. Abraham (eds.), *Metaheuristics for Scheduling in Industrial and Manufacturing Applications*, vol. 128, pp. 255–276. Springer, Berlin.

55. Wang, P., and Q.-P. Zhang. 2019. "Train Delay Analysis and Prediction Based on Big Data Fusion." *Transportation Safety and Environment* 1 (1): 79–88.
56. Wang, M., L. Wang, X. Xu, Y. Qin, and L. Qin. 2019. "Genetic Algorithm-Based Particle Swarm Optimization Approach to Reschedule High-Speed Railway Timetables: A Case Study in China." *Journal of Advanced Transportation* 2019 (March): 1–12.
57. Liu, Y., Z. Liu, and R. Jia. 2019. "DeepPF: A Deep Learning Based Architecture for Metro Passenger Flow Prediction." *Transportation Research Part C: Emerging Technologies* 101 (April): 18–34.

10 Design and Manufacturing of Railway Track

Smitirupa Pradhan and Prajapati Naik

10.1 INTRODUCTION

The dynamics of the railway vehicles depend upon both the wheel geometry of the railway vehicle/rolling stock and rail profile of the track. The interface between the wheel tread and the rail head, is a complex geometry. The dynamic behaviour such as dynamic response and dynamic stability of the rolling stock (as depicted in Figure 10.1) depends upon the different operating parameters as well as external input. The track is a crucial element that is responsible for the noise/vibration. Over time, unevenness/roughness appears on both the wheel and rail surfaces leading to generating noise with frequency $f = V/\lambda$, where V represents the speed of the vehicle and λ denotes the wavelength present on the rail. The irregularities are distributed continuously through the track which leads to noise, oscillation or vibration of the train. Railway vehicles exhibit instability, derailment, and hunting due to self-excitation in the lateral direction. Derailment occurs due to several reasons such as failures of different bogie components, inferior quality of track structure, and improper track design parameters.

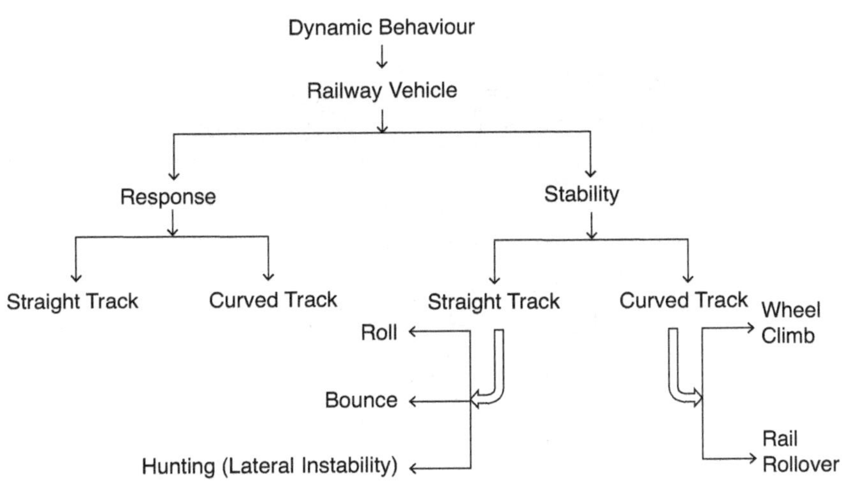

FIGURE 10.1 Dynamic behaviour of railway vehicles.

DOI: 10.1201/9781003659983-10

This chapter deals with different features of track dynamics and wheel-rail interaction which includes design parameters, track components, modelling of the track with the track irregularities.

10.2 COMPONENTS OF THE TRACK

A railway track system has two sub-systems, namely substructure and superstructure. The superstructure comprised up rail, sleepers, rubber pad, fasteners, ballast, and sub-ballast [1] as shown in Figure 10.2a and b; whereas the substructure comprises subgrade and foundation. The mechanism of transmission of the axle load

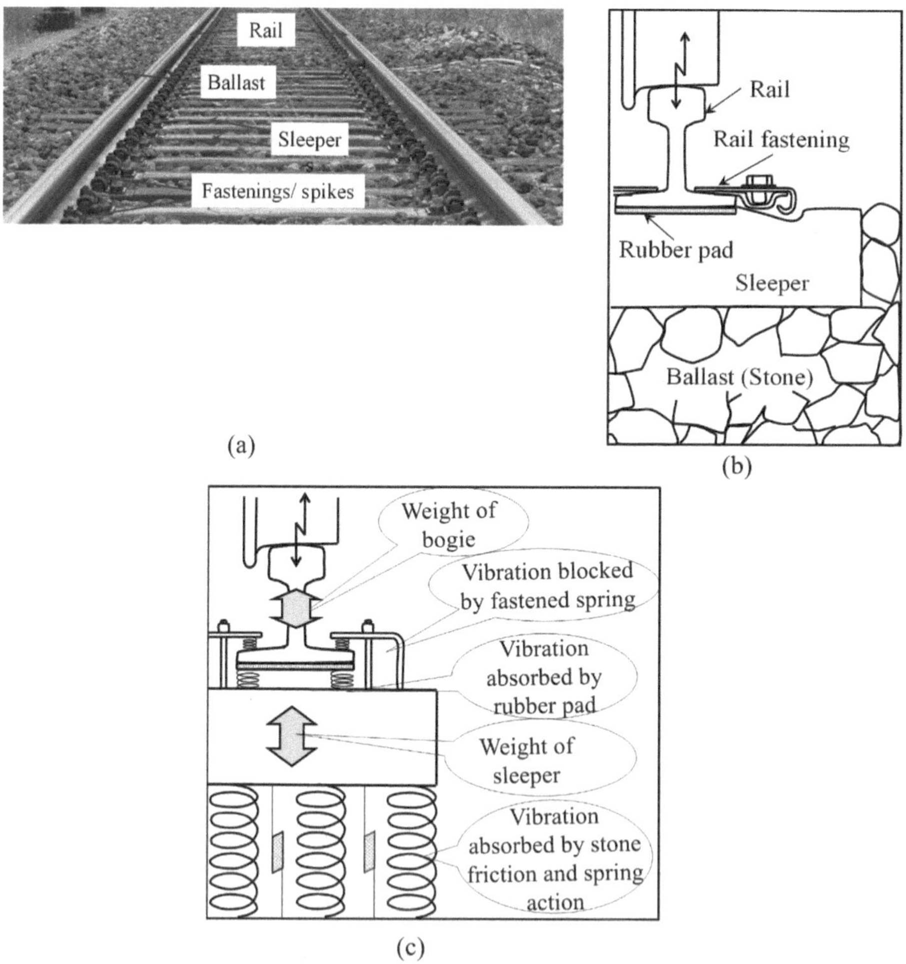

FIGURE 10.2 (a) Different components of track system and (b) Schematic diagram of a track system and (c) Schematic diagram of transmission of wheel load and vibration in the track.

through wheel and vibration from the wheel to the ground through the components of the track is presented in Figure 10.2c. The descriptions and purposes of different elements of the ballasted track are given below.

10.2.1 Rails

The shape of the modern steel rail is the I-section. The top portion of the rail acts as rail head as presented in Figure 10.3. Different countries such as America, China, British, Europe, Japan, India, and Australia use different standard rail profiles throughout the world. The rail profile UIC 60 is generally used in Europe where 60 stands for the mass of the rail/metre (kg).

The following functions are as follows:

- Rail bears axle load from wheel to sleeper through rail pad.
- It guides the wheelset and provides a smooth surface for running the vehicle.
- Longitudinal forces are generated due to braking and traction and are also transmitted to the sleepers.
- Behaves as an electrical conductor for the signals.

Various types of rails are used with distinct properties and benefits in the railway lines.

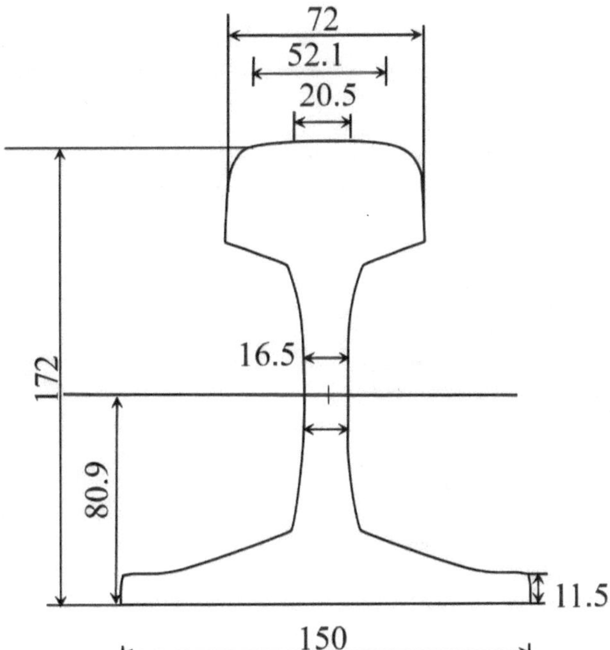

FIGURE 10.3 Rail profile UIC 60 (All dimensions are in mm).

10.2.2 Sleepers

Railway sleepers are also called crossties or railroad ties. These are placed between rail and ballast to maintain gauge. Depending upon the track gauge such as broad gauge, narrow gauge, standard gauge, etc. as well as sleeper material, the dimensions of the sleeper vary. The dimension of the concrete sleeper is depicted in Figure 10.4. The functions of different sleepers are mentioned in Table 10.1.

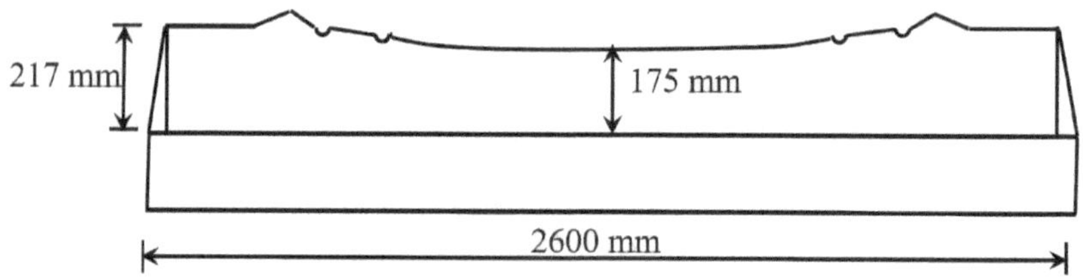

FIGURE 10.4 Cross sections of B70 prestressed concrete sleepers [2].

TABLE 10.1
Different Types of Sleepers Such as Wooden, Concrete, Steel, Composite Sleepers, etc.

Types of Sleepers	Advantages	Disadvantages
Wooden	• More economical than competitors and easy to manufacture. • The product is lightweight, making it easy to transport and handle during installation. • It's easy to put fasteners on wooden sleepers. • Rails are highly protected since they are made of good insulators. • Wooden sleepers can be used to maintain any type of gauge. • Applicable to rail sections of any kind. • Ideal for coastal areas' trails.	• In comparison to others, it has a shorter life cycle. • Poorly resistant to fire. • Susceptible to high levels of humidity, which can lead to the development of dry rot, wet rot, and other related issues. • Attacks by vermin are a possibility. • Poor creep resistance. • More maintenance is required.
Concrete sleepers	• It is heavier than all the others and makes the rails stable. • A long lifespan is economically advantageous. • High level of fire resistance. • Corrosion does not develop in concrete sleepers. • Incidents of vermin infestation and degradation have not been seen. Therefore, they are appropriate for any kind of soil and moisture conditions.	• The weight is considerable and makes handling challenging. • Inappropriate for use on bridges and at crossings, as there is a risk of damage during transportation.

(Continued)

TABLE 10.1 (*Continued*)
Different Types of Sleepers Such as Wooden, Concrete, Steel, Composite Sleepers, etc.

Types of Sleepers	Advantages	Disadvantages
	• Bucking strength is more. • An excellent insulator that is particularly well-suited for use in circuit tracks. • Provides robust track retention and ensures consistent gauge.	
Steel sleepers	• Easy transport and installation due to lightweight. • Good scrap value as they are recyclable. • Life expectancy is significantly longer, typically around 30 years. • Excellent fire and creep resistance • Designed for high-speed rails and heavy loads. • Gripping the rail firmly and establishing a connection between the rail and the sleeper	• Highly susceptible to chemical substances. • Demands extensive upkeep. • Not universally compatible with all varieties of ballast, which serve as a foundation for sleepers. • If a derailment occurs, the trains are severely damaged and cannot be reused. • Not universally compatible with all rail sections and gauges.
Composite sleepers	• Longer service life of almost 50 years. • Eco-friendly sleepers are made of composite materials. • Despite their low weight, they are incredibly strong. • Recyclable nature contributes to their high scrap value. • Easily scalable, much like hardwood sleepers. Thus, they are suitable for all kinds of rail sections. • Effectively reduces vibrations from railroads.	• Sleepers are not very fire resistant. • Sleeper costs are high.

The functions of sleepers are as follows:

- Hold firmly the rail to ensure proper alignment and distribute the weight of the rail to the ballast.
- Prevent the rail movement both in the lateral and longitudinal directions.
- Act as an insulator between two rails.

10.2.3 Rail Pads

On a railroad track, rail pads are positioned in between concrete sleepers and steel rail. They help to keep them safe from damage and wear due to cyclic impact loads and behave as an insulator of the rail. The rail cushions facilitate the

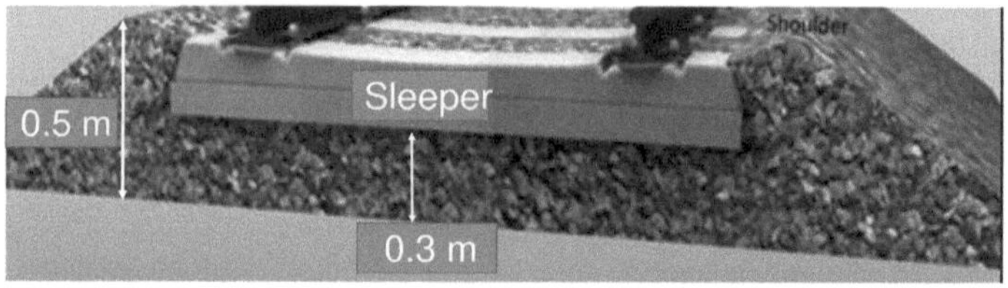

FIGURE 10.5 Ballast bed.

flexion of the rail, ensuring that the axle force is evenly distributed over the sleepers. High frequency is isolated or suppressed by soft rail pads, whereas vibration is allowed to transmit through the stiffer rail pad to the sleepers and then into the ballast.

10.2.4 Ballast and Sub-Ballast

Ballast means coarse stones which are generally triangular, made up of hard stone and rocks as shown in Figure 10.5. This bed supports both the rail and sleepers and carries vertical and lateral forces, produced by railway vehicles. Tightly compacted ballast helps to align and level the track properly. The depth of ballast is standardized, i.e. 0.3 m, whereas 0.5 m is around the ends of the sleepers to ensure lateral stability. Ballast is elastic and the range for modulus of elasticity varies from 200 to 280 MPa and 105–245 MPa by several researchers [3, 4].

Sub-ballast lies between the ballast and acts as the upper layer and a fine-graded lower layer. It prevents mutual penetration of ballast and subgrade. Gravels and sand materials may be used as sub-ballast.

10.2.5 Subgrade

Subgrade is the surface of the earth where ballast and sub-ballast are placed. Proper development of subgrade leads to avoiding inadequate track conditions and track failure. Generally, maintenance practice is very difficult in the existing track. Once the track is laid, very little effort may be required to improve its characteristics/properties, once the track is placed.

10.2.6 Fasteners

Fasteners are an integral part of the railway track that is responsible for holding the rail. It helps to grip the rail in the proper position and prevent movement. Different types of fasteners along with the functions are shown in Table 10.2.

10.3 MANUFACTURING AND HEAT TREATMENT PROCESS FOR RAIL STEEL

10.3.1 Manufacturing Process

Generally, rail is mainly subjected to fatigue and wear failure criteria which cause degradation of rail profiles as well as material properties. Hence, compositions of

TABLE 10.2
Different Types of Fasteners

Name of Fasteners	Figures	Description
Hey back		Used to firmly hold rails to sleepers or ties in train tracks
Rail spike		A rail spike/cut spike/crampon is a big nail with an offset head that is used to fasten base plates and rails to railroad ties (sleepers) in the track.
Elastic spike with plates		Used for flat-bottomed rail, base plates, and wooden sleepers, to keep the rail in place and prevent tipping,
K-fastening		It is used on wooden sleepers throughout the world.

(*Continued*)

TABLE 10.2 (*Continued*)
Different Types of Fasteners

Name of Fasteners	Figures	Description
Pandrol-e-clip		The assembly comprises a pair of elastic rail clips, a pair of rail shoulders, a single rail pad, and a pair of rail insulators.
Pandrol Fastclip	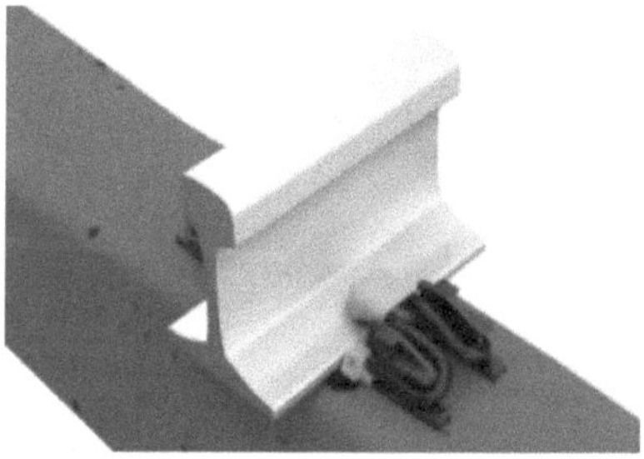	Fast clip reduces wear by enabling independent rail movement in various orientations.
W-fastening		Ideal for high-speed and conventional rail systems with concrete sleepers on ballasted tracks.

the steel and heat treatment process are important for the rail manufacturing process. Rail lines are subjected to extreme cold and hot climates throughout the world depending upon the geographical locations. The extreme climate conditions affect the mechanical properties and stiffness of the rail. Different countries use different grades of steel such as R260 and R350HT in European countries, U75V in Middle Eastern countries, JIS E 1101-2001 in Japan and IRS-T-12-96 in India as per usage and climate conditions as mentioned in Table 10.3.

TABLE 10.3
Different Compositions of the Rail Steel

Elements Standards	C	Mn	Si	S	P	V	Al	References
R260	0.62–0.8	0.65–1.25	0.13–0.6	≤0.03	≤0.03	0.03		[5]
R350HT	0.72–0.8	0.65–1.25	0.13–0.6	≤0.030	≤0.025	0.01		
U75V	0.71–0.8	0/95–1.05	0.5–0.8	≤0.025	≤0.025	0.04–0.12		
IRS-T-12-96	0.6–0.8	0.8–1.3	0.5–1.3	0.035	0.035		0.015	[6]
EN13674-1:2011	0.72–0.8	0.7–1.2	0.15–0.58	0.25	0.02			[7]
JIS E 1101-2001 (40N, 50N, 60N)	0.63–0.75	0.7–1.1	0.15–0.3	0.025	0.03			

Several countries use different manufacturing techniques to ensure good services. The rail steel may be manufactured in four stages:

i. Steel is manufactured by using an oxygen/electric arc process
ii. Continuous casting of blooms
iii. Rail rolling process with controlled cooling
iv. Rail finishing with eddy current testing, ultrasonic testing followed with finishing work.

Different countries implement different standards such as EN 13674 in European countries for manufacturing rail steel. The flow chart of the whole manufacturing process of rail steel [7] is shown in Figure 10.6.

10.3.2 Heat Treatment Process

Researchers throughout the world have focussed on the improvement of the performance of rail steel to meet the requirements according to train speed and load. Different countries such as Japan, the UK, Europe, and America have developed hypereutectoid pearlite steel, martensite steel, and bainite steel, respectively. The performance of hypereutectoid steel can be improved by accumulating more ferrite. The wear resistance martensitic steel (low carbon) is almost the same as heat-treated steel, whereas the toughness is much better than the heat-treated rail steel. Bainite steel has high wear resistance, hardened and tensile strength along with toughness is two to five times better than pearlite steel. The operating conditions and environmental conditions vary throughout the world and different countries develop different compositions of rail steel followed by different heat treatment approaches. In this chapter, R350HT rail steel is considered for the heat treatment process as a case study. The sample of R350HT is exposed to the isothermal heat treatment, then it is attained to bainitic microstructure (R360B). The R360B sample was then subjected

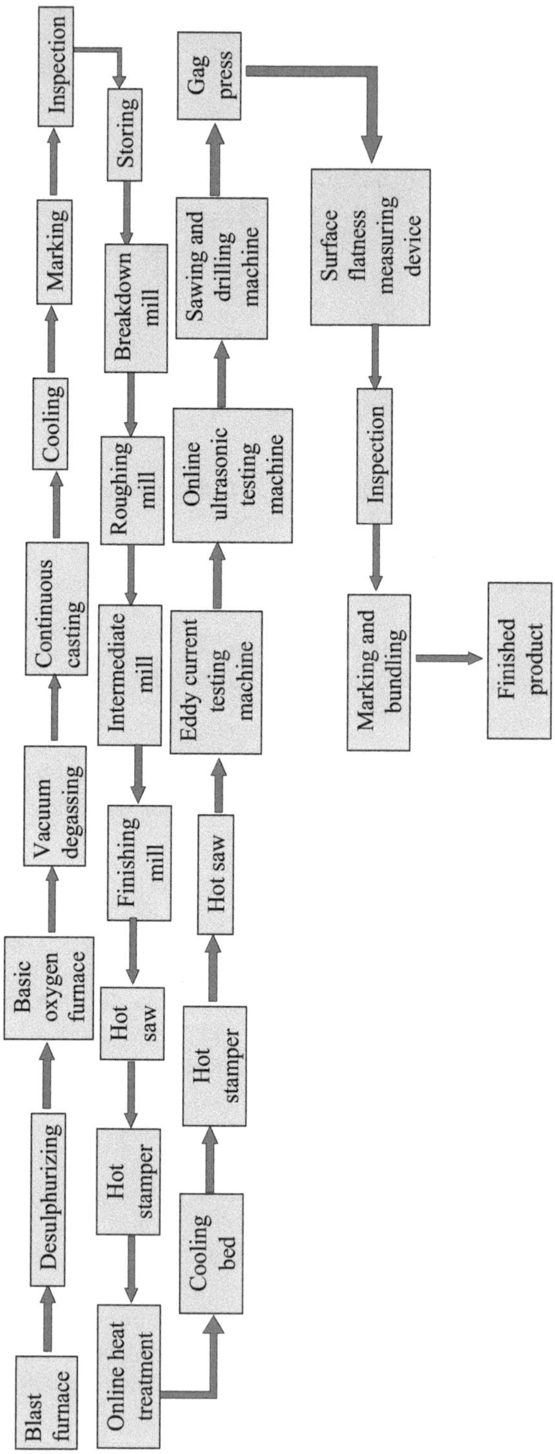

FIGURE 10.6 Flow chart of rail steel manufacturing process from iron ore [7].

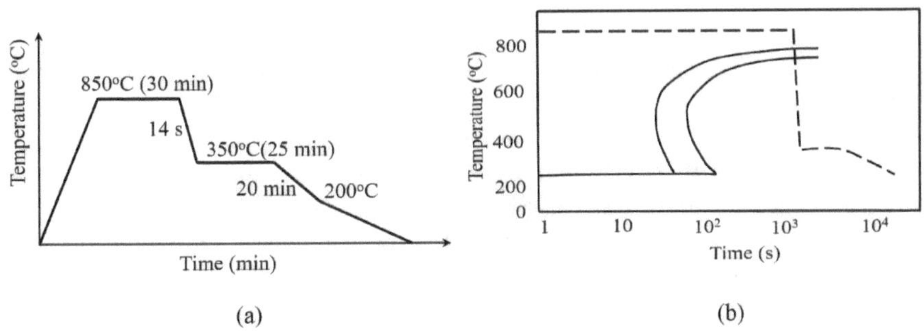

FIGURE 10.7 Schematic diagram (a) heat treatment of R350HT and (b) TTT curve (continuous cooling) of R350HT steel in isothermal condition [8].

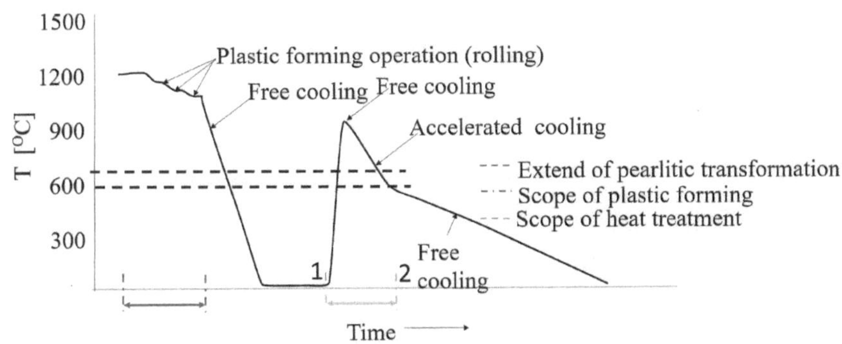

FIGURE 10.8 Heat treatment of rail steel R350HT [10].

to austenitizing temperature, i.e. 850°C for 30 minutes under isothermal conditions, subsequently, the sample was cooled to 350°C for 14 seconds and at this condition the sample was annealed for 25 minutes. Finally, the sample is cooled to 200°C in the presence of air as shown in Figure 10.7a. The bainitic microstructure is obtained by following the continuous cooling curve as shown in Figure 10.7b.

Nowadays, rails are mainly manufactured from different grades of pearlitic steel as per standard EN16273. The cooling rate is a crucial parameter for obtaining different microstructure and hardness [9]. The thermoplastic treatment of rail R350HT is shown in Figure 10.8, where 1–2 portion is scheduled for the heat treatment process. The cooling rate differs from the head/top portion (3.68°C/s) to inside the rail portion (2.63°C/s) of the rail.

10.4 TRACK GEOMETRY

Monitoring track design parameters, such as track gauge, super-elevation/cant height or cant angle, radius of curvature, rail inclination, and transition length of the curve, is essential for ensuring the dynamic stability of the vehicle.

10.4.1 Track Gauge

The space between the inner faces of both the rails (left and right) is termed a track gauge as shown in Figure 10.9. Different countries use different types of gauges as mentioned in Table 10.4. A vehicle is more stable when the gauge is large. However, it becomes difficult to turn with a large gauge. To study the vehicle–track interaction and dynamic behaviour of the vehicle, passengers' comfort, etc., some analytical descriptions of track design parameters and track irregularities are necessary. The whole track is the summation of straight/tangent, transition/entry or exit curves and curved tracks.

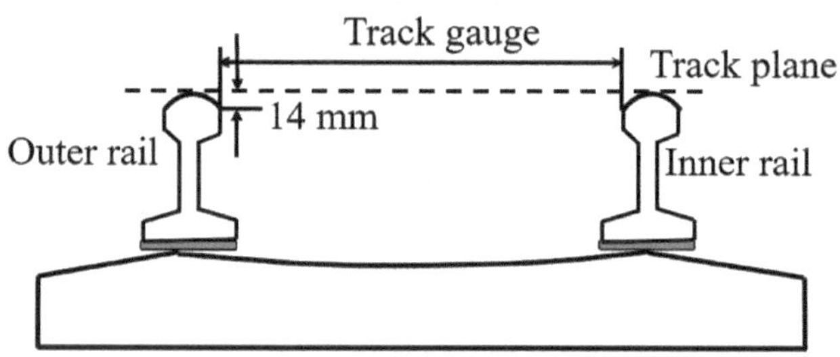

FIGURE 10.9 Track gauge.

TABLE 10.4
Countries Use Different Track Gauge

Types of Track Gauge	Countries	Values (mm)
Standard gauge	All high-speed rail lines except those in Russia, Finland, and Uzbekistan.	1,435
Broad gauge	Russia	1,520
	United State	1,575, 1,581, 1,588, and 1,638.
	England	1,727
	India	1,676
	Russia	1,520
	France	1,750
Narrow gauge	United States	508 and 578
	United Kingdom	508, 578 and 660
	Germany	520, 560, 860 and 1,100
	Japan	838 and 1,067
Metre gauge	France, Britain, and Germany	1,000

10.4.2 Tangent/Straight Tracks

Common terminologies such as vertical profile, cross-level, gauge and alignment are used to represent the track geometry. The average of the position of both left (y_l) and right rail (y_r) in the lateral direction is termed alignment (y_a), which is expressed as $y_a = (y_l + y_r)/2$ (Figure 10.10a). Cross-level (z_c) is outlined as the deviation between left and right rail elevation, i.e. $z_c = z_l - z_r$. Vertical profile (z_v) is the average of left and right rail elevation, which is expressed as $z_v = (z_l + z_r)/2$ (as illustrated in Figure 10.10b), where y and z designate for the lateral direction and elevation, respectively and suffices r and l, respectively, signify the right and left side of the rail [11]. The deviations of the above-mentioned parameters are termed irregularities such as rail profile error, rail alignment error, gauge error, and warp/twist in a straight track. Geometrical track irregularities, otherwise called static irregularities, generally arise from initial installation errors and degradation due to usage. The main sources of lateral vibration in railway vehicles are cross-level and alignment. Gauge is also responsible for the lateral stability of the rolling stocks. Static irregularities are the external sources whereas the dynamic irregularities of the railway track depend upon track flexibility. Dynamic irregularities are generated during the vehicle passing over the track as shown in Figure 10.10c.

Static irregularities are classified as periodic, aperiodic or random.

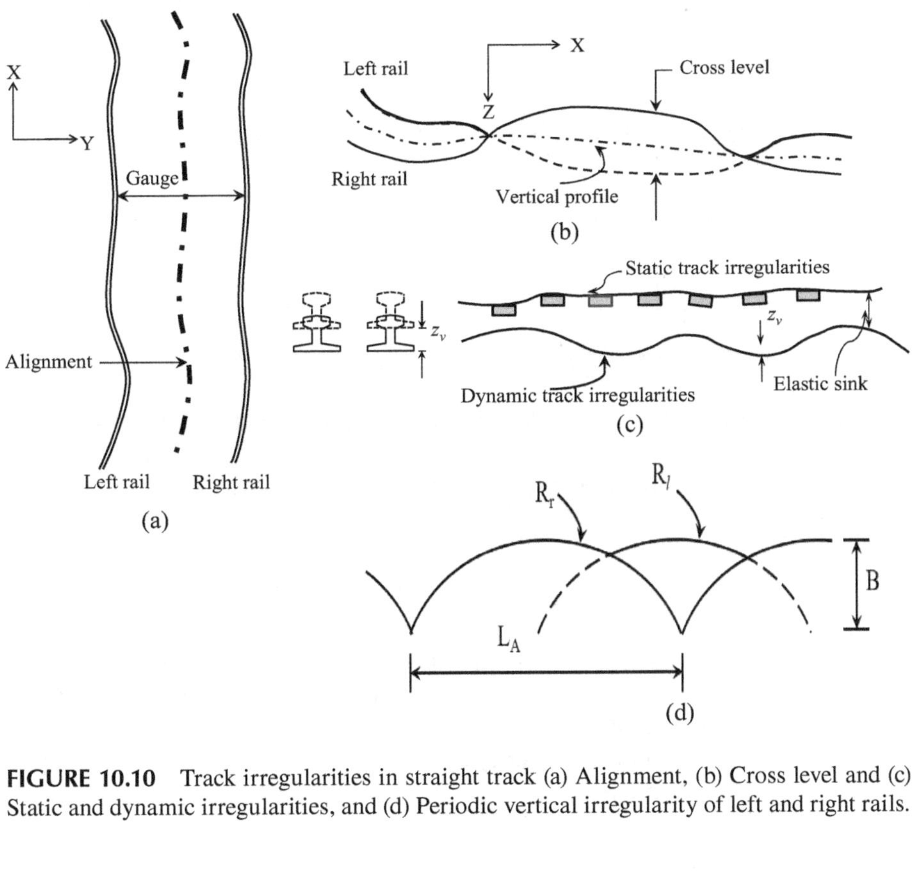

FIGURE 10.10 Track irregularities in straight track (a) Alignment, (b) Cross level and (c) Static and dynamic irregularities, and (d) Periodic vertical irregularity of left and right rails.

10.4.2.1 Periodic Irregularities

This type of irregularities appears due to the joining of rails. Each rail segment ends with a flattened end and the roughness of each rail may be considered a rectified cosine wave (RCW) (as shown in Figure 10.10d approximately [11]), i.e.

$$Z_r = B|\cos u| \quad \text{and} \quad Z_l = B\left|\cos\left(u - \frac{\pi}{2}\right)\right|, \tag{10.1}$$

where B is the amplitude of RCW.

$$u = \Omega_p\left(x/2\right) \quad \text{and} \quad \Omega_p = 2\pi/L_A, \tag{10.2}$$

where L_A and x represent for the rail length (m) and the track distance (m), respectively and Ω_p is the wave number (rad/m). The vertical profile variations are given by

$$Z_v = \frac{(Z_r + Z_l)}{2} = \sum_n A_n \cos n\Omega_p x, \quad n = 2, 4, 6, \ldots, \tag{10.3}$$

where $A_n = \dfrac{-4B}{\pi\left(4n^2 - 1\right)}$.

Similarly, the cross-level irregularities can be determined by

$$Z_c = Z_l - Z_r = -2\sum_n A_n \cos n\,\Omega_p x, \quad n = 1, 3, 5, \ldots. \tag{10.4}$$

The alignment irregularities are

$$Y_a = Y_a - \frac{2C}{\pi} = \frac{4C}{\pi}\left[\frac{1}{1\times 3}\cos\Omega_p x - \frac{1}{3\times 5}\cos 2\Omega_p x + \frac{1}{5\times 7}\cos 3\Omega_p x - \ldots.\right], \tag{10.5}$$

where B is the amplitude of the RCW. The corresponding spectral density is

$$S_a(\Omega) = \frac{\pi}{2}\sum_n B_n^2\,\delta\left(\Omega - n\Omega_p\right), \quad n = 1, 2, 3, \ldots;$$

$$\text{where } B_n = \frac{4C}{\pi\left(4n^2 - 1\right)}. \tag{10.6}$$

The amplitudes of these types of irregularities may vary over the track, however, wavelength remains constant over several cycles. The dynamics of the vehicle–track interface might be impacted by these kinds of modifications. Resonance occurs in large amplitude as the natural frequency of the irregularities equals or nearly equal to the vehicle's natural frequency.

10.4.2.2 Aperiodic Irregularities

Railway track structures have switches and frogs which can modelled by considering track geometry as input. Several aperiodic irregularities such as cusp, jog, bump, plateau, damped sinusoid, and tough sine cycle. The shapes and mathematical functions are used to describe the above-mentioned irregularities as shown in Table 10.5. The analytical representation of aperiodic irregularities is represented by two parameters such as duration-related parameter (k) and amplitude (B). The limitations of the B and k are found in the track geometry as represented in Table 10.6.

TABLE 10.5
Different Types of Aperiodic Irregularities [11]

Name of Aperiodic Irregularities	Shape	Analytical Representation
CUSP	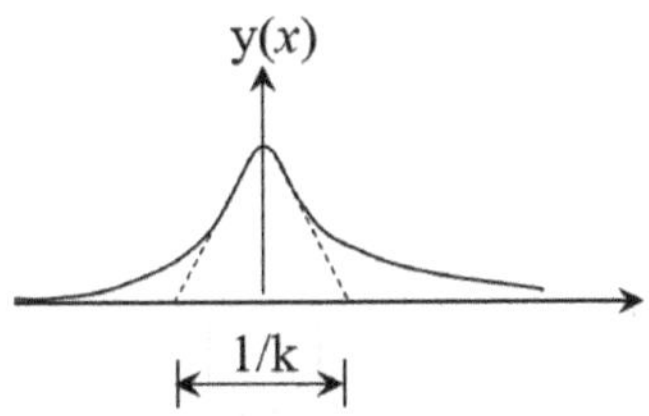	$y = Ae^{-k\|x\|}$
BUMP	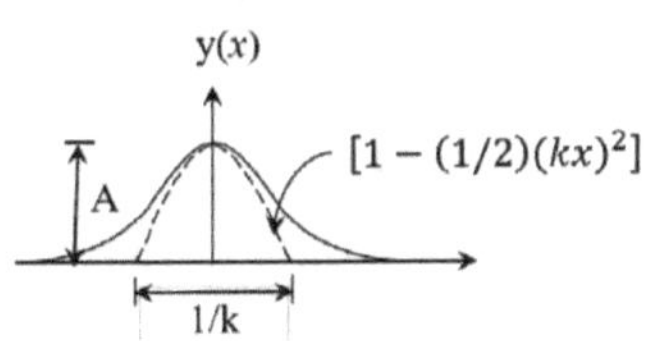	$y = Ae^{(-1/2)(kx)^2}$
JOG	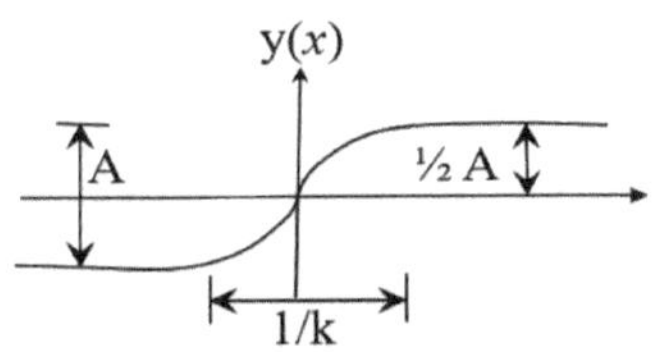	$y = \frac{Akx}{(1+4k^2x^2)^{1/2}}$
PLATEAU	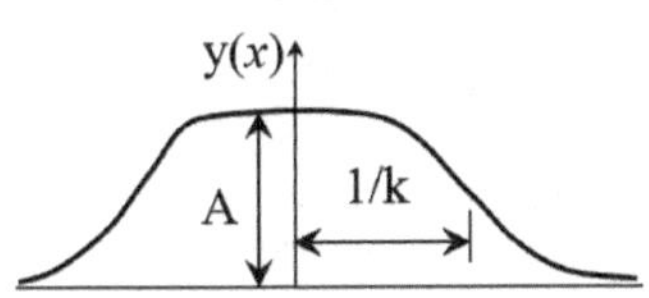	$y = \left(\frac{A^2}{1+(kx)^8}\right)^{1/2}$
TROUGH	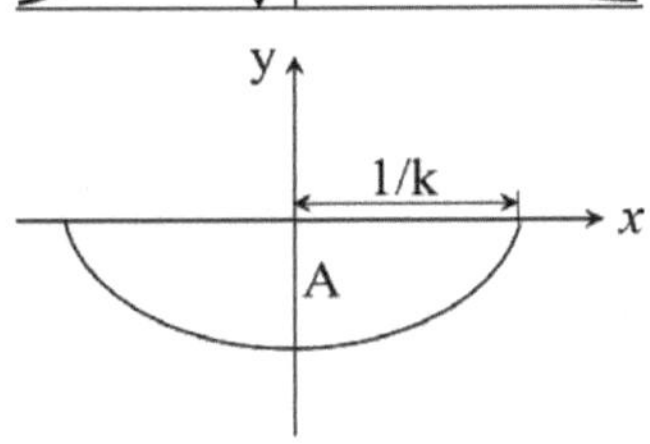	$y = Ak\left[\left(\frac{1}{k}\right)^2 - x^2\right]^{1/2}$

(Continued)

TABLE 10.5 (*Continued*)
Different Types of Aperiodic Irregularities [11]

Name of Aperiodic Irregularities	Shape	Analytical Representation
SINUSOID	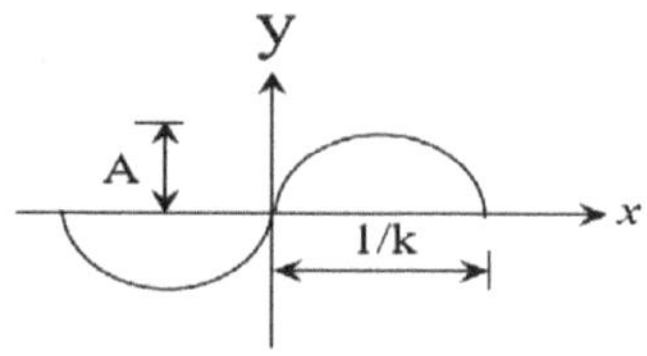	$y = A\sin(\pi kx)$
DAMPED SINUSOID	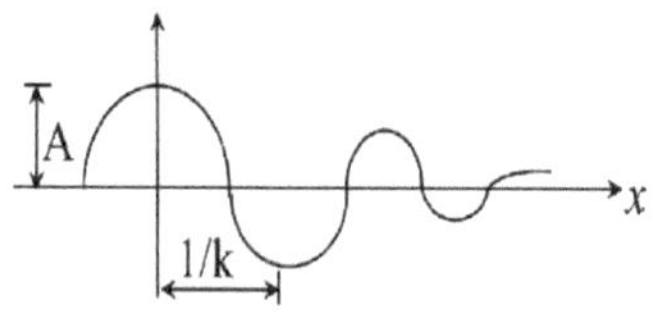	$y = Ae^{-kx}\cos(\pi kx)$

TABLE 10.6
Analytical Representation of the Parameters of *B* and *k* Values for Different Irregularities [12]

	Gauge		Alignment		Cross Level		Vertical Profile	
Signature	***B*** **(in)**	***k*** **(ft^{-1})**	***B*** **(in)**	***k*** **(ft^{-1})**	***B*** **(in)**	***k*** **(ft^{-1})**	***B*** **(in)**	***k*** **(ft^{-1})**
Cusp	0.8–1.4	0.016–0.061	0.5–3.0	0.011–0.103	0.9–3.0	0.031–0.095	0.9–3.0	0.016–0.095
Bump	0.8–1.4	0.031–0.04	0.5–2.8	0.009–0.083	1.0–3.0	0.017–0.831	0.5–4.0	0.013–0.065
Jog	–	–	0.5–3.3	0.006–0.025	1.6–2.8	0.020–0.050	0.5–5.0	0.008–0.045
Plateau	0.8–1.3	0.029–0.08	1.2–1.6	0.025–0.027	0.6–1.0	0.026–0.040	0.9–3.0	0.009–0.033
Tough	–	–	1.4–2.2	0.013–0.029	–	–	0.7–2.0	0.020–0.025
Sinusoid	–	–	0.8–1.2	0.033–0.020	–	–	1.0–1.5	0.020–0.025
Damped sinusoid	0.5–1.0	–	1.0–2.2	0.013–0.019	0.9–1.2	0.051–0.061	–	–

Cusps in the cross-level irregularities appear due to the presence of consecutive low joints, observed in the gauge and single side alignment in the curve. Periodic undulations and sinusoidal patterns normally perform the role of alignment irregularities on the bridges.

10.4.2.3 Random Irregularities

Power spectral densities (PSDs) or spectral densities are termed for stationary random track irregularities, which is a function of frequency or wave number. Road and track surfaces can be measured by spectral density which is represented as $S(\Omega) = A/\Omega^2$, where A is the type of surface measured and Ω is the wave number. Different spectral density functions are used for different types of anomalies that take place such as alignment, vertical profile, gauge and cross-level irregularities.

For alignment irregularities in terms of PSD (cm² m/rad or in² ft/rad): $S_a = \frac{A_a\Omega_c^2}{\Omega^2\left(\Omega^2+\Omega_c^2\right)}$, where A_a is a roughness parameter, and Ω_c is break wave number [13]. No adjustment is required for the above-mentioned irregularities for the shorter wavelength which is less than 3.048 m.

For the vertical profile irregularities: $S_v(\Omega) = \frac{A_v\Omega_c^2}{\Omega^2\left(\Omega^2+\Omega_c^2\right)}$, where A_v is roughness parameter.

The cross-level spectral density (S_c) and gauge spectral density (S_g) are independent of both the alignment (S_a) and vertical profile spectral densities (S_v) are represented by

$$S_g(\Omega) = S_c(\Omega) = \frac{4A_v\Omega_c^2}{\left(\Omega^2+\Omega_s^2\right)\left(\Omega^2+\Omega_c^2\right)}, \tag{10.7}$$

where Ω_s is a break wave number. Track PSDs are used for the frequency domain analysis of the rail vehicles during the calculating of rail deviations, forces exerted at the rail-wheel interface, vibration analysis of the vehicle and relative displacement between different components of the railway vehicles.

10.4.3 Curved Track

The curved track geometry comprises straight, transition, and steady curve segments as depicted in Figure 10.11a. The key elements of the curved track are super-elevation/cant, curve radius (as presented in Figure 10.11b) and transition/entry or exit

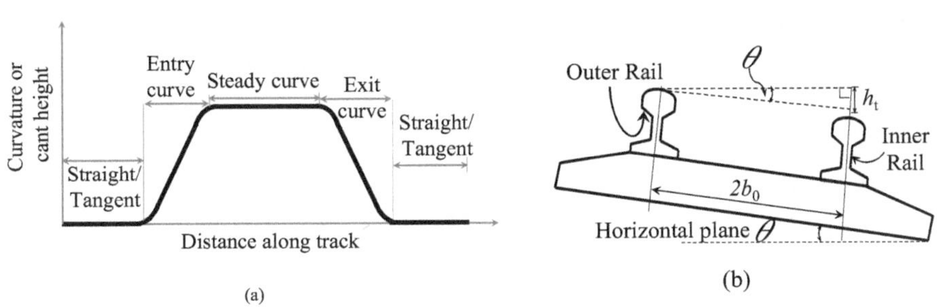

FIGURE 10.11 Designed parameters for curved track: (a) Different segments of a real railway track and (b) Track cant/super-elevation.

curve. It is crucial to understand the behaviour of the railway vehicle throughout different segments, such as the transition/entry or exit curve, during dynamic analysis.

10.4.3.1 Track Cant

The elevation between two inner and outer rails of the track in the curved path is called track cant Figure 10.11b. The cant angle can be calculated as $\theta_c = \sin^{-1}\left(h_t/2b_0\right)$ where h_t is the super-elevation/cant height (mm) and $2b_0$ is the distance between the centre of the inner and outer rail (mm). Cant deficiency (θ_d) is defined by the curvature ($1/R_{\text{curve}}$), track cant and vehicle speed as

$$\theta_d = \frac{V^2}{gR_{\text{curve}}} - \theta_c, \tag{10.8}$$

where g gravitational acceleration. The velocity at which the unbalanced lateral acceleration is nullified yields zero cant deficiency.

10.4.3.2 Curvature

Generally, the curvature of the curved track is measured in terms of radius. However, the curvature in some countries like the United States and India, etc. is expressed as a degree of curvature. The speed limit is lower in the curved path to prevent overturning of the train due to lateral centrifugal force. Cant may be used to improve the speed limit over the same curve.

10.4.3.3 Transition Part of the Curve

A transition curve is the change in curvature at the start and finish of a curved path. Generally, it is a clothoid or spiral type in nature, which varies linearly with the distance. It is used to join two similar portions (straight to straight or curve to curve) and two dissimilar portions (straight to curved and vice versa) of the track. The transition curve is represented as clothoid type which is written as $\rho(s) = \rho_0 + s/A_{\text{curve}}^2$, where $s = s_0 = 0$ at the entry of the transition curve, ρ ($1/R_{\text{curve}}$) is the curvature at any distance s and A_{curve} is the clothoid parameter. If the clothoid starts from the straight and ends with a radius of curvature (R_{curve}) with transition length (L_t), $A_{\text{curve}}^2 = L_t R_{\text{curve}}$ [14].

10.4.3.4 Steady Curve

The curvature and super-elevation are constant in a steady curve as shown in Figure 10.11.

10.5 RAILWAY TRACK DESIGN

10.5.1 Forces Exerted on the Rail

Generally, the design wheel load is expressed in terms of static wheel load, i.e. $P = \Phi P_s$, where Φ is dimensional impact factor and P_s is static load. The impact factor is calculated by using different empirical formulae based throughout the world [15].

$$\left.\begin{aligned}
&\Phi = 1+\frac{V}{58.14(K)^{0.5}} && \text{for Indian standard}\\
&\left.\begin{aligned}
&\Phi = 1+\frac{V^2}{3\times 10^4} && \text{when } V \text{ is up to 100 km/h}\\
&\Phi = 1+\frac{4.5V^2}{10^5}-\frac{1.5V^3}{10^7} && \text{when } V>100
\end{aligned}\right\}\text{German formula}\\
&\Phi = 1+4.92\frac{V}{D} && \text{for African standard}\\
&\Phi = \left(1+3.68\times 10^{-5}V^2\right)^{0.67} && \text{for Washington Metropalitan Transit Authority}
\end{aligned}\right\}. \tag{10.9}$$

Rail joints are considered as discrete irregularities which affect vehicle velocity and unsprung mass. British main rail line consists of 54 kg/m rail (continuously weld rail) with sleeper spacing 760 mm. As per the British Railway (BR) standard, the dynamic load exerted due to discrete load [16] (welded rail) is

$$P = P_s + 8.784\,(\alpha_1+\alpha_2)\left[\frac{D_j P_u}{g'}\right]^{0.5} V, \tag{10.10}$$

where P_s stands for static load, P_u is unsprung mass, D_j is track stiffness at the joint and $\alpha_1+\alpha_2$ is total rail joint dip angle.

The amplitude of the lateral guiding force exerted on the rail head depends on velocity of the vehicle, track curvature (1/track curve radius) and wheelbase, etc. The estimated lateral force H (kN) exerted on the rail head as per International Union of Railway (ORE) is expressed as

$$H = 35+\frac{7,400}{R}, \tag{10.11}$$

where R is the curve radius of the track.

As per Swedish Railways, mean lateral force H_{mean} (kN) is estimated from series of field tests as

$$H_{\text{mean}} = 17+\frac{V}{27.6}. \tag{10.12}$$

10.5.2 Sleeper Analysis

The performance such as longitudinal and lateral loading of a sleeper depends upon surface geometry, weight, shape and size and spacing between adjacent sleepers. Effective length of sleeper can be calculated as $L=(l-g')\left[1-\frac{(l-g')}{125\,t^{0.75}}\right]$, where l is the total sleeper length, g' stands for gap between central line of rail sheet (mm) and t is sleeper thickness. Sleeper length is approximately equal to $L=L/3$.

Sleeper support area (effective) (As in mm^2) for both concrete and timber sleepers is $As = B(l - g')$ where B is the breadth of the sleeper.

Maximum load on rail seat (q_r in kN) is calculated by assuming rail beam is on the continuous elastic foundation model, $q_r = Sky_m F_1$, S is distance between two consecutive sleepers, y_m is maximum rail deflection and F_1 is a factor which includes track support and maintenance. Contact pressure between sleeper and ballast (P_a) $= \frac{q_r}{\text{BL}} F_2$. By AREA approach, $P_a = \frac{2q_r}{\text{BL}} F_2$. The maximum bending stress and bending moment occurs at the rail seat and sleeper centre. Maximum BM at the rail seat is $M_r = q_r \left(\frac{l - g'}{2}\right)$.

10.5.3 Design Parameters of the Foundations

The total deflection (y) of the rail at the support is the sum of deflections due to rail at every sleeper location. Rail foundation is assumed as a continuous linear elastic foundation (Winkler's method)

D (kN/mm) is the foundation spring constant which is estimated as $D = \frac{P_s}{y}$, where P_s is the static load of the wheel exerted on the rail (kN). Track modulus k (MPa) is determined as force applied on the rail/unit deflection/unit length which is expressed as per American Railroad Engineering Association (AREA) approach

$$k = \frac{P_s}{y \times s \times 10^3}, \tag{10.13}$$

where s is the sleeper spacing (mm). By assuming constant sleeper distance, the effective rail support area (As in mm^2)$=B(l-g)$, where l and B are the length and breadth of the sleeper, g stands for span between the rail seats. Equivalent breadth (B') of longitudinal sleeper$=A_s/s$ as per German Railway. Coefficient of the subgrade (C in kN/mm^3) for a single rail is determined as

$$C = \frac{P_s}{ysB'}. \tag{10.14}$$

Track modulus is a static parameter which depends upon bending moment of the rail at its base and ballast conditions. The track modulus varies as sleeper size and track condition varies as shown in Table 10.7.

The minimum rail resistance (R_0) in longitudinal direction due to rail fastener, i.e.

$$R_{0(\min)} = \frac{sF^2 10^3}{2\delta_{r\max} A_r E}, \tag{10.15}$$

where δ_r is the end movement of the rail (maximum) in mm which is calculated as $\delta_r = \frac{(F - J_r)^2 s\ 10^3}{2R_0 A_r E}$, where F stands for total force required to check the rail

TABLE 10.7
Track Modulus for Different Ballast Conditions [17]

Rail (kg/m)	Size of Rail (mm)	Conditions of the Ballast	k (MPa)
42	178×229×2,590	150 mm fine cinder-sized ballast (Poor)	3.7
		150 mm fine cinder-sized ballast (fair condition)	5.2
42	152×203×2,440	150 mm limestone (good before tamping)	6.7
		300 mm limestone after tamping	7.5
42	178×229×2,440	300 mm limestone before and after tamping	7.3
64	178×229×2,590	600 mm gravel ballast and 200 mm heavy limestone with well-packed subgrade	20–20.7
55	178×229×2,590	Limestone ballast on wide stable roadbed	25.5

TABLE 10.8
Rail Joint Gap for Different Rail Length as per German Conditions [18]

Rail Length (l_r) (m)	Rail Joint Gap (l_g) (mm)
15	5
30	9
45	11
60	14

movement, i.e. $F = A_r E\ \sigma_t (t_1 - t_0) 10^3\ (kN)$. σ_t is thermal expansion coefficient of rail steel = 1.15×10^{-5}, t_1 and t_0 are the maximum deviation from ambient temperature at base and ambient temperature at base, J_r deals with resistance force of the rail joint, R_0 is average sleeper resistance against rail displacement in the longitudinal direction.

10.5.3.1 Estimation of the Longitudinal Stresses (German Approach)

Maximum stress occurs at the centre of the rail, which is calculated as

$$\sigma_t = \frac{2F_0 + W_o l_r}{2 \times 10^3 A_r}, \tag{10.16}$$

where W_o is the track resistance against the longitudinal displacement, l_r length of rail between two fishplates (m) and A_r stands for rail cross-sectional area.

For long rail track, maximum compressive stress in longitudinal direction at rail centre

$$\sigma_t = \frac{W_o l_r}{4 \times 10^3 A_r} + (t_1 - t_0)\sigma_t E - \frac{l_g E}{l_r}, \tag{10.17}$$

where l_g stands for gap between the rail joints (mm) (Table 10.8).

TABLE 10.9
P/D Ratios for Different Railway Operators [18]

Railway Operators	P/D Value (kN/m)
BR railways operator (British)	105.9–122.6
SNCF Railways (French)	88.3
JNR Railways (Japanese)	90.3–98.1
SAR Railways (SAR)	111.8
Australian state railway	90.3–105.9
AAR (USA)	144.2

Rail breadth for long rail can be estimated as

$$l_b = \left[\frac{A_r\, Eg}{2\times 10^6 W_Q}\right]^{0.5}. \tag{10.18}$$

10.5.3.2 Static Wheel Load (Gross) (P) to Wheel Diameter (D)

The quotient of the static wheel load to wheel diameter is the indication of contact stress between wheel and rail head. *P/D* ratio used by railway operators is given in Table 10.9.

10.5.4 Rail Wear Life

Three major approaches such as Illinois method, AREA approach and Couard's method are there to estimate rail wear theory [15].

10.5.4.1 Rail Wear by University of Illinois

Worn out of the rail is due to abrasive, which is estimated as

$$w_a = w_t\left(1+K_w D_c\right)D_A, \tag{10.19}$$

where, w_a is abrasive rail head wear (mm²/year), w_t stands for average wear of the rail head, D_c is degree of the curve in °, D_A deals with annual gross tonnage (Mt gross/year) and K_w stands for wear factor which depends on degree of curve.

For 57 kg/m plain carbon CWR: $w_a = 0.42\left(1+K_w D_c\right)D_A$ and for 66 kg/m plain carbon CWR: $w_a = 0.413\left(1+K_w D_c\right)D_A$.

Expected rail life (T_y) in year, $T_y = \frac{\theta_A}{W_a}$, where θ_A stands permissible limit of rail head wear. Expected rail life T (M_t gross) is estimated by $T = T_y\, D_A$.

10.5.4.2 Rail Wear by AREA Standard

For the tangent rail with fish plate where average velocity is 80 km/h, then the estimated rail life (T) is $T = 1.839K_T R_{wt}\left(1.102D_A\right)^{0.565}$, where K_T is the composite constant for track condition. A generalized formula for the tangent track to estimate rail life (mt. gross) by adopting track and traffic conditions,

$$T = 1.839 \frac{K_c K_G K_R R_{wt} (1.102 D_A)^{1.565}}{\sum_{i=1}^{n} \left[\frac{1.102\ D_i}{K_{Vi} K_{Ai} K_{Si}} \right]}, \tag{10.20}$$

where R_{wt} stands rail weight (kg/m), K_c stands for track curvature, K_G is track gradient, K_R denotes rail factor including joint type, rail weight and rail strength, n stands for different types of traffics, K_{Vi}, K_{Ai} and K_{Si} are speed class factor, wheel load class factor and service type factor, respectively.

10.5.4.3 Couard's Approach for Rail Wear

For estimating the wear rate annually, the subsequent equations [19]:

a. Rail wear in the vertical direction for tangent track

$$W_v = 8.7 \times 10^{-5} Q_1 V D_A \left(1 + 0.23 G^{1.7}\right) + 0.0635, \tag{10.21}$$

b. Rail wear in the vertical direction on high rail in a curved track is

$$W_v = 9.57 \times 10^{-5} Q_1 V D_A \left(1 + \frac{U_B}{254} + 0.23 G^{1.7}\right) + 0.0635(3.105),$$

c. Side rail wear in the vertical direction on high rail on curved track

$$W_s = 1.39 \times 10^{-4} Q_2 V D_A D_C \left(1 + \frac{U_B}{254} + 0.23 G^{1.7}\right) + 0.0635, \tag{10.22}$$

d. Side rail wear on high rail on curved track due to lubrication

$$W_{sl} = 0.7\left(W_s - 0.0635\right) + 0.0635, \tag{10.23}$$

where W_v, W_s and W_{sl} represent vertical wear at head, side rail wear and side rail wear with the presence of lubrications annually, respectively, D_A stands for traffic density, U_B is unbalanced cant height (mm), Q_1 stands for the ratio of the width of rail head to the 69.5 kg/m and Q_2 stands for the ratio of the depth of rail head to the 69.5 kg/m.

Generally, with increase in rail section size, maximum bending stress and deflection increases. For the analysis two different standards such as Victorian Railways (VicRail) with locomotive X-class and Western Australian Govt. Railways (Westrail) with two locomotives with L class have been adopted. The maximum bending stress and deflection are estimated by considering the rail as beam on elastic foundation. The rail with cross section 60 kg/m is used with track modulus value 15 MPa and corresponding bending stress and maximum deflection is given in Table 10.10. The effect of sleeper length and loading condition on the estimation of contact pressure is tabulated in Table 10.11.

TABLE 10.10
Maximum Rail Bending and Rail Deflection by Adopting Track Modulus 15 MPa [15]

		Maximum Bending Stress (MPa)		Maximum Rail Deflection (mm)	
Rail Section	Speed (Km/h)	VicRail X-Locomotive	WestRail L Locomotive	VicRail X-Locomotive	WestRail L Locomotive
60 kg/m	80	80.1	94.8	4.9	6.3
	100	85.9	101.7	5.3	6.8
	120	91.7	108.5	5.6	7.2

TABLE 10.11
Estimation of Ballast Contact Pressure by Change in Sleeper Length and Loading Condition in American Railroad Engineering Association (AREA) Approach [15]

	Sleeper to Contact Pressure (AREA Approach)	
Sleeper Length (m)	VicRail X-Locomotive	WestRail L Locomotive
2.44	470	554
2.54	452	534
2.64	436	514

10.5.5 Ballast Analysis

The material properties, depth of ballast and grading of ballast are significant parameters for present design approach.

10.5.5.1 Determination of Ballast Depth

The key function of ballast is the transmission of load exerted by the wheel to the subgrade; hence it is important to estimate the vertical pressure (stress in vertical direction) due to sleeper loading. The stress in the vertical direction (σ_z) at any distance (z) in kPa under the load is estimated by assuming a semi-infinite elastic medium [20].

$$\sigma_z = P_a \left[1 - \frac{z^3}{\left(a^2 + z^2\right)^{1.5}} \right], \tag{10.24}$$

and horizontal stress in both x and y directions are

$$\sigma_x = \sigma_y = \frac{P_a}{2} \left[(1 + 2\upsilon) - \frac{2(1+\upsilon)z}{\left(a^2 + z^2\right)^{0.5}} + \frac{z^3}{\left(a^2 + z^2\right)^{1.5}} \right]. \tag{10.25}$$

The horizontal and vertical stresses are considered as minor and major stresses, respectively.

Maximum shear stress (τ_r) in kPa after applying unit load

$$\tau_{z\max} = \frac{\sigma_z - \sigma_x}{2} = P_a \left[\frac{(1-2\upsilon)}{4} + \frac{(1+\upsilon)z}{\left(a^2+z^2\right)^{0.5}} - \frac{3z^3}{4\left(a^2+z^2\right)^{1.5}} \right], \quad (10.26)$$

where P_a is average uniform pressure over the area where load is applied, z = vertical depth beneath the surface (m), Q_0 is the point load (kN) and r is the horizontal distance (m) from vertical to the particular position, υ is the Poisson's ratio of the material and a is the radius of the loaded area, i.e. $a = \left[\frac{A_s}{\pi}\right]^{0.5} = \left[\frac{BL}{\pi}\right]^{0.5}$, where B is the sleeper breadth (m) and L is length of sleeper support (m).

Different countries have adopted different empirical relation.

i. As per Japanese National Railway (JNR) [21]

$$\sigma_z = P_a \left(\frac{58}{10+(100z)^{1.35}} \right), \quad (10.27)$$

where z is the depth of ballast (m).

For broken stone, $\sigma_z = P_a \left(\frac{125}{50+(100z)^{1.6}} \right)$ and for gravel ballast, $\sigma_z = P_a \left(\frac{125}{50+(100z)^{1.5}} \right)$.

Different countries adopt different standards such as EN 13803-1 in European countries and STN73 6360-1 in Slovak Republic, etc. to design the modern track. When the velocity of the train increases to a new higher limit, i.e. $V \leq 300$ km/h, the STN73 6360 modified the design criteria by considering other limits (RP5 for $V \leq 230$h and $V \leq 300$ km/h) [22]. For designing the railway line, it is very difficult to satisfy the parameter criteria in industrial parks and most populated area without reducing the passenger comfort. When the railway line is considered for the passenger trains only within the speed limit ($230 < V \leq 300$ km/h), the parameters are designed as per EN13803-1 standard. For example, the proposed curve radius (r) is in between $r_{\min}$ and $r_{\max}$ and calculated as per Equation (10.28).

$$\frac{11.8}{h_t + h_d} V_{\max}^2 \leq r \leq \frac{11.8}{h_t - h_e} V_{\min}^2, \quad (10.28)$$

where h_t, h_d and h_e represent cant height, cant deficiency and cant excess, respectively.

TABLE 10.12
Different Eq. of Transition Curve for Different Speed Limit [11]

Types of Transition Curve	Speed Limit	Eq. of Transition Curve
Cubic parabola with linear grade line	0> $V \leq 300$ km/h	$y = \frac{x^3}{6rl_p}$
Spiral with non-linear grade line	80> $V \leq 300$ km/h	$y = \frac{x^4}{4l_p^2 r} - \frac{x^5}{10l_p^3 r}$
Spiral with non-linear grade line	200> $V \leq 300$ km/h	$y = \frac{x^2}{4r} - \frac{l_p^2}{2r\pi^2}\left(\cos\frac{\pi x}{l_p} - 1\right)$

While designing the track, the designer should consider the weight of the vehicle along with their pay load. However, reconstructing of the existing track is necessary to improve the velocity. For the mountain or hill side route area, the velocity is reduced, then the proposed radius is also reduced. While designing the curve, the transition curve is an important parameter for increasing speed as mentioned in Table 10.12 (as per STN 73 6360-1).

10.6 SUMMARY

Track design and manufacturing are the essential elements in the railway track system that are responsible for improving the safety, efficiency and longevity of the track. The design practice includes the optimization of track parameters under different operating and environmental conditions. Rails, sleepers, fastening systems, and ballast are crucial components as they provide stability, effective load distribution which helps to reduce wear and tear. The life cycle of the rail depends upon the quality of the steel. The hot rolling and heat treatment process enhances hardness, strength and fatigue resistance. The development of the track is essential to accommodate heavy haul, high speed and urban transit rolling stocks and reduces the maintenance and environmental impacts. The incorporation of manufacturing and design proficiency is vital for reliable and resilient railway track. Different countries use different manufacturing processes and heat treatment processes, depending upon the climatic conditions and uses. This chapter gives a basic idea regarding the different compositions of rail steel, the manufacturing process and heat treatment process for further studies. Researchers focussed on the simulation of the heat treatment process of different rail steel materials and studied the microstructures and all the material behaviour before implementing the whole process, which is not covered.

REFERENCES

1. Iwnicki, S. *Handbook of Railway Vehicle Dynamics*. CRC Press, Boca Raton, FL, 2006.
2. Soltanian, H., A. Firouzi, and S. Mohammadzadeh. "Time Dependent Reliability Analysis of Railway Sleepers Subjected to Corrosion." *Structural Concrete* 19, no. 5 (2018): 1409–1418.
3. Ahlbeck, D. R., H. C. Meacham, and R. H. Prause. "The Development of Analytical Models for Railroad Track Dynamics." In *Railroad Track Mechanics and Technology*, pp. 239–263. Pergamon, Bergama, Turkey, 1978.
4. Bowles, J. E., and Y. Guo. *Foundation Analysis and Design*, Vol. 5. McGraw-Hill, New York, 1996.
5. EN 16273. *Railway Applications—Track*. European Committee for Standardization, Brussels, Belgium, 2014.
6. Chandra, S. *Railway Engineering*. Oxford University, Oxford, 2007.
7. JFE Steel Corporation. "Steel Shapes and Sections." https://www.jfe-steel.co.jp/en/works/west/process.html. Accessed October 10, 2023.
8. Dopierała, K., J. Ćwiek, K. Labisz, and R. Konieczny. "Application of Isothermal Heat Treatment for Perlite Rail-Steel." *Inżynieria Materiałowa* 38, no. 2 (2017): 82–87.
9. Ferreira, V. M., M. G. Mecozzi, R. H. Petrov, and J. Sietsma. "Microstructure Development of Pearlitic Railway Steels Subjected to Fast Heating." *Materials & Design* 221 (2022): 110989.
10. Jabłońska, M., F. Lewandowski, B. Chmiela, and Z. Gronostajski. "Advanced Heat Treatment of Pearlitic Rail Steel." *Materials* 16, no. 19 (2023): 6430.
11. Dukkipati, R. V., and J. R. Amyot. "Computer-Aided Simulation in Railway Dynamics." *Mechanical Engineering* 1988 (1988): 61.
12. Garg, V. *Dynamics of Railway Vehicle Systems*. Elsevier, Amsterdam, Netherlands, 2012.
13. Anis, Z., and J. K. Hedrick. *Characterization of Rail Track Irregularities*. Final Report; Contract No. DOT-TSC-1206. US DOT, Boston, MA, 1977.
14. Lindahl, M. *Track Geometry for High-Speed Railways*. TRITA-FKT Report 54 (2001).
15. Doyle, N. F. *Railway Track Design: A Review of Current Practice*. Australia, Bureau of Transport Economics, Canberra, 1980.
16. Jenkins, H. H., J. E. Stephenson, G. A. Clayton, G. W. Morland, and D. Lyon. "The Effect of Track and Vehicle Parameters on Wheel/Rail Vertical Dynamic Forces." *Railway Engineering Journal* 3, no. 1 (1974): 1–12.
17. Hay, W. W. *Railroad Engineering*, Vol. 1. John Wiley & Sons, New York, 1991.
18. Schramm, G. *Permanent Way Technique and Permanent Way Economy*. Otto Elsner Verlagsgesellschaft, Berlin, 1961.
19. Prause, R. H., and H. C. Meacham. *Assessment of Design Tools and Criteria for Urban Rail Track Structures: Volume I. At-Grade Tie-Ballast Track*. Columbus Laboratories, Columbus, 1974.
20. Boussinesq, J., and M. A. Caquot. *Application des Potentiels: À l'Étude de l'Équilibre et du Mouvement des Solides Élastiques*. Gauthier-Villars, France, 1885.
21. Okabe, Z. "Laboratory Investigations of Railroad Ballast." *Permanent Way* 11 (1961): 37–46.
22. Hodas, S. "Design of Railway Track for Speed and High-Speed Railways." *Procedia Engineering* 91 (2014): 256–261.

11 Formulation of a Metamodel-Based Multi-Objective Optimization Problem for a Rail Vehicle's Dynamic Response

Manish Pandey

11.1 INTRODUCTION

A Rail vehicle moving on the railway track is a complicated dynamic system in context to the rail vehicle dynamics [1–6]. The two rails guide the wheels in lateral and longitudinal directions apart from supporting the rail vehicle and transmitting the dynamic loads. The track irregularities provide excitation to the running rail vehicle through the wheels. The excitation is forwarded to the carbody of the vehicle through the bogies. In turn, the dynamic loads are transferred from the vehicle to the rails through the wheels. There are three subsystems- The vehicle sub-system, the wheel–rail interaction subsystem and the track subsystem in a rail vehicle dynamics system as demonstrated in Figure 11.1.

The rail vehicle subsystem consists of several bodies like carbody, bolsters, side frames, wheelsets, etc. These bodies are connected through different types of suspension/connection elements to each other. The rail vehicle and track interaction system can be represented as a multibody dynamic system. As far as rail vehicle dynamics is considered, the bogie (along with the suspension systems) should be able to carry the vehicle stably on the curved and tangent track geometries and that too in both the vehicle empty and vehicle-loaded conditions. Moreover, the bogie should generate low dynamic forces so that the generation of track irregularities is minimal. If the track conditions and the inertia parameters of the rail vehicle are given, proper selection of suspension parameters may result in optimised running behaviour of the bogie suspension system.

The dynamic considerations in the suspension design of a rail vehicle can be segregated into three parts: Running behaviour (carbody vertical and lateral accelerations), safety (lateral frame force, vertical carbody acceleration, and lateral carbody acceleration), and track fatigue (lateral frame force, vertical track force, and lateral

DOI: 10.1201/9781003659983-11

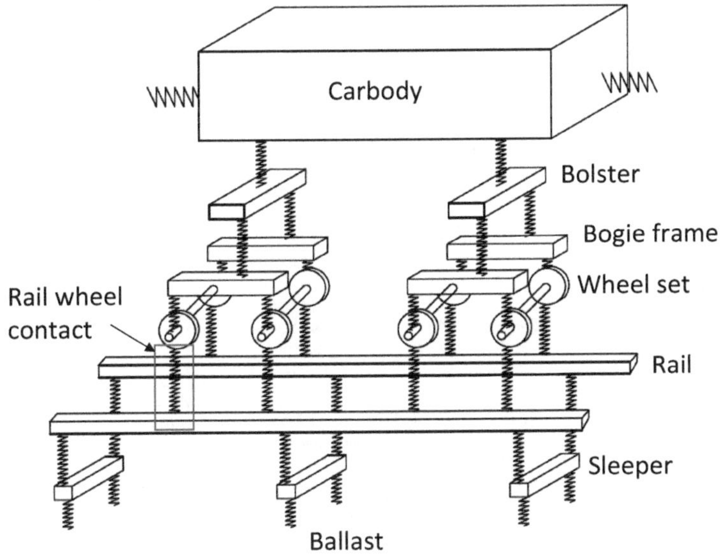

FIGURE 11.1 The vehicle sub-system, the wheel-rail interaction subsystem and the track subsystem.

track force). The running behaviour of a vehicle is to be examined on the curved and tangent track geometries and in the vehicle's empty and vehicle-loaded conditions. The above dynamic performance parameters of a vehicle depend on its vertical and lateral suspension parameters and the inertia values of its components.

Initially, specific dynamic models were made focusing on individual areas of dynamic response like the hunting of a wheelset, hunting of a bogie and curving behaviour of bogies, etc. Individual codes were written to construct the subject models. This posed a serious limitation in making a single mathematical that could take care of all the aspects of rail vehicle dynamic response. Since the advent of specific sophisticated rail vehicle dynamic software, it has become possible to assess a of aspects regarding the rail vehicle's dynamic response.

Similarly, the advancement in the domain of decision sciences using optimization algorithms has also made it possible to attempt for a comprehensive optimization of a large number of aspects of the dynamic response of a rail vehicle under the framework of multi-objective optimization (MOO). It is highlighted here that in such cases, the objective functions cannot be characterised by definitive equations but are estimated computationally through simulations run on the rail vehicle. The conjunction between the rail vehicle dynamic simulations and the optimization algorithms is a challenging task wherein parallel computing, co-simulation and metamodelling are some of the techniques used. The typical structure followed in applying the MOO in the field of Rail vehicle dynamic simulation is shown in Figure 11.2.

Since the process starts with creating a dynamic multibody model for the subject vehicle, Section 11.2 of this chapter will present the mathematical formulations for the multi-body system. In Section 11.3, the implementation of the MBS system in the Rail vehicle simulation software will be delved into. Discussion will also be made in

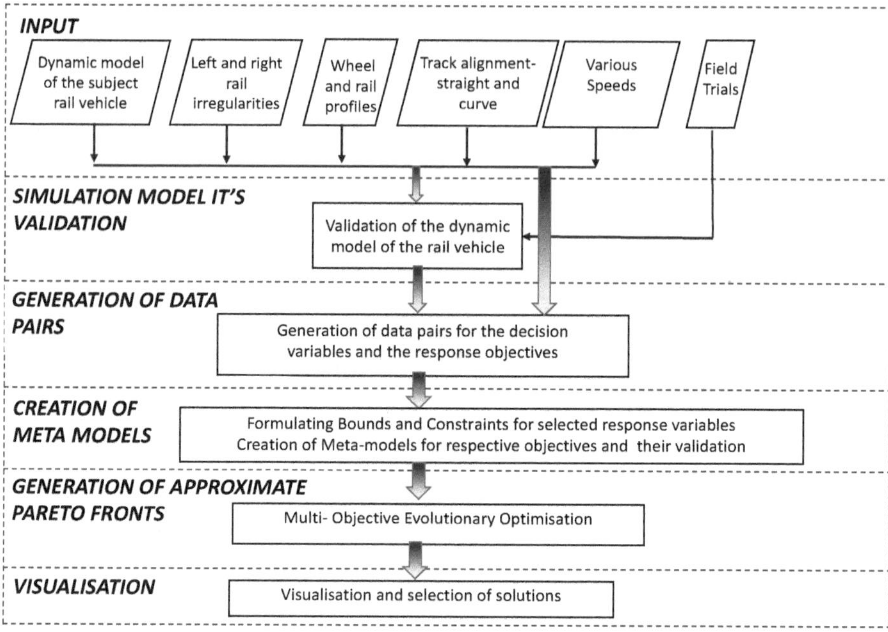

FIGURE 11.2 The typical flow chart followed in applying MOO to rail vehicle dynamic simulation.

this section regarding the methodology used for the validation of the dynamic simulation models. In Section 11.4, the current state of research in the dynamic response analysis of rail vehicles will be discussed. In Section 11.5, state of research in the field of metamodelling and subsequent MOO for the dynamic response of rail vehicles will be discussed. The intricacies and the challenges in the metamodelling and subsequent MOO as applied to the rail vehicle dynamics will also be discussed in Section 11.5.

11.2 MATHEMATICAL FORMULATION FOR THE MULTI-BODY SYSTEM

11.2.1 Reference Frames

Usually, moving reference frames (apart from an inertial fixed reference frame) are used in Rail vehicle dynamic analysis. Two moving reference systems used are the track-following reference system and the body-following reference system (Figure 11.3).

The track following system ($\hat{e}1$, $\hat{e}2$, $\hat{e}3$) moves along the nominal track centre line (longitudinal coordinate s). The motion of its origin says T0 with respect to the inertial reference frame ($\mathbf{e}_1, \mathbf{e}_2, \mathbf{e}_3$) is described by its position vector, velocity, and acceleration with respect to the inertial reference frame. The degrees of freedom of a rigid body i.e., longitudinal, lateral and vertical motions (say x, y and z) and the small relative rotations roll (ϕ), pitch (θ) and yaw (ψ) are defined in the body

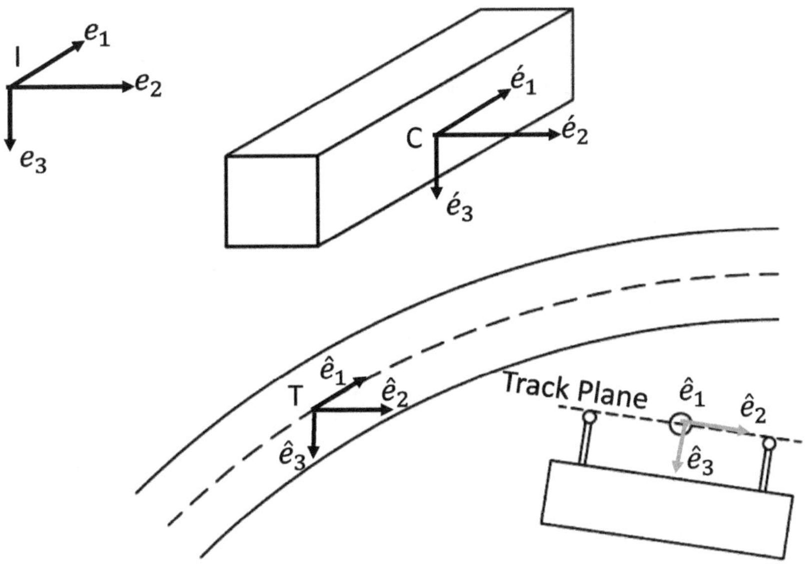

FIGURE 11.3 The three reference frames- one fixed and two moving.

following reference system. The governing force and moment equilibrium equations are formed in the track following coordinate system. However, the creepages and subsequently creep forces are calculated in the Contact plane reference system and then the creep forces are resolved in the track following reference system.

The Euler moment equations are first formed in the body following reference system (the Moment vector and the body mass of inertia are also in the body following reference system. The moment equations are then extended to the track following reference system using inertia tensors and rotational matrices. Firstly, creep velocities are expressed in terms of wheel set's roll angle, yaw angle, contact angle and their derivatives along with the tread radius of wheel and half the distance between the contact points, in the track following coordinate system (XYZ). Then the creep velocities are converted to longitudinal and lateral creep velocities in the contact point coordinate system (ξ, η, ζ) as shown in Figure 11.4. However, spin creepages are directly expressed in the contact point coordinate system (ξ, η, ζ). From these creep velocities, creep forces are calculated in the contact point coordinate system. Then the creep forces are resolved in the track following reference system.

11.2.2 Formulation of MBS System

Shabana [7] and Shabana et al. [8] have presented the details for multibody formulations. As shown in Figure 11.5, the global position of a random point '*P*' on a body, denoted by the body '*i*'

$$\mathbf{r}_p^i = \mathbf{R}^i + \mathbf{A}^i \overline{\mathbf{u}}^i. \tag{11.1}$$

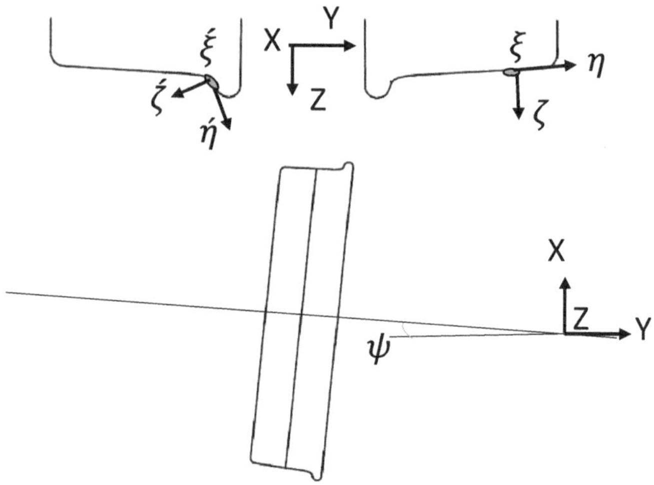

FIGURE 11.4 Contact plane reference system.

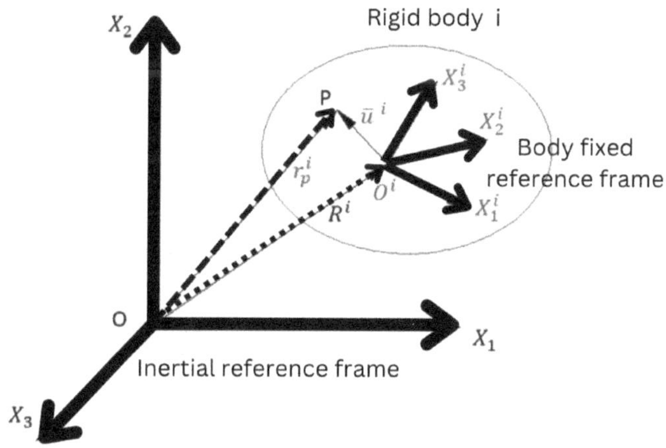

FIGURE 11.5 Global position of a random point.

The origin of the selected body reference $\mathbf{X}_1^i\mathbf{X}_2^i\mathbf{X}_3^i$ is $\mathbf{O}^i$ whose position vector is $\mathbf{R}^i$

In order to transform a coordinate from the body coordinate system ($\mathbf{X}_1^i\mathbf{X}_2^i\mathbf{X}_3^i$) to the global coordinate system ($\mathbf{X}_1\mathbf{X}_2\mathbf{X}_3$), the transformation matrix $\mathbf{A}^i$ is used. The global coordinate system $\mathbf{X}_1\mathbf{X}_2\mathbf{X}_3$ can be taken as track following coordinate system in case of a vehicle is running on a straight track.

The local position vector of the random point '*P*' measured with respect to the body coordinate system is $\bar{\mathbf{u}}^i$.

$\mathbf{A}^i$ is a function of the set of rotational coordinates $\boldsymbol{\theta}^i$. $\boldsymbol{\theta}^i$ has three or four elements in the 3D analysis and only one element in the 2D or Planer analysis. Techniques like Euler angles, Euler Parameters and Rodriguez parameters are used to ascertain the $\boldsymbol{\theta}^i$.

Generally, Euler Angles are used for orientation parameters. To define the orientation of a rigid body, three serialised rotations are used.

$$\mathbf{q}_r^i = \left[\mathbf{R}^{i^T} \boldsymbol{\theta}^{i^T}\right]. \tag{11.2}$$

For planer analysis

$$\mathbf{q}_r^i = \left[R_1^i \; R_2^i \; \theta^i\right]^T. \tag{11.3}$$

In 3-D analysis,

$$\mathbf{q}_r^i = \left[R_1^i \; R_2^i \; R_3^i \; \theta^{i^T}\right]^T. \tag{11.4}$$

θ^{i^T} contains three elements $[\theta_1^i, \theta_2^i, \theta_3^i]$.

Bar shows with respect to body reference frame, $\overline{\omega} = \overline{\mathbf{G}}\dot{\theta}$ and $\overline{\alpha} = \overline{\mathbf{G}}\,\ddot{\theta}$ are the absolute angular velocity vector and absolute angular acceleration defined in the body coordinate system.

The acceleration vector in the body coordinate system can be written as

$$\ddot{\mathbf{r}} = \ddot{\mathbf{R}} + \mathbf{A}^i\left\{\overline{\alpha} \times \overline{\mathbf{u}}^i + \overline{\omega} \times \left(\overline{\omega} \times \overline{\mathbf{u}}^i\right)\right\}. \tag{11.5}$$

Matrix $\mathbf{A}^i$ is used to convert body coordinates to global reference coordinates. Matrix $\mathbf{G}^i$ is used to convert time derivatives of the generalised coordinates into angular velocity and angular acceleration vectors and to convert the external forces/moments to generalised forces/moments. The column vectors of $\mathbf{G}^i$ represent unit vectors defined in the global coordinates about which the three serialised rotations have been performed. The column vectors of $\overline{\mathbf{G}^i}$ represent the unit vectors in the body coordinate system along which the Euler angles are performed.

$$\mathbf{G}^i = \mathbf{A}^i\overline{\mathbf{G}^i}. \tag{11.6}$$

11.2.3 Newton-Euler Equations for a Body

In this section, Newton-Euler equations of motion will be presented for a rigid body in the multibody system [8]. Absolute acceleration and external forces in the inertial reference frame which is the same as the track following the reference frame for the straight track. The Newtonian equations of motion are formed with respect to the inertial reference frame but can be expressed in (referred to) the body-fixed reference frame. The Euler moment equations are generally formed with respect to a body-fixed reference frame with a time-invariant inertia matrix.

$$\mathbf{m}_{RR}\ddot{\mathbf{R}} = \mathbf{F}_e, \tag{11.7}$$

$$\overline{\mathbf{I}}_{\theta\theta}\left(\overline{\mathbf{G}}\,\ddot{\theta}\right) + \omega_{\text{axis}} \times \left\{\overline{\mathbf{I}}_{\theta\theta}\,.(\overline{\mathbf{G}}\,\dot{\theta}\right\} = \overline{\mathbf{M}}_e = \overline{\mathbf{F}}_\theta. \tag{11.8}$$

$\bar{\omega}$, $\bar{\alpha}$ and $\bar{\mathbf{M}}_e$ are absolute angular velocity, angular acceleration, and external moment vectors.

$$\bar{\alpha} = \bar{\mathbf{G}}\ddot{\theta},\ \bar{\omega} = \bar{\mathbf{G}}\ \dot{\theta}. \tag{11.9}$$

$\bar{\mathbf{F}}_e$ represents the vector of external forces
$\bar{\mathbf{F}}_\theta$ is the vector of external moments acting on the body
$\bar{\mathbf{I}}_{\theta\theta}$ is the inertia matrix.

In matrix form

$$\begin{bmatrix} m\mathbf{I} & 0 \\ 0 & \bar{\mathbf{I}}_{\theta\theta} \end{bmatrix} \begin{bmatrix} \ddot{\mathbf{R}} \\ \bar{\alpha} \end{bmatrix} = \begin{bmatrix} \mathbf{F}_e \\ \bar{\mathbf{M}}_e - \omega_{\mathbf{axis}} \times \left\{ \bar{\mathbf{I}}_{\theta\theta}\ .(\bar{\mathbf{G}}\ \dot{\theta}) \right\} \end{bmatrix}. \tag{11.10}$$

The Newton-Euler equations are written in terms of generalised coordinates as follows:

$$\begin{bmatrix} \mathbf{m}_{RR} & 0 \\ 0 & \mathbf{m}_{\theta\theta} \end{bmatrix} \begin{bmatrix} \ddot{\mathbf{R}} \\ \ddot{\theta} \end{bmatrix} = \begin{bmatrix} (\mathbf{Q}_e)_R \\ (\mathbf{Q}_e)_\theta + (\mathbf{Q}_v)_\theta \end{bmatrix}. \tag{11.11}$$

$\mathbf{m}_{RR} = m\mathbf{I}$ and $\mathbf{m}_{\theta\theta} = \bar{\mathbf{G}}^T \bar{\mathbf{I}}_{\theta\theta} \bar{\mathbf{G}}$ are the generalised mass matrices with respect to R and θ respectively.

$$(\mathbf{Q}_v)_\theta = -\bar{\mathbf{G}}^T \left[\bar{\omega} \times \left(\bar{\mathbf{I}}_{\theta\theta}\ .\bar{\omega} \right) + \bar{\mathbf{I}}_{\theta\theta}\ \dot{\bar{\mathbf{G}}}\dot{\theta} \right], \tag{11.12}$$

$$(\mathbf{Q}_e)_R = \mathbf{F}_e, \tag{11.13}$$

$$(\mathbf{Q}_e)_\theta = \bar{\mathbf{G}}^T \bar{\mathbf{M}}_e. \tag{11.14}$$

The vectors for external force and moment can also be written in the global coordinate as follows:

$$\mathbf{F}_e = A\bar{\mathbf{F}}_e, \tag{11.15}$$

$$\mathbf{M}_e = A\bar{\mathbf{M}}_e, \tag{11.16}$$

$$\text{Also } (\mathbf{Q}_e)_\theta = \bar{\mathbf{G}}^T\ \bar{\mathbf{M}}_e = \mathbf{G}^T \mathbf{M}_e. \tag{11.17}$$

We can write the system equations for the motion of the rigid body 'i' as

$$\mathbf{M}^i \ddot{\mathbf{q}}^i = \mathbf{Q}_e^i + \mathbf{Q}_v^i. \tag{11.18}$$

where

$$\mathbf{M}^i = \begin{bmatrix} \mathbf{m}_{RR} & 0 \\ 0 & \mathbf{m}_{\theta\theta} \end{bmatrix}, \ddot{\mathbf{q}}^i = \begin{bmatrix} \ddot{\mathbf{R}} \\ \ddot{\theta} \end{bmatrix}, \mathbf{Q}_e^i = \begin{bmatrix} (\mathbf{Q}_e)_R \\ (\mathbf{Q}_e)_\theta \end{bmatrix}, \mathbf{Q}_v^i = \begin{bmatrix} 0 \\ (\mathbf{Q}_v)_\theta \end{bmatrix}. \quad (11.19)$$

$\mathbf{Q}_e^i$ represents the vector of externally applied forces
$\mathbf{Q}_v^i$ represents the quadratic velocity vector.

Augmented Formulation for incorporating constraints (dependent coordinates also)

$$\mathbf{M}^i\ddot{\mathbf{q}}^i + \mathbf{C}_q^{iT}\boldsymbol{\lambda} = \mathbf{Q}_e^i + \mathbf{Q}_v^i; \quad i = 1, 2, \ldots n_b. \quad (11.20)$$

In matrix form

$$M\ddot{\mathbf{q}} + \mathbf{C}_q^T\boldsymbol{\lambda} = \mathbf{Q}_e + \mathbf{Q}_v. \quad (11.21)$$

The above formulation is valid for the tangent track with no twist i.e., track following reference frame and inertial frame are same.

11.2.4 Mathematical Formulation for a Flexible Body

Structural flexibility means the body deformations due to the forces acting on the body. In many railway applications involving high speeds and lightweight rolling stocks, it is necessary to consider the structural flexibility of the bodies/components in vehicle track interaction simulations. In the rolling stock, different components like bogies and car bodies can have significant structural flexibilities, particularly in bending and torsion (Figure 11.6).

Wasfy and Noor [9] presented the current status in computational modelling of multibody systems. The incorporation of modal analysis techniques results in a significant reduction in the number of the physical finite element nodal coordinates while using the floating reference frame. Escalona et al. [10] paid attention to the floating frame of reference formulation.

11.2.5 Mathematical Formulation of the Flexible Multibody System

For a flexible body, the generalised equation of motion in a floating frame of reference

$$\mathbf{M}^c\ddot{\mathbf{q}}^c + \mathbf{K}^c\mathbf{q}^c = \mathbf{Q}_e^c + \mathbf{Q}_v^c + \mathbf{Q}_r^c, \quad (11.22)$$

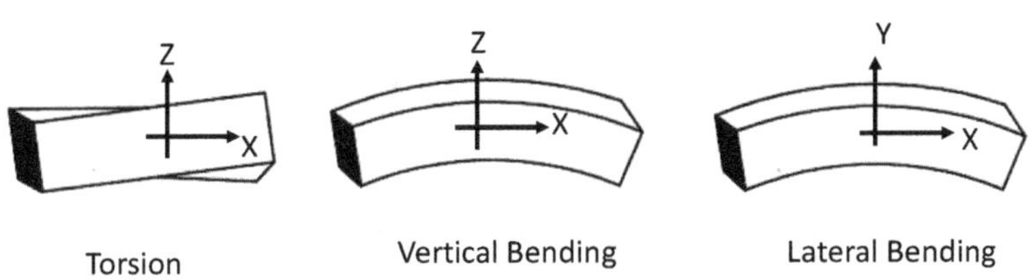

FIGURE 11.6 Flexible body torsion, vertical bending, and lateral bending modes.

where $\mathbf{M}^c$ and $\mathbf{K}^c$ are the coordinate dependent mass matrix and stiffness matrix, respectively.

$\mathbf{Q}_v^c$ represents the generalised quadratic – velocity inertia vector
$\mathbf{Q}_e^c$ represents the generalised applied force vector
$\mathbf{Q}_R^c$ represents the generalised reaction force vector, respectively.

For the carbody, the generalised coordinates vector can be written as

$$\mathbf{q}^c = \left[\mathbf{q}_r^{c^T} \mathbf{q}_f^{c^T} \right] T. \tag{11.23}$$

where $\mathbf{q}_r^c = \left[\mathbf{R}^{c^T} \boldsymbol{\theta}^{c^T} \right] T$ is the vector of reference coordinates and $\mathbf{q}_f^c$ represents the vector of nodal coordinates.

So, the equation of motion becomes

$$\begin{bmatrix} \mathbf{m}_{rr}^c & \mathbf{m}_{rf}^c \\ \mathbf{m}_{fr}^c & \mathbf{m}_{ff}^c \end{bmatrix} \begin{bmatrix} \ddot{\mathbf{q}}_r^c \\ \ddot{\mathbf{q}}_f^c \end{bmatrix} + \begin{bmatrix} 0 & 0 \\ 0 & \mathbf{K}_{ff}^c \end{bmatrix} \begin{bmatrix} \mathbf{q}_r^c \\ \mathbf{q}_f^c \end{bmatrix} = \begin{bmatrix} (\mathbf{Q}_e^c)_r \\ (\mathbf{Q}_e^c)_f \end{bmatrix} + \begin{bmatrix} (\mathbf{Q}_v^c)_r \\ (\mathbf{Q}_v^c)_f \end{bmatrix} + \begin{bmatrix} (\mathbf{Q}_R^c)_r \\ (\mathbf{Q}_R^c)_f \end{bmatrix}. \tag{11.24}$$

In the expanded forms

$$\mathbf{m}_{rr}^c \ddot{\mathbf{q}}_r^c + \mathbf{m}_{rf}^c \ddot{\mathbf{q}}_f^c = (\mathbf{Q}_e^c)_r + (\mathbf{Q}_v^c)_r + (\mathbf{Q}_R^c)_r = \mathbf{Q}_r, \tag{11.25}$$

$$\mathbf{m}_{fr}^c \ddot{\mathbf{q}}_r^c + \mathbf{m}_{ff}^c \ddot{\mathbf{q}}_f^c + \mathbf{K}_{ff}^c \ddot{\mathbf{q}}_f^c = (\mathbf{Q}_e^c)_f + (\mathbf{Q}_v^c)_f + (\mathbf{Q}_R^c)_f = \mathbf{Q}_f. \tag{11.26}$$

Furthermore, the Linear Theory of Elastodynamics [7, 11] can be used here, and the equations for rigid body motion and elastic deformation are solved separately while ignoring the nonlinear geometric effect. The term $\mathbf{m}_{rf}^c \ddot{\mathbf{q}}_f^c$ may also be ignored for further simplification. The matrix $\mathbf{m}_{rr}^c$ and the vector $\mathbf{Q}_r$ are assumed to be independent of the elastic deformation of the carbody. Therefore, Equations (11.25 and 11.26) can be simplified as

$$\mathbf{m}_{rr}^c \ddot{\mathbf{q}}_r^c = \mathbf{Q}_r, \tag{11.27}$$

$$\mathbf{m}_{ff}^c \ddot{\mathbf{q}}_f^c + \mathbf{K}_{ff}^c \ddot{\mathbf{q}}_f^c = \mathbf{Q}_f - \mathbf{m}_{fr}^c \ddot{\mathbf{q}}_r^c. \tag{11.28}$$

11.3 MATHEMATICAL MODELLING OF THE RAIL VEHICLE AND ITS COMPONENTS

Most of the dynamic mathematical models catered to the lateral stability and curving models of wheelsets. Investigations using wheelset models were presented by various researchers [12–19]. Curving and hunting response analysis of railway trucks have also been conducted by some researchers [20, 21]. Mathematical models of complete rail vehicles can be found in the literature [22–30]. However, the models were targeted to specific purposes like lateral stability, vertical response, lateral response, curving behaviour, etc.

11.3.1 Implementation in Simulation Software

Three-dimensional models of a rail vehicle system including the three subsystems i.e. rail vehicle, railway track and the rail–wheel contact can be created which can solve any of the above-mentioned problems at a time. Multibody formulations are the foundation stones for the Rail vehicle dynamic simulations. The formulation methodologies, modelling and numerical procedures have been dealt in detail [31–33].

The multibody system dynamics problem for a railroad vehicle consists of two sub-problems- the rail wheel contact problem and the dynamics problem. The rail wheel interaction is modelled using Special force elements or kinematic constraint elements. Shabana and Sany [34], Shabana et al. [35], Netter et al. [36], and Schupp et al. [37] presented methods for including the rail wheel contact problem into the computational multi software (MBS) algorithms. Shabana et al. [38] presented an elastic force model and compared with the results obtained using the constraint approach. Shabana [33] dealt with the various issues of rail vehicle multibody dynamic formulations comprehensively, including dealing with flexible bodies.

The nonlinear dynamic analysis of Rail vehicle multibody systems [8] uses three approaches- Ordinary differential equations (ODE's), Differential and algebraic equations (DAEs) with minimal coordinates and DAEs with reference coordinates. All the above approaches have their relative advantages and disadvantages. A typical rail vehicle dynamic simulation software is used for static analysis, quasi-static analysis, modal analysis, frequency response analysis, power spectral density analysis, and time history analysis. Kortüm [39] has compared different MBS rail vehicle dynamics software with respect to the formalisms and methodologies used. A summary is presented in Table 11.1.

In the Manchester Benchmark [40], various contact methods currently employed in MBS were compared. Weidmann [41] stated that there are general-purpose multi-body dynamic software like ADAMS (Europe), A'GEM (Canada), SIMPACK (Germany) and Universal Mechanism (Russia), etc. which can be utilised for Rail vehicle dynamic analysis also. There are certain specialised Rail vehicle dynamic software like GENSYS (Sweden), MEDYNA (Germany), NUCARS (USA), VAMPIRE (UK), and VOCO (France).

11.3.2 Modelling of Suspension and Connection Elements

The reliability and consistency of a dynamic model depend upon the correct modelling of the different types of connection elements like helical springs, rubber springs, dampers, friction wedges and air springs, etc. Eickhoff et al. [42] and Bruni et al. [43] reviewed the existing models for railway vehicle suspension components. In case of a three-piece bogie fitted in freight wagons, the modelling of the friction wedges is intricate. Kaiser et al. [44], Ballew et al. [45], Xia [46], and Wu et al. [47] have made significant insights in the modelling of friction wedges.

TABLE 11.1
Comparison of Rail Vehicle Dynamic Software

Sr. No.	Item	SIMPACK	ADAMS	NUCARS
1.	Joints/Constraints	Yes	Yes	No. Hence, every degree of freedom for a body must be supported by a force restraint
2.	Closed Loop (The topology of constraints is checked after every step)	Yes	Yes	No. But it is suitable for small displacements and rotations
3.	Coordinates Absolute-Degrees of freedom defined with respect to the inertial frame Relative- Relative motion of contiguous bodies is specified	Both Absolute and Relative Coordinates	Both Absolute and Relative Coordinates	Relative coordinates only
4.	Minimal Coordinates	Yes	Yes	No
5.	Moving Reference Frame Degrees of freedom are defined relative to moving reference frames.	Yes	Yes	Yes
6.	Kinematics	Nonlinear Kinematics	Linear and Nonlinear Kinematics	Linear Kinematics
7.	Type of Equations ODE or DAE	Both ODE and DAE	Both ODE and DAE	Only ODE
8.	Constraint Forces	Yes	Yes	No

11.3.3 Numerical Methods for Solving the Governing Equations

The second-order DAE's equations derived in Sections 11.2.8 and 11.3 are complicated in nature and are usually solved numerically as an initial value problem. The differential equations are generally nonlinear due to the non-linearity in the rail wheel contact, non-linearity in the stiffness and friction, etc. Three explicit and 04 implicit direct integration schemes were compared by [2] with reference to stability, CPU times and time steps.

11.3.4 Validation of Simulation Models

Evans and Berg [48] has discussed various challenges in the fields of modelling and rail vehicle dynamic simulation. The on-track tests have been used majorly for the approval of rolling stocks with respect to dynamic behaviour. The application of simulations for rolling stock approval was proposed in the European standard UIC 518:2009 [49]. But the methodology was not concrete.

The Australian Standards and Rail Corporation, New South Wales [50], attempted to use dynamic simulations instead of field experimentations. Other countries, such as Japan, China, and Korea, allow the use of simulation for some parts of the vehicle approval process [51]. Polach and Böttcher [52], Polach et al. [53], and Polach and Evans [54] proposed a standard deviation-based criterion for model validation. The latest version of European standard EN14363:2016 [55] includes a comprehensive methodology for the validation of simulation models to be used instead of on-track tests.

11.4 DYNAMIC CONSIDERATIONS IN THE SUSPENSION DESIGN

The dynamic considerations in the suspension design of a rail vehicle can be segregated into three parts:

1. Safety
   Lateral frame force
2. Guidance and Stability
   Hunting Stability on Straight Tracks
   Curving Behaviour
3. Response/Riding
   Vertical Acceleration/Vertical Ride Index
   Lateral Acceleration/Lateral Ride Index

Dukkipatti and Garg [2] showed that the vertical and lateral accelerations of the carbody can be considered as primary or intermediate response variables for most of the dynamic performance parameters of a Rail vehicle. The carbody accelerations have also been included in the European standard UIC-518 [56] as the dynamic performance parameters for the purpose of approval of railway rolling stock.

The car body dynamic response analysis has been used for passenger and freight vehicles. Rosenberger et al. [57] modeled Amtrak's F-40 non-powered control units and performed a parametric study considering the vehicle acceleration response in pitching and bouncing modes. Suarez et al. [58] investigated about how the dynamic behavior of a passenger vehicle is affected by the elastic properties of the suspensions. Pandey and Bhattacharya [59] attempted to investigate the dynamic response of a car body of a freight vehicle fitted with three-piece bogies, and the influence of central suspension parameters on various dynamic modes. Multibody models in NUCARS (New and Untried Cars Analysis Regime Simulation) consisting of eleven bodies and seventy connections were created for the study. The empty and loaded wagon models were validated using the standard deviation-based methodology provided by the EN 14363:2016 standard.

Pandey and Bhattacharya [60] presented a parametric study regarding a freight vehicle fitted with three-piece bogies, the effect of various central suspension parameters on the lateral frame force caused by the vehicle.

11.4.1 Flexible Car Bodies

The carbody flexible multibody analysis generally uses a combinatorial technique using the carbody's finite element model (FEM) and a multibody model of the rail vehicle. The modal parameters for a flexible carbody may be assessed using FEM or Experimental modal analysis [EMA] [61, 62]. Knothe and Grassie [56] presented that curving, stability, and passenger comfort are phenomena that happen at comparatively low frequencies below 20 Hz.

Carlbom [63, 64] studied a flexible passenger-carrying vehicle, and the simulation results were compared with the experimental (field results) data. studied the influence of carbody vertical stiffness on the vehicle ride quality were studied by [65, 66] analysed the influence of accounting for the structural flexibility of the carbody of open wagon on its hunting speed. Shi and Wu [67] studied the flexible carbody vibration of carbody in high-speed Electric multi-units. Dumitriu and Camil [68] have presented certain aspects of the modelling of flexible car bodies. Ling et al. [69] investigated the flexible vibrations in high-speed train carriages. The interactions caused between the high-speed lightweight passenger vehicles and the train bridge have also been studied [70, 71].

The effect of structural vertical vibrations in high-speed trains has been studied by researchers and methods like introducing control techniques in suspension systems [72], using the underfloor mounted equipment [73] or the bogie as a dynamic vibration absorber [74] and using piezoelectric actuators [75, 76]. Shust and Iler [77] studied the flexible modes of railway freight cars and presented the band of natural frequencies for them. Pandey [78] analysed the effect of including the structural flexibility of the carbody of an open freight car in the analysis of its dynamic response.

11.5 MULTI-OBJECTIVE OPTIMIZATION FOR THE RAIL VEHICLE DYNAMIC PERFORMANCE

Optimization algorithms are mainly divided as deterministic and non-gradient-based stochastic algorithms. Deterministic or 'Hill-climbing' algorithms may be further classified into direct search (e.g., Simplex) and indirect search (e.g., SQP algorithm) algorithms. Direct search algorithms and stochastic algorithms are mostly heuristic in nature and the derivative evaluations are not needed. In case of suspension optimization of rail vehicles with respect to the dynamic performance, the objective functions cannot be generally expressed explicitly in terms of the design variables. The optimization problem is to be solved as a discrete optimization problem using some non-gradient-based algorithm.

The deterministic optimization algorithms like sequential quadratic programming (SQP) methods, though offer rapid converging, have the tendency to stop at local minima, far away from the global optimum. Stochastic optimization algorithms offer a way to perform a global optimization, but they usually require many more iterations [79]. For multibody dynamics problems where criteria evaluations are performed using time-consuming numerical integration of ODEs or DAEs, it can be prohibitive to perform a large number of criteria evaluations. Advanced techniques such as quenching or reannealing may be applied to get satisfying results with a reasonable number of iterations [80].

If more than one objective is to be optimised, then the problem is called a multi-objective, multicriteria, or vector optimization problem. However, in some cases scalarisation of a vector optimization problem may be done by combining the objectives and solving the problem as a single objective optimization problem. In MOO, the Pareto fronts [81] can be obtained by using metaheuristic optimization methods like genetic algorithms [82], simulated annealing [83] and swarm optimization [84, 85]. Wu et al. [86] surveyed the application of Particle swarm optimization (PSO) in railway-related research. There is no single winner within the methods of Evolutionary algorithms and the results will depend upon the problem on hand as well as the algorithm settings.

To reduce the computational efforts and time, several approaches have been followed. Parallel computing of multibody dynamics software and the optimization algorithm is one of the approaches to reduce computational time. A summary of the efforts available in literature for optimising the rail vehicle dynamic response using various approaches is presented in Table 11.2.

11.5.1 Metamodel-Based Optimization of Rail Vehicle Dynamic Behaviour

As can be seen from Table 11.2, some of the recent research regarding multi-objective optimization (MOO) problems related to the suspension parameters of rail vehicles have started using meta models or virtual models for approximating the objective functions. Since the rail vehicle dynamic simulations/experiments are costly and time-consuming, metamodels may be formed from the detailed simulation models for applying the optimization algorithm. The detailed simulation models are represented by simplified meta models with smooth gradients and are formed using the results of several detailed simulation runs. There are several techniques for forming metamodels such as polynomial regression, Kriging, artificial neural networks, multivariate adaptive regression splines, radial basis functions, radial basis function networks, support vector regression, etc [98]. Jin et al. [99] compared Polynomial regression, kriging, MARS, and radial basis functions (RBF) models and opined that RBF models worked the best for him. Performing rail vehicle dynamic simulations to generate sample data points is a very effort-intensive and computationally expensive procedure. For such applications, metamodels such as kriging models, RBFs, and linear and quadratic models may be used to generate additional data points for input to optimization algorithms [100].

It cannot be said conclusively which of the meta modelling technique is completely superior to the other ones. Furthermore, the results of the used software may vary substantially depending on the tuning of the parameters. The performance may vary depending upon various factors like the size of data set, nonlinearity in the problem, scale of the problem (whether it is a large-scale problem or a small-scale problem) and the noise in the data, etc. For large sample of data sets, MARS, RBF and KG are very close in average accuracy and robustness. Radial Basis function is best for small and scarce data set. However, some of the distinguishing features of the four most used metamodelling techniques are given in Table 11.3:

It is again iterated that there is no universal model that always is the best choice. Instead, the choice of the metamodel will depend upon the problem to be tackled. All the above-mentioned modelling techniques have their unique features. Hence, a normal trend is there in research to have a combination of different models so that

TABLE 11.2
Summary of Research Works for Optimising the Rail Vehicle Dynamic Response

Sr. No.	Reference	Modelling of Dynamic System	Design Goals	Design Variables	Optimization Algorithm	Implementation of Integrated Optimization Algorithm	Remarks
1	[79]	17 DOF rail vehicle model using A'GEM software. Calculation of the 'critical speed' using SQP	Maximising Critical speed	The suspension stiffness and damping coefficients, the inertial property parameters, and the geometric parameters	• Genetic algorithm • Nelder-Mead's Simplex algorithm (non-gradient based)	A two-loop method using GA as the outer loop algorithm, and SQP (with A'GEM) as the interior loop algorithm	Parallel computing
2	[87]	MOO using three models in A'GEM • A lateral stability model having 17 degrees of freedom (DOF) • A vertical ride quality model using 36 DOF • A 21 DOF nonlinear dynamic curving performance model.	The conflicting requirements for dynamic curving, lateral stability, and vertical ride quality	The suspension stiffness and damping coefficients, the inertial property parameters, and the geometric parameters	Genetic algorithm	Hybrid MDO Combinative method using IDF and A-i-O	Vector optimization problem reduced to Scaler problem by combining the objectives.
3	[88]	First, a complete three-car model for optimising the complete vehicle dynamics. Second, a linearised half-car model having eight degrees of freedom is used	Comfort and safety	Primary and secondary damping in the lateral and longitudinal suspension.	Multi-objective evolutionary algorithm (MOEA)	MATLAB-GENSYS Environment	Parallel computing Co-simulation

(Continued)

TABLE 11.2 (*Continued*)
Summary of Research Works for Optimising the Rail Vehicle Dynamic Response

Sr. No.	Reference	Modelling of Dynamic System	Design Goals	Design Variables	Optimization Algorithm	Implementation of Integrated Optimization Algorithm	Remarks
4	[89]	A model of 10 DOF system	Vertical acceleration using weighting function, transfer function, and input PSD.	Primary and secondary suspension	BFGS (Broyden-Fletcher-Goldfarb-Shanno)		Parallel computing
5	[90]	21 DOF model for lateral dynamics	Safety (derailment risk) and comfort (noise level)	The suspension stiffness and damping coefficients, the inertial property parameters, and the geometric parameters	A combination of Genetic Algorithms and Monte Carlo Simulation		Parallel Computing
6	[91]	50 DOF model in SIMPACK	Running stability, risk of derailment and the track force		Co-simulation using MATLAB/ SIMPACK And Genetic Algorithm		Co-simulation Parallel computing

(*Continued*)

TABLE 11.2 (*Continued*)
Summary of Research Works for Optimising the Rail Vehicle Dynamic Response

Sr. No.	Reference	Modelling of Dynamic System	Design Goals	Design Variables	Optimization Algorithm	Implementation of Integrated Optimization Algorithm	Remarks
7	[92]		A combination of three objectives. First is a Dynamic Performance. Second objective is the non-compensated acceleration. The third objective is a penalty associated with cases where the vehicle presents an unacceptable dynamic behaviour		GLODS i.e. global and local optimization using direct search.		Scalarising Strategy.
8	[93]	A 1/8th wagon model using FORTRAN	The wheel/ rail contact forces and carbody accelerations	wedge angle, toe angle, and spring stiffness	A combinatorial approach using parallel multi-objective Genetic Algorithm and Particle Swarm Optimization.		Parallel computing

(*Continued*)

TABLE 11.2 (*Continued*)
Summary of Research Works for Optimising the Rail Vehicle Dynamic Response

Sr. No.	Reference	Modelling of Dynamic System	Design Goals	Design Variables	Optimization Algorithm	Implementation of Integrated Optimization Algorithm	Remarks
9	[94]	A 42 DOF model in SIMPACK	There are three objectives. First objective is the lateral ride quality index. The second objective is the reduction ratio of the wheel.	Six suspension parameters	NSGA II		Meta modelling technique using a virtual prototype (model) of a rail vehicle and three Kriging models for the three objectives.
10	[95]	A 14 DOF model	Three objectives. The first two objectives are vehicle speed robustness and suspension parameter robustness, while the third is wheel-rail contact equivalent conicity robustness.	The primary longitudinal and lateral stiffness, the yaw damping, secondary lateral damping, the yaw damper series stiffness and the secondary lateral damper series stiffness.	Cluster analysis of Pareto frontier		

(Continued)

TABLE 11.2 (*Continued*)
Summary of Research Works for Optimising the Rail Vehicle Dynamic Response

Sr. No.	Reference	Modelling of Dynamic System	Design Goals	Design Variables	Optimization Algorithm	Implementation of Integrated Optimization Algorithm	Remarks
11	[96]	A 52 DOF model of wagon in NUCARS	There are three objectives. First is the vertical acceleration of empty wagon on a tangent track. Second is the lateral acceleration of empty wagon on 2° curve. Third objective is the lateral acceleration of a loaded wagon on 2° curve.	Six suspension parameters of the central suspension.	NSGA II		Non- linear regression equations formed on the basis of experimental data.
12	[97]	A 52 DOF model of wagon in NUCARS 52 DOF model	Five objective functions.	Six suspension parameters	NSGA II	Radial basis functions	Meta modelling technique using Radial Basis Functions (RBFs)

TABLE 11.3
Comparison of the Four Meta Modelling Techniques

Sr. No.	Metamodelling Method	Features
1.	Polynomial regression (PR)	One of the best for the data having noise since it's a non- interpolating method. Best for low order nonlinear and small scale problems. Mainly used for screening purposes and in iterative optimization procedures.
2.	Kriging (KG)	Second best in terms of prediction accuracy and robustness. Hold joint second position in computational efficiency along with the RBF method. Best for low order non-linear and large scale problems. Having good flexibility.
3.	Radial basis functions (RBF)	Second in terms of computational efficiency. Highly suitable for high-order nonlinear problems
4	Multivariate Adaptive Regression Splines (MARS)	Best in terms of computational efficiency. One of the best for model interpretations.

the combined model performs at least as good as the best constituent model while guarding against the worst-performing individual model. Once a meta modelling technique is selected and accurate meta models are constructed, then evolutionary algorithm is applied to a large number of random vectors created through surrogate models, for identification of optimal Pareto fronts.

11.5.1.1 Validation of the Meta Models

Before applying the optimization algorithm on a metamodel, the accuracy of the metamodel needs to be ascertained. The accuracy of a metamodel depends upon the type of metamodel itself as well as on the quality and quantity of the data set on which the model has been built. Both the overfitting as well as the underfitting should not be present in the model. The accuracy of a metamodel can be determined in two ways- ***Error measures*** i.e. by evaluating the residuals and variability, or by ***Cross Validating*** with other models. The features of various methods for validating the accuracy of meta models are given in Table 11.4.

If a regression model is made, its accuracy can be ascertained with reference to maximum R-square adjusted (a penalised version of R-square: penalty is based on number of variables in the model), or minimum S (this is a measure of spread of random noise) [101] or maximum R-square predicted (this is a cross-validation measure: measures how the model does to any new prediction) or a combination of all the above three criterions [102] with a good Mallows Cp (indicates unbiases of the model) [103]. If a Kriging or RBF model is made, its accuracy can be ascertained calculating the Root Mean Square Error (RMSE) from Leave One Out (LOO) Cross Validation or by calculating the coefficient of variation (CV) of RMSE of the metamodel [104, 105].

TABLE 11.4
Features of the Methods for Validating the Accuracy of the Meta Models

Sr. No.	Validation Method	Features
1.	Estimating Residuals	The difference between simulated and predicted values is computed. RMSE (root mean squared error) is the most commonly used measure. MAPE (mean absolute percentage error) is a better method to avoid bias in the data because it is based on relative error.
2.	Variability	Coefficient of Determination (R^2) is the most commonly used metric. Doesn't work well with an over-fitted model and also for interpolating models. One improvisation to use it for interpolating models is to estimate the predicted values at unknown design points.
3.	Cross-validation	Comparing interpolating meta models with approximate ones. It uses the same dataset for fitting the model. Leave-one-out (LOO) Cross Validation is used for RBFs and low order polynomials meta models. Leave – k -out Cross Validation is used for Kriging models with $k = 0.1n$ or $\sqrt{n}$, where n is the number of input-output pairs in the dataset. It is more computationally expensive than the LOO Cross Validation method.

11.5.1.2 Visualisation Methods

If a comprehensive optimization problem is to be solved for optimising the dynamic response of Rail vehicle, then there are often more than three objectives (some of which may be conflicting to each other). Visualisation of the approximate Pareto fronts in more than three dimensions is a challenge. The objective space as well as the decision space are to be suitably visualised. The existing methods used in the visualisation of MOO are Scatter plot (after dimensional reduction), bubble chart, parallel coordinate plots, heat maps, radial coordinate visualisation, Sammon mapping, neuroscale and self-organising maps [106–108]. The points of discussion may be about what insights a visualisation method provides into the shape and distribution of a solution set. And what information is provided in terms of the relation among the objectives etc.

There are relative merits and demerits of the methods used in visualisation of all objective information like scatter plot (SP), Bubble chart (BP), parallel coordinate plots (PCP) and heat maps (HP). The frequently used visualisation methods in evolutionary multi-objective algorithms are PCP and HP. PCP is specifically more advantageous in representing the interdependencies of the objectives. In fact, both the PCP and HP techniques are being continuously improvised to reduce the lacunae. And there are some new visualisation techniques being developed via dimensional reduction and via transferring the coordinates.

11.6 SUMMARY

The optimization problem for the optimization of dynamic response of a rail vehicle can be solved under the framework of MOO. Coupling of optimization algorithm and the rail vehicle dynamics simulation software is a very arduous task. When the field of rail vehicle dynamic simulation is itself very challenging, linking the optimization algorithms through co-simulation, parallel computing or metamodelling, etc. adds more complexity to the job. The following can be concluded for discussions made in this chapter:

- Using meta models and approximating objective functions is a suitable approach to avoid many simulations in rail vehicle dynamics for MOO.
- The area of simulation in rail vehicle dynamics is itself very challenging with regard to the validation and verification of simulation models. The inclusion of meta models infuses more complexity into the problem since the metamodel will have its own errors and variability. Collaboration of dynamic experts in the fields of rail vehicle dynamics, multibody dynamic simulation, meta modelling and optimization is required to meaning research that could contribute to actual field results.
- The interpolative meta models interpolate the data points (i.e. zero error at the sample points) and it will tend to be accurate in regions close to these sample points. However, since we normally have small data points in the search space in case of rail vehicle dynamic simulations, the metamodel might not be accurate in regions that are far from the sample points, especially if the objective functions are highly nonlinear.
- Most of the earlier reported MOO studies were concentrated on passenger vehicles. However, with the rise of high-speed freight trains, optimising the dynamic behaviour of freight vehicles gaining importance.
- More than three objectives have not been taken a time so far. However, there would be challenges in visualisation of more than three objectives. Appropriate visualisation techniques need to be developed depending on the problem being tackled. Suitable modifications in the existing PCP method and heat pump methods are required. Or new visualisation algorithms need to be developed using dimensional reduction or transferring of coordinates.
- Despite of above challenges, the area is very promising and needs research efforts into. Once, reliable and accurate meta models are established for different sets of problem, a large number of solutions can be generated.

REFERENCES

1. Koffman, J. L. "Vibrational Aspects of Bogie Design." *Journal of the Institution of Locomotive Engineers* 47, no. 60 (1957): 549–686.
2. Dukkipatti, R., and Garg, V. K. *Dynamics of Railway Vehicle Systems*. London: Academic Press, 1987.
3. Spiryagin, M., Cole, C., Sun, Y. Q., McClanachan, M., Spiryagin, V., and McSweeney, T. *Design and Simulation of Rail Vehicle*. Boca Raton, FL: CRC Press, 2014.
4. Wickens, A. H. *Fundamentals of Rail Vehicle Dynamics: Guidance and Stability*. Lisse, Netherlands: Swets & Zeitlinger Publishers, 2003.

5. Iwnicki, S., Spiryagin, M., Cole, C., and McSweeney, T. (Eds.). *Handbook of Railway Vehicle Dynamics*. 2nd ed. CRC Press, 2019. doi: 10.1201/9780429469398.
6. Andersson, E., Berg, M., and Stichel, S. *Rail Vehicle Dynamics*. Stockholm: Railway Group KTH, 2014.
7. Shabana, A. A. *Dynamics of Multibody Systems*. 4th ed. New York: Cambridge University Press, 2013.
8. Shabana, A. A., Zaazaa, K. E., and Sugiyama, H. *Railroad Vehicle Dynamics: A Computational Approach*. Boca Raton, FL: CRC Press, 2008.
9. Wasfy, T. M., and Noor, A. K. "Computational Strategies for Flexible Multibody Systems." *Applied Mechanics Reviews* 56, no. 6 (2003): 553–613.
10. Escalona, J. L., Sugiyama, H., and Shabana, A. A. "Modeling of Structural Flexibility in Multibody Railroad Vehicle Systems." *Vehicle System Dynamics* 51, no. 7 (2015): 1027–58.
11. Shabana, A. A., and Sany, J. R. "A Survey of Rail Vehicle Track Simulations and Flexible Multibody Dynamics." *Nonlinear Dynamics* 26, no. 2 (2001): 179–210.
12. De Pater, A. D. "The General Theory of the Motion of a Single Wheelset Moving Through a Curve with Constant Radius and Cant." *Journal of Applied Mathematics and Mechanics (ZAMM)* 61, no. 5 (1981): 277–92.
13. Boocock, D. *The Equations of Motion of an Elastically Restrained Wheelset on Curved Track and Preliminary Study of Steady-State Motion*. Report no. DYN/35. British Railways Research Department, Derby, England, August 1966.
14. Law, E. H. "Nonlinear Wheelset Dynamic Response to Random Lateral Rail Irregularities." ASME Paper no. 73-WA/RT-3, November 1973.
15. Wickens, A. H. "The Dynamic Stability of Railway Vehicle Wheelsets and Bogies Having Profiled Wheels." *International Journal of Solids and Structures* 1, no. 3 (1965): 319–41.
16. Wickens, A. H. "Static and Dynamic Stability of Railway Vehicle Wheelsets Having Profiled and Coned Wheels." Technical Note DYN/90. British Railways Research Department, Derby, England, April 1968.
17. Law, E. H., and Brand, R. S. "Analysis of the Nonlinear Dynamics of a Railway Vehicle Wheelset." *Journal of Dynamic Systems, Measurement, and Control* 95, no. 1 (1973): 28–35.
18. Hobbs, A. E. W. *The Response of a Restrained Wheelset to Variations in the Alignment of an Ideally Straight Track*. British Railways Research Department Report no. E542, 1964.
19. True, H. "Dynamics of a Rolling Wheelset." *Applied Mechanics Reviews* 46, no. 8 (1993): 438–44.
20. Clark, J. W., and Law, E. H. *Investigation of the Truck Hunting Instability Problem of High-Speed Trains*. ASME Paper no. 67-Tran-17, August 1967.
21. Cooperrider, N. K. "The Hunting Behavior of Conventional Railway Trucks." *Journal of Engineering for Industry, Transactions of the ASME, Series B* 94, no. 2 (1972): 752–62.
22. Matsudaira, T. "Hunting Problem of High-Speed Railway Vehicles with Special Reference to Bogie Design for the New Tokaido Line." *Interaction between Vehicle and Track, Proceedings of the Institution of Mechanical Engineers, London* 180, no. 3F (1966): 58–66.
23. Wickens, A. H. "The Dynamic Stability of a Simplified Four-Wheeled Railway Vehicle Having Profiled Wheels." *International Journal of Solids and Structures* 1 (1965): 385–406.
24. Wickens, A. H. "The Dynamics of Railway Vehicles on Straight Track: Fundamental Considerations of Lateral Stability." *Interaction between Vehicle and Track, Proceedings of the Institution of Mechanical Engineers, London* 180, no. 3F (1966): 29–44.

25. Blader, F. B., and Kurtz Jr., E. F. *Dynamic Stability of Cars in Long Freight Trains.* ASME Paper no. 73-WA/RT-2, November 1973.
26. Gilchrist, A. O., Hobbs, A. E. W., King, B. L., and Washby, V. "The Riding of Two Particular Designs of Four-Wheeled Railway Vehicle." *Interaction between Vehicle and Track, Proceedings of the Institution of Mechanical Engineers, London* 180, no. 3F (1966): 99–113.
27. Tse, Y. H., Garg, V. K., and Lo, A. *Validation of the Freight Car Hunting (Non-Linear/Linear) Model.* Research Report R-324. Chicago: Association of American Railroads, May 1979.
28. Hedrick, J. K., and Krolewski, S. M. *Freight Car Dynamic Curving Model, Volume 1: Technical Documentation.* Research Report R-557. Chicago: Association of American Railroads, prepared by the Mechanical Engineering Department, Massachusetts Institute of Technology, Cambridge, July 1983.
29. Law, E. H., and Cooperrider, N. K. "Nonlinear Dynamic and Steady-State Curving of Rail Vehicles." Paper presented at the 1980 American Society of Mechanical Engineers Winter Annual Meeting, San Francisco, California, December 1981.
30. Tse, Y. H., and Martin, G. C. *Flexible Body Railroad Freight Car Model.* Research Report R-199. Chicago: Association of American Railroads, 1976.
31. Schiehlen, W. *Technische Dynamik.* Stuttgart: Teubner, 1986.
32. Huston, R. L. "Multibody Dynamics—Modeling and Analysis Methods." *Applied Mechanics Reviews* 44, no. 3 (1991): 109–17.
33. Shabana, A. A. *Dynamics of Multibody Systems.* 4th ed. New York: Cambridge University Press, 2013.
34. Shabana, A. A., and Sany, J. R. "An Augmented Formulation for Mechanical Systems with Non-Generalized Coordinates: Application to Rigid Body Contact Problems." *Nonlinear Dynamics* 24, no. 2 (2001): 183–204.
35. Shabana, A. A., Berzeri, M., and Sany, J. R. "Numerical Procedure for the Simulation of Wheel/Rail Contact Dynamics." *Transactions of the ASME* 123, no. 2 (2001): 168.
36. Netter, H., Schupp, G., Rulka, W., and Schroeder, K. "New Aspects of Contact Modelling and Validation within Multibody System Simulation of Railway Vehicles." *Vehicle System Dynamics* 29 (1998): 246–69.
37. Schupp, J., Weidemann, C., and Mauer, L. "Modelling the Contact between Wheel and Rail within Multibody System Simulation." *Vehicle System Dynamics* 45 (2004): 349–64.
38. Shabana, A. A., Zaazaa, K. E., Escalona, J. L., and Sany, J. R. "Development of Elastic Force Model for Wheel/Rail Contact Problems." *Journal of Sound and Vibration* 269, no. 3 (2004): 295–325.
39. Kortüm, W. "Review of Multibody Computer Codes for Vehicle System Dynamics." *Vehicle System Dynamics* 22, Suppl. 1 (2007): 280–300.
40. Shackleton, P., and Iwnicki, S. "Comparison of Wheel–Rail Contact Codes for Railway Vehicle Simulation: An Introduction to the Manchester Contact Benchmark and Initial Results." *Vehicle System Dynamics* 46, nos. 1–2 (2008): 129–49.
41. Weidmann, C. "State of the Art in Railway Vehicle Design with Multi-Body Simulation." *Journal of Mechanical Systems for Transportation and Logistics* 3, no. 1 (2010): 12–26.
42. Eickhoff, B. M., Evans, J. R., and Minnis, A. J. "A Review of Modeling Methods for Railway Vehicle Suspension Components." *Vehicle System Dynamics* 24, nos. 6–7 (1995): 469–96.
43. Bruni, S., Vinolas, J., Berg, M., Polach, O., and Stichel, S. "Modelling of Suspension Components in a Rail Vehicle Dynamics Context." *Vehicle System Dynamics* 49, no. 7 (2011): 1021–72.
44. Kaiser, A. B., Cusumano, J. P., and Gardner, J. F. "Modeling and Dynamics of Friction Wedge Dampers in Railroad Freight Trucks." *Vehicle System Dynamics* 38, no. 1 (2002): 55–82.

45. Ballew, B., Chan, B. J., and Sandu, C. "Multibody Dynamics Modelling of the Freight Train Bogie System." *Vehicle System Dynamics* 49, no. 4 (2011): 501–26.
46. Xia, F. "Modelling of Wedge Dampers in the Presence of Two-Dimensional Dry Friction." *Vehicle System Dynamics* 37 (2003): 565–78.
47. Wu, Q., Cole, C., Spiryagin, M., and Sun, Y. Q. "A Review of Dynamic Modelling of Friction Wedge Suspensions." *Vehicle System Dynamics* 52, no. 11 (2014): 1389–1415.
48. Evans, J., and Berg, M. "Challenges in Simulation of Rail Vehicle Dynamics." *Vehicle System Dynamics* 47, no. 8 (2009): 1023–48.
49. International Union of Railways (UIC). *UIC 518, Testing and Approval of Railway Vehicles from the Point View of Their Dynamic Behaviour-Safety-Track Fatigue-Running Behaviour.* 4th ed. Paris: International Union of Railways, 2009.
50. Spiryagin, M., George, A., Sun, Y. Q., Cole, C., McSweeney, T., and Simson, S. "Investigation of Locomotive Multibody Modelling Issues and Results Assessment Based on the Locomotive Model Acceptance Procedure." *Proceedings of the Institution of Mechanical Engineers, Part F: Journal of Rail and Rapid Transit* 227, no. 5 (2013): 453–68.
51. Wilson, N., Fries, R., Witte, M., Haigermoser, A., Wrang, M., Evans, J., and Orlova, A. "Assessment of Safety Against Derailment Using Simulations and Vehicle Acceptance Tests: A Comparison of the State-of-the-Art Worldwide." *Vehicle System Dynamics* 49, no. 7 (2011): 1113–57.
52. Polach, O., and Böttcher, A. "A New Approach to Define Criteria for Rail Vehicle Model Validation." *Vehicle System Dynamics* 52 (2014): 125–41.
53. Polach, O., Böttcher, A., Vannucci, D., Sima, J., Schelle, H., Chollet, H., Götz, G., Prada, M. G., Nicklisch, D., Mazzola, L., Berg, M., and Osman, M. "Validation of Multi-Body Models for Simulations in Authorisation of Rail Vehicles." In *9th International Conference on Railway Bogies and Running Gears*, Budapest, Hungary, September 9–12, 2013.
54. Polach, O., and Evans, J. "Simulation of Running Dynamics for Vehicle Acceptance, Application and Validation." In *2nd International Conference on Railway Technology: Research Development and Maintenance (RAILWAYS 2014)*, Ajaccio, Corsica, April 8–11, 2014.
55. European Committee for Electrotechnical Standardisation (CENELEC). *Railway Applications—Testing and Simulation for the Acceptance of Running Characteristics of Railway Vehicles—Running Behaviour and Stationary Tests.* Standard No. EN 14363:2016. Brussels: CENELEC, 2016.
56. Knothe, K. L., and Grassie, S. L. "Modelling of Railway Track and Vehicle/Track Interaction at High Frequencies." *Vehicle System Dynamics* 22, nos. 3–4 (2018): 209–62.
57. Rosenberger, G. W., Klauser, P. E., George P. Binns, G. P., and Wolf, G. P. "Using Parametric Simulation to Optimize Suspension Design." In *Proceedings of IMECE'03*, Washington, DC.
58. Suarez, B., Mera, J. M., Martinez, M. L., and Chover, J. A. "Assessment of the Influence of the Elastic Properties of Rail Vehicle Suspensions on Safety, Ride Quality and Track Fatigue." *Vehicle System Dynamics* 51, no. 13 (2013): 280–300.
59. Pandey, M., and Bhattacharya, B. "A Parametric Study Analyzing the Effect of Bolster Suspension Parameters on the Carbody Dynamic Response in a Freight Wagon Fitted with Three-Piece Bogie." *Journal of Vibration Engineering & Technologies* 8 (2019): 1–17.
60. Pandey, M., and Bhattacharya, B. "Effect of Bolster Suspension Parameters of Three-Piece Freight Bogie on the Lateral Frame Force." *International Journal of Rail Transportation* 8, no. 1 (2019): 45–65.

61. Sun, W., Zhou, J., Gong, D., and You, T. "Analysis of Modal Frequency Optimization of Railway Vehicle Car Body." *Advances in Mechanical Engineering* 8, no. 4 (2016): 1–12.
62. Mallik, A. K. *Principles of Vibration Control.* New Delhi, India: East-West Press Private Limited, 2010.
63. Carlbom, P. F. "Combining MBS with FEM for Rail Vehicle Dynamics Analysis." *Multibody System Dynamics* 6 (2001): 291–300.
64. Carlbom, P. F. *Car Body and Passengers in Rail Vehicle Dynamics.* Doctoral Thesis, KTH, Stockholm, 2000.
65. Zhou J., Goodall R., Ren L., and Zhang, H. "Influences of Car Body Vertical Flexibility on Ride Quality of Passenger Railway Vehicles." *Proceedings of the Institution of Mechanical Engineers, Part F: Journal of Rail and Rapid Transit.* 223, no.5 (2009): 461–71. doi: 10.1243/09544097JRRT272
66. Casas, W. J., and Viganico, C. E. "Comparison between Rigid and Flexible Car Body in the Dynamics of Railway Freight Vehicles." In *Proceedings of the XVII International Symposium on Dynamic Problems of Mechanics (DINAME)*, Natal, Brazil, February
67. Shi, H. L., and Wu, P. "Flexible Vibration Analysis for Carbody of High-Speed EMU." *Journal of Mechanical Science and Technology* 30, no. 1 (2016): 55–66.
68. Dumitriu, M., and Camil, C. "Modelling of Structural Flexibility of the Railway Vehicle Carbody." *MATEC Web of Conferences* 112 (2017): 07007.
69. Ling, L., Zhang, Q., Xiao, X., Wen, Z., and Jin, X. "Integration of Car-Body Flexibility into Train–Track Coupling System Dynamics Analysis." *Vehicle System Dynamics* 56, no. 4 (2017): 485–505.
70. Shin, Y. J., You, W. H., Hur, H. M., and Park, J. H. "Semi-Active Control to Reduce Carbody Vibration of Railway Vehicle by Using Scaled Roller Rig." *Journal of Mechanical Science and Technology* 26, no. 11 (2012): 3423–31.
71. Wang, K., Xia, H., Xu, M., and Guo, W. "Dynamic Analysis of Train-Bridge Interaction System with Flexible Car-Body." *Journal of Mechanical Science and Technology* 29, no. 9 (2015): 3571–80.
72. Sugahara, Y., Kazato, A., Tomioka, T., and Sampei, M. "Vertical Vibration Suppression of Railway Vehicle by Primary Suspension Damping Force Control (Results of Running Tests Using Shinkansen Vehicle)." *Transactions of the Japan Society of Mechanical Engineers* 74, no. 741(C) (2008): 1222–30.
73. Huang, C., Zeng, J., Luo, G., and Shi, H. "Numerical and Experimental Studies on the Car Body Flexible Vibration Reduction Due to the Effect of Car Body-Mounted Equipment." *Proceedings of the Institution of Mechanical Engineers, Part F: Journal of Rail and Rapid Transit* 232, no. 1 (2016): 1–18.
74. Tomioka, T., and Takigami, T. "Reduction of Bending Vibration in Railway Vehicle Carbodies Using Carbody–Bogie Dynamic Interaction." *Vehicle System Dynamics* 48, sup. 1 (2010): 467–86.
75. Orvanäs, A. *Methods for Reducing Vertical Carbody Vibrations of a Rail Vehicle.* Stockholm: Report in Railway Technology, 2010.
76. Dumitriu, M., and Cruceanu, C. "Approaches for Reducing Structural Vibration of the Carbody Railway Vehicles." *MATEC Web of Conferences* 112 (2017): 07006.
77. Shust, W. C., and Iler, D. "Variability in Natural Frequencies of Railroad Freight Car Components." In *Structural Dynamics, Volume 3: Proceedings of the 28th IMAC, A Conference on Structural Dynamics, 2010*, pp. 1273–86. New York: Springer, 2011.
78. Pandey, M. "Effect of Carbody Flexibility on the Dynamic Performance of an Empty Freight Wagon." *Proceedings of the Institution of Mechanical Engineers, Part F: Journal of Rail and Rapid Transit* 236, no. 5 (2021): 1–11.
79. Yuping, H., and McPhee, J. "Optimisation of the Lateral Stability of Rail Vehicles." *Vehicle System Dynamics* 38, no. 5 (2002): 361–90.
80. Eberhard, P., Schiehlen, W., and Bestle, D. "Some Advantages of Stochastic Optimization of Multibody Systems." *Archive of Applied Mechanics* 69 (1999): 543–54.

81. Arora, J. S. *Introduction to Optimum Design*. Boston: Academic Press, 2012.
82. Sivanandam, S. N., and Deepa, S. N. *Introduction to Genetic Algorithms*. Berlin: Springer, 2007.
83. Kirkpatrick, S., Gelatt, C. D., and Vecchi, M. P. "Optimization by Simulated Annealing." *Science* 220 (1983): 671–80.
84. Essiet, I. O., Yanxia, S., and Zenghui, W. "Improved Genetic Algorithm Based on Particle Swarm Optimization-Inspired Reference Point Placement." *Engineering Optimization* 51, no. 7 (2019): 1097–114.
85. Gazi, V., and Passino, K. M. *Swarm Stability and Optimization*. Berlin: Springer, 2011.
86. Wu, Q., C. Cole, and T. McSweeney. "Applications of Particle Swarm Optimization in the Railway Domain." *International Journal of Rail Transportation* 4, no. 3 (2016): 167–90.
87. He, Y., and McPhee, J. "Multidisciplinary Optimisation of Multibody Systems with Application to the Design of Rail Vehicles." *Multibody System Dynamics* 14, no. 1 (2005): 11–35.
88. Johnsson, A., Berbyuk, V., and Enelund, M. "Pareto Optimization of Railway Bogie Suspension Damping to Enhance Safety and Comfort." *Vehicle System Dynamics* 50, no. 9 (2012): 1379–407.
89. Baek, S. G., Shin, B., Lee, S. W., Choi, Y. -S., Kim, J., and Koo, J. C. "Optimisation of High-Speed EMU Suspension Parameters for Vibration Reduction." *Journal of Mechanical Science and Technology* 27, no. 2 (2013): 305–311
90. Nejlaoui, M., Houidi, A., and Affiz, A. "Multiobjective Robust Design Optimization of Rail Vehicle Moving in Short Radius Curved Tracks Based on the Safety and Comfort Criterion." *Simulation Modelling Practice and Theory* 30 (2013): 21–34.
91. Mousavi-Bideleh, S. M., and Berbyuk, V. V. "Multiobjective Optimisation of Bogie Suspension to Boost Speed on Curves." *Vehicle System Dynamics* 54, no. 1 (2016): 58–85.
92. Magalhães, H., Madeira, J. F. A., Ambrósio, J., and Pombo, J. "Railway Vehicle Performance Optimisation Using Virtual Homologation." *Vehicle System Dynamics* 54, no. 9 (2016): 1177–1207.
93. Wu, Q., Sun, Y., Spiryagin, M., and Cole, C. "Methodology to Optimize Wedge Suspensions of Three-Piece Bogies of Railway Vehicles." *Journal of Vibration Control* 24, no. 3 (2016): 565–581.
94. Yang, Y., Zeng, W., Wang, T., and Liao, Y. "Multi-Objective Optimization of Suspension Parameters for Rail Vehicles Based on a Virtual Prototype Surrogate Model." *Journal of the Chinese Institute of Engineers* 39, no. 6 (2016): 746–54.
95. Yao, Y., Li, G., Wu, G., Zhang, Z., and Tang, J. "Suspension Parameters Optimum of High-Speed Train Bogie for Hunting Stability Robustness." *International Journal of Rail Transportation* 8, no. 3 (2019): 195–214.
96. Pandey, M., Datta, R., Dey, R., and Bhattacharya, B. "Multi-objective Optimisation of Dynamic Responses for a Rail Freight Wagon using Regression Models," In *2019 IEEE 18th International Conference on Cognitive Informatics & Cognitive Computing (ICCI&CC)*, Milan, Italy, pp. 361–65, 2019. doi: 10.1109/ICCICC46617.2019.9146044.
97. Pandey, M., Regis, R. G., Datta, R., and Bhattacharya, B."Surrogate-Assisted Multi-Objective Optimization of the Dynamic Response of a Freight Wagon Fitted with Three-Piece Bogies." *International Journal of Rail Transportation* 9, no. 3 (2020): 290–309. https://doi.org/10.1080/23248378.2020.1792808
98. Ryberg, A. B., Backryd, R.D., and Nilsson, L. *Metamodel-Based Multidisciplinary Design Optimisation for Automotive Applications*. Technical Report LIU-IEI-R-12/003. Division of Solid Mechanics, Linköping University, 2012.
99. Jin, R., Chen, W., and Simpson, T. W. "Comparative Studies of Metamodeling Techniques under Multiple Modelling Criteria." *Structural and Multidisciplinary Optimization* 23 (2001): 1–13.

100. Won, K. S., and Ray, T. "A Framework for Design Optimization Using Surrogates." *Engineering Optimization* 37, no. 7 (2005): 685–703.
101. Allen, D. M. "The Relationship between Variable Selection and Data Augmentation and a Method for Prediction." *Technometrics* 16 (1984): 125–27.
102. Mallows, C. L. "Some Comments on CP." *Technometrics* 15, no. 4 (1973): 661–75.
103. Gilmour, S. G. "The Interpretation of Mallows's Cp-Statistic." *Journal of the Royal Statistical Society, Series* D 45, no. 1 (1996): 49–56.
104. Aman, S., Simmhan, Y., and Prasanna, V. K. "Holistic Measures for Evaluating Prediction Models in Smart Grids." *IEEE Transactions on Knowledge and Data Engineering* 27 (2015): 475–88.
105. Ga, C. H., Markus, F., and Martin, K. "Modelling the Physical Multiphase Interactions of HNO_3 between Snow and Air on the Antarctic Plateau (Dome C) and Coast (Halley)." *Atmospheric Chemistry and Physics* 18 (2018): 1507–34.
106. Tušar, T., and Filipič, B. "Visualization of Pareto Front Approximations in Evolutionary Multiobjective Optimization: A Critical Review and the Prosection Method." *IEEE Transactions on Evolutionary Computation* 19 (2015): 225–45.
107. Ibrahim, A. "3D-RadVis: Visualization of Pareto Front in Many-Objective Optimization." In *2016 IEEE Congress on Evolutionary Computation (CEC)*, pp. 736–45. New York: IEEE, 2016.
108. Li, M., Zhen, L., and Yao, X. "How to Read Many-Objective Solution Sets in Parallel Coordinates." *IEEE Computational Intelligence Magazine* 12, no. 4 (2017): 88–100.

12 Railway Track System and Dynamics

Smitirupa Pradhan and Prajapati Naik

12.1 INTRODUCTION

Railway track is an essential infrastructure for the smooth and stable running of the train. The track system consists of sleepers, rail, fasteners, ballast, and subgrade. The role of each component is critical for stability, reliability, safety, etc. Railway dynamics deals with the lateral and normal forces acting on the wheel and rail when the train is running both straight and curved path. Track dynamics comprises train–track interaction, load distribution, study of vibration both time and frequency domain and track deformation due to transmission of dynamic loads. The study of railway dynamic is important for reducing vibration and improving ride comfort for the passengers, predict the wear pattern for maintenance, and optimize the track design parameters to improve the performance, safety and stability.

Track irregularities can affect passengers' comfort, performance and safety adversely. These irregularities occurred due to manufacturing errors, material degradation, issues during design and external forces. Different types of track irregularities such as vertical and lateral irregularities, gauge irregularities, cross-level (super-elevation) irregularities, twist and surface irregularities are found in the railway track over the time which lead increase the risk of derailment, enhance wear and tear of wheel and other track components, reduce passengers' comfort and increase the maintenance cost. This chapter focusses briefly all the aspects which affect the dynamics of the track system.

12.2 INTERACTION BETWEEN TRACK AND RAILWAY VEHICLE

12.2.1 Track–Vehicle Interaction

The forces at the wheel-rail interface and the interaction between the wheelset and the track are influenced by the coupling between the train and the track. The dynamics of the vehicles are influenced by several factors which include axle load, train speed, wheel profiles, axle box, primary and secondary suspension system, curve radius, cant, and rail inclination in the track design, and flexible track components like rail profiles, types and dimensions of sleepers, rail pads, fasteners, and ballast. In addition to all the above-mentioned factors, sliding, cyclic loading, and extra forces generated during braking and traction are also significant.

As the vehicle passes steadily on the curved path, gravitational and centrifugal forces exert on the vehicle in lateral and vertical directions, respectively. The

DOI: 10.1201/9781003659983-12

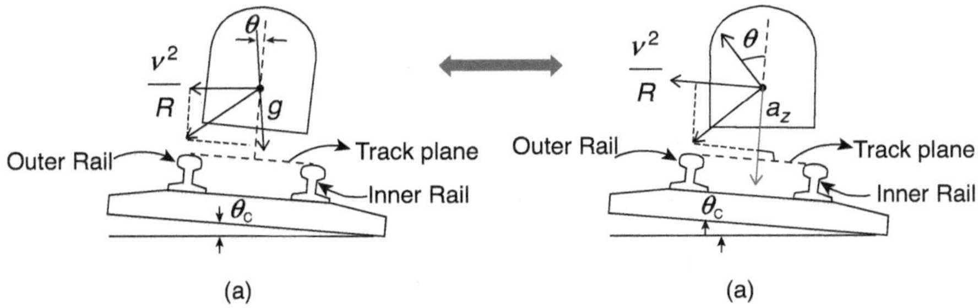

FIGURE 12.1 Acceleration on the car body during curving.

resultant acceleration due to the forces acting can be split into the parallel component (a_y) and perpendicular component to the track (a_z) as depicted in Figure 12.1b. It is assumed that differences in the load distribution on the left and right rail can be ignored for low velocities. Due to static equilibrium (as per Figure 12.1), the accelerations in lateral and vertical directions are given as

$$a_y = \frac{V^2}{R_{\text{curve}}}\cos\theta_c - g\sin\theta_c = \frac{V^2}{R_{\text{curve}}}\cos\theta_c - g\frac{h_t}{2b_0}, \tag{12.1}$$

$$a_z = \frac{V^2}{R_{\text{curve}}}\sin\theta_c + g\cos\theta_c. \tag{12.2}$$

Here, g is the gravitational acceleration.

For small cant angle θ_c, Equations (12.1 and 12.2) may be written as

$$a_y = \frac{V^2}{R_{\text{curve}}} - g\frac{h_t}{2b_0}, \tag{12.3}$$

$$a_z = g. \tag{12.4}$$

For a particular speed and radius, when $a_{y=}0$, the corresponding cant is termed as equilibrium cant. The equilibrium cant is given by $h_{\text{eq}} = (2b_0/g)(V^2/R_{\text{curve}})$. Equilibrium speed, i.e. the speed when $a_{y=}0$, for a given radius and cant, is given by $V_{\text{eq}} = \sqrt{(R_{\text{curve}}gh_t)/2b_0}$. Note, it is very difficult to maintain the desired speed, at a given radius and curvature which gives zero acceleration in the lateral direction. Hence, cant excess and cant deficiency arise depending upon the operating speed. When the operating speed of the vehicle is less than the designed speed of the track, then cant excess is caused which is given by $h_{e=}h_t{-}h_{\text{eq}}$. As per Equation 12.5, cant deficiency is defined as

$$h_d = h_{\text{eq}} - h_t = \frac{2b_0}{g}\frac{V^2}{R_{\text{curve}}} - h_t. \tag{12.5}$$

Cant/super-elevation is useful to a certain extent to minimize the acceleration and creep force in the lateral direction in the curved track.

12.2.2 Normal and Tangential Forces Due to Wheel-Rail Interaction

Both the wheel and rail deform elastically when a wheel passes on a rail and contact patch is elliptical in shape. The contact pressure on the contact patch is the summation of normal force and tangential forces due to traction and braking. When a wheel set is rolling on a straight track, the lateral suspension force F_{sy}, vertical force F_z which consists of constant axle load and variable vertical suspension force and spin-creep moment as shown in Figure 12.2. The normal pressure and tangential forces must be ascertained using Hertz theory and Kalker's theory, respectively, in order to investigate the dynamic behaviour of the railway vehicle.

12.2.2.1 Normal Contact

The contact patch at the wheel-rail interface is an elliptical shape with major and minor axes are a and b, respectively and deflection between centres of the wheel and rail is δ as calculated as:

If $a > b$

$$a = m\left(\frac{3}{2}N\frac{1-\upsilon^2}{E}\frac{1}{A+B}\right)^{1/3},$$

$$b = n\left(\frac{3}{2}N\frac{1-\upsilon^2}{E}\frac{1}{A+B}\right)^{1/3},$$

$$\delta = r\left(\left(\frac{3}{2}N\frac{1-\upsilon^2}{E}\right)^2(A+B)\right)^{1/3}. \tag{12.6}$$

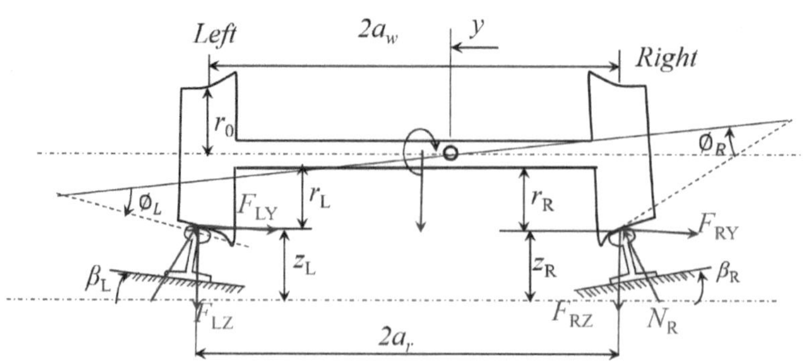

FIGURE 12.2 Geometry parameters and forces acting on wheel–rail interface [1].

where N is normal load, by assuming both rail and wheel are the same material, E is Young's modulus, υ is the Poisson's ratio. The non-dimensional parameters are m, n and r which is the function of $g = n/m$. The values of m, n and r are given in Table 12.1.

The pressure distribution for elliptical contact patch is elliptical shape, i.e. $\sigma = N / \pi ab$ and the maximum pressure is

$$\sigma_{\max} = 1.5N / \pi ab. \tag{12.7}$$

12.2.2.2 Tangential Force Calculation

Partly elastic and partly frictional at the contact between rail and wheel is termed as creep, where both adhesion and slip occur simultaneously. At the creep region, there is the transition from pure rolling to pure sliding. To understand the dynamic behaviour of the vehicle, it is necessary to calculate the lateral and longitudinal creep forces and moment (spin) at the wheel-rail interface as presented in Figure 12.3.

TABLE 12.1
Values for Hertz Coefficients ([a]$A/B<1$) [2]

θ^o	90	80	70	60	50	40	30	20	10	0
$G = n/m$	1	0.7916	0.6225	0.4828	0.3652	0.2656	0.1806	0.1080	0.047	0
M	1	1.128	1.285	1.486	1.754	2.136	2.731	3.816	6.612	∞
N	1	0.8927	0.8	0.7171	0.6407	0.5673	0.4931	0.4122	0.311	0
R	1	0.9932	0.9726	0.9376	0.8867	0.8177	0.7263	0.6038	0.428	0

[a] $A = \frac{1}{2r_n}$ and $B = \frac{1}{2}\left(\frac{1}{R_{wx}} + \frac{1}{R_{rx}}\right)$, where r_n and R are normal radius of the wheel and curvature, respectively, w and r stands for wheel and rail and $\cos\theta = \frac{|B - A|}{B + A}$.

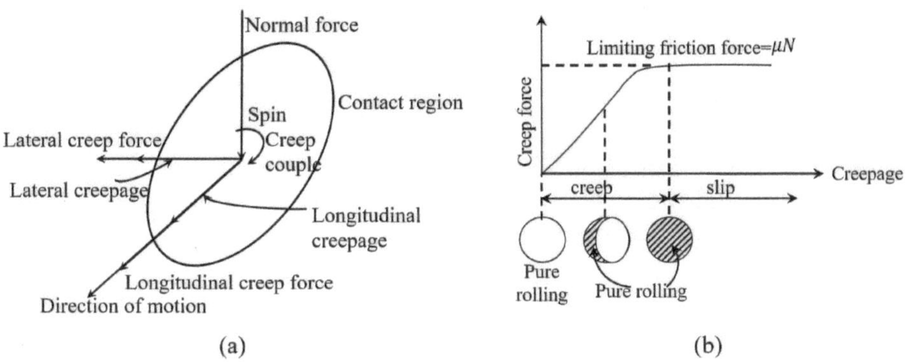

FIGURE 12.3 (a) Creep force and creepage in the contact region and (b) Relation between creepage and creep force [3].

The deviation from pure rolling is termed as creepages [3]. Longitudinal creepage: $\xi_x = \left(v_x - v_x^{'}\right) / v_x^{'}$.

In quasi-static condition, for left wheel $\xi_x = (v_x - r\omega) / \frac{1}{2}(v_x + r\omega)$, Right wheel will be with opposite sign to the left wheel [4]. v_x and $v_x^{'}$ are actual and pure rolling velocity, respectively in the forward direction,

$$\text{Lateral creepage} : \xi_y = \left(v_y - v_y'\right) / v_x',$$

where v_y and $v_y^{'}$ are actual and pure rolling velocity, respectively in the lateral direction. In quasi-static condition, lateral creepage is equal to yaw angle with small creepage, common to both wheels.

$$\text{Spin creepage} : \xi_{sp} = (\Omega - \Omega') / v_x',$$

where Ω and $\Omega'$are angular velocities of the wheel and rail about the normal to the contact surface. $\Omega' = 0$, as rail speed is equal to zero and the general expression is simplified to $\xi_{sp} = sin\varnothing / r_0$. In quasi-static conditions with small creepages, $v_x \approx r_0\,\omega$. $\varnothing$ is an algebraic value, i.e. difference between two wheels.

Spin has an important role during flange contact with a larger value. Moment and creep forces are functions of creepage that are nonlinear. Creep force is zero when creepage is zero and it varies linearly up to μN at maximum creepage as shown in Figure 12.3; where μ represents friction coefficient and N is the normal force. F_x and F_y are the longitudinal and lateral creep forces in the contact plane. For small creepage, a linear relationship exists between creepage and creep force (as shown in Figure 12.3b). The proportionality constant is called the creep coefficient [5]. For Hertzian contact, creep forces are the function of creepage. The general expressions of creep forces are expressed by assuming the linear theory of Kalker [5, 6].

$$F_x = -Gabc_{11}v_x,$$

$$\begin{cases} F_{y\,\text{yaw}} = -Gabc_{22}v_y \\ F_{y\,\text{spin}} = -Gabc_{23}c\varphi \end{cases}, \tag{12.8}$$

where G represents the shear modulus of wheel and rail material, c_{ij} refers to Kalker's coefficient. The transversal force caused by the spin φ, being separated from the force due to the lateral creepage v_y. The overall transversal force is

$$F_y = F_{y\,\text{yaw}} + F_{y\,\text{spin}}. \tag{12.9}$$

The c_{ij} are functions of *b*/*a* ratio of the ellipse (as in Table 12.2). C_{11} and C_{22} values are taken for Poisson's ratio for the steel is about to 0.27 and interpolated to propose the polynomial fit as:

TABLE 12.2
Kalker's Coefficient [7]

		C_{11}			C_{22}			$C_{23}=-C_{32}$			C_{33}		
b/a		$\sigma=0$	1/4	1/2	$\sigma=0$	1/4	1/2	$\sigma=0$	1/4	1/2	$\sigma=0$	1/4	1/2
	$g=0$	$\pi^2/4(1-\sigma)$			$\pi^2/4$			$\pi/\sqrt{g/3}$	—	—	$\pi^2/16(1-\sigma)g$		
	1	3.40	4.12	5.20	3.40	3.67	3.98	1.33	1.47	1.63	1.21	1.19	1.16
	0.9	3.51	4.22	5.30	3.51	3.81	4.16	1.44	1.59	1.77	1.16	1.11	1.06
	0.8	3.65	4.36	5.42	3.65	3.99	4.39	1.58	1.75	1.94	1.10	1.04	0.954
	0.7	3.82	4.54	5.58	3.82	4.21	4.67	1.76	1.95	2.18	1.05	0.965	0.852
	0.6	4.06	4.78	5.8	4.06	4.50	5.04	2.01	2.23	2.5	1.01	0.892	0.751
	0.5	4.37	5.10	6.11	4.37	4.90	5.56	2.35	2.62	2.96	0.958	0.819	0.650
	0.4	4.84	5.57	5.57	4.84	5.48	6.31	2.88	3.24	3.7	0.912	0.747	0.549
	0.3	5.57	6.34	7.34	5.57	6.40	7.51	3.79	4.32	5.01	0.868	0.674	0.446
	0.2	6.96	7.78	8.82	6.96	8.14	9.79	5.72	6.63	7.89	0.828	0.601	0.341
	0.1	10.7	11.7	12.9	10.7	12.8	16.0	12.2	14.6	18.0	0.795	0.526	0.228
a\b	0.1	2.51	3.31	4.85	2.51	2.52	2.53	0.334	0.473	0.731	6.42	8.28	11.7
	0.2	2.59	3.37	4.81	2.59	2.63	2.66	0.483	0.603	0.809	3.46	4.27	5.66
	0.3	2.68	3.44	4.80	2.68	2.75	2.81	0.607	0.715	0.889	2.49	2.96	3.72
	0.4	2.78	3.53	4.82	2.78	2.88	2.98	0.720	0.823	0.977	2.02	2.32	2.77
	0.5	2.88	3.62	4.83	2.88	3.01	3.14	0.827	0.929	1.07	1.74	1.93	2.22
	0.6	2.98	3.72	4.91	2.98	3.14	3.31	0.930	1.03	1.18	1.56	1.68	1.86
	0.7	3.09	3.81	4.97	3.09	3.28	3.48	1.03	1.14	1.29	1.43	1.50	1.60
	0.8	3.19	3.91	5.05	3.19	3.41	3.65	1.113	1.25	1.40	1.34	1.37	1.42
	0.9	3.29	4.01	5.12	3.29	3.54	3.82	1.23	1.36	1.51	1.27	1.27	1.27

$$c_{11} = 3.2893 + \frac{0.975}{b/a} - \frac{0.012}{(b/a)^2},$$

$$c_{22} = 2.4014 + \frac{1.3179}{b/a} - \frac{0.02}{(b/a)^2}, \tag{12.10}$$

$$c_{23} = 0.4147 + \frac{1.0184}{b/a} + \frac{0.0565}{(b/a)^2} - \frac{0.0013}{(b/a)^3}.$$

12.3 TRACK DYNAMICS

12.3.1 Dynamic Properties of the Track

Receptance is the quotient obtained by dividing the track deflection by the force applied on the track, i.e. deformation (*m*) per applied load (*N*); the inverse of the track stiffness. Track stiffness depends upon the frequency of the load and preload of the track [8]. Generally, resonance occurs due to properties of the track structure, i.e. when rails and sleepers vibrate on the ballast bed, i.e. soft ground or hard ground. It is noticed that resonance may be seen in the frequency range of 20–40 Hz for the soft ground. The rail track system behaves as a spring–mass system, where rail/sleeper and ballast behave as mass and spring, respectively. Resonance is well-damped due to the good damping properties of the ballast. The rail pad also behaves as a spring which is inserted between the rail and sleepers. Rail bounces on the rail pads and due to this resonance may be found in the frequency 200–600 Hz. The highest resonance frequency, i.e. pinned-pinned resonance frequency is obtained at about 1,000 Hz. The resonance occurs when the wavelength of the bending waves of the rail doubles the distance between two sleepers. The nodes of the bending waves are placed at the sleepers and act as a support. This vibration is lightly damped due to the rail material i.e. steel. The pinned-pinned frequency can be determined by assuming the rail is a simply supported beam. Applying Bernoulli's beam theory, developed by Euler, yields the fundamental frequency as

$$f = \frac{\omega}{2\pi} = \frac{1}{2\pi}\sqrt{\frac{EI}{mL^4}}, \tag{12.11}$$

where m is the mass of the rail per unit length and EI is the rail's bending stiffness. The sleeper spacing is denoted by L.

12.3.1.1 Excitation Sources of Track and Track Vibration

The different components such as ballast, sleepers, rail pads and rail behave as springs and masses which are responsible for the dynamics of the train. On the other hand, due to several components of the train such as wheelsets, suspensions, bogie frame, etc. several resonance frequencies will appear. When wheelset vibrates on the track, the wheel and track act as mass and spring, respectively. The resonance frequency of the wheel and track system falls in the frequency range of 30–100 Hz. Again, the wheel-rail interface is considered as a spring and the elastic deformation at the interface is similar to the nonlinear spring deformation. The generation of the vibration in the ground transmits to the buildings along with the railway lines which brings passengers' discomfort [9]. Vibration and noise are generated in the train, track and surroundings due to the excitation source such as track irregularities. Corrugation (short wavelength irregularities) creates vibrations with high frequencies which are developed on the rail head due to heavy traffic.

Due to wear, periodic irregularities appeared at the tread of the wheel. This type of irregularities may be either short wavelength or long wavelength. The long wavelength can be otherwise named as polygonalization which contains one to five wavelengths on the surface of the wheel. Another source of vibration excitation is due to uneven stiffness on the track can be the source of vibration excitation. The track exhibits greater rigidity at the sleeper locations and increased flexibility between the sleepers. Thus, rail deflects more between two sleepers rather than on the sleeper. The excitation frequency due to the above-mentioned condition is known as sleeper passage frequency.

12.3.1.2 Excitation Frequencies

When the train moves with velocity V (m/s) on the track with periodic (sinusoidal) irregularities; the corresponding excitation frequency f (Hz) is given by $f = V/\lambda$, where λ is the wavelength of the sinusoidal irregularities. Longer wavelength provides lower frequencies and vice-versa. Noise and vibrations in the track and train are induced due to excitation. The amplitude of the vibration will magnify at the resonance frequencies.

12.3.1.3 Corrugation/Short Wavelength Irregularities at Railhead

Sometimes, periodic irregularities with different wavelengths are developed on the wheel tread and railhead. The progress of the irregularities on the rail head arises due to the dynamics of the track-train system with resonance. The periodic irregularities with a wavelength in the range of 30–300 mm are named corrugation [10]. The dynamic forces due to wheel-rail interaction will be amplified on the corrugated rail head which will help to enhance the rate of deterioration of the track.

12.3.1.4 Long Wavelength Irregularities

Even if there are no irregularities on the rail, the wheel is excited with the excitation frequency, i.e. $f = V/\lambda$ due to sleeper spacing. The stiffness of the track varies from one place to another, i.e. stiffness is more on the sleepers and less in between the sleepers. Due to wear, the wheels go out-of-roundness and the train-track will excite with one or more excitation frequencies; correspond to the wavelengths of the wheel out-of-roundness. An oval-shaped wheel induces two times the frequencies. Non-periodic irregularities, i.e. short wavelength corrugation may be visible on the wheel tread due to the block brake.

12.3.1.5 Rail Manufacturing

Rail irregularities with wavelengths one to several metres are induced during the rail manufacturing process. These kinds of errors will cause the train to be excited at a low frequency.

12.3.1.6 Track Stiffness Irregularities

Due to the presence of switches and turnouts, the stiffness of the track varies from location to location. At switches, both left and right rails have different stiffness and the rails with the presence of the switches become stiffer as longer sleepers support that rail. The change in stiffness occurs due to the presence of a frog in the rail switch which induces high-frequency vibration in the train-track system. Sometimes, the substructure of the rail track system is not homogeneous. The variation of stiffness is there due to non-homogeneous distribution of the substructure/subgrade of the rail track system which causes low frequency random oscillation of the train.

12.3.1.7 Track Embankment

When a load-carrying train passes on the railway track, the ballast and sub-ground may experience plastic deformations. Ballast and subgrade will not come to the original position after passing the trains. After several train passages, small plastic deformations will accumulate and various sections of the track will shift separately and result in a new track position. This notation of the above track is called differential track settlement. Plastic deformation in the subgrade leads to the track irregularities with a wavelength in the order of metre to thousands of metres. The uneven track causes low-frequency vibration of the train.

12.3.2 Impact Loads

Impact loads come from different sources such as the rail joints, flatness of the wheel, and frog (switch crossing) to the track which is discussed in this section.

12.3.2.1 Wheel Flats

When brakes at the wheels are not properly working, i.e. due to limited adhesion, the wheel slides along the rail instead of rolling. Some materials are removed from the wheel surface and form wheel flatness. At every evolution, this flat surface hits the rail and the impact load comes from the wheel. This type of wheel is the source of high-frequency vibration of the track and creates noise. Fermér and Nielsen have examined the effect of flat wheel [11]. Bending stress is developed in the sleeper due to the impact load, coming from the wheel flat. This wheel flat contributes large contact forces and stress to affect the dynamics of the train.

12.3.2.2 Rail Joints

A small difference between the levels of the left and right rails appears due to rail joints. When a wheel climbs on the different levels of the rail, high dynamic forces arise. Due to this type of irregularities, Variations of stiffness are due to rail joint.

12.3.2.3 Switches

Due to the existence of switches in the rail track, variations in the stiffness will be there as mentioned above. The wheel-rail contact patch will shift abruptly on the wheel tread laterally when a wheel passes on the switchblade and crossing. This leads to inducing longitudinal force both on the wheel and track and enhances the yaw movement of the wheelset. Due to the presence of wing rail and nose rail (portions of switches), the difference in height gives a vertical impact on the wheel [12].

12.3.3 Mathematical Modelling of Track System

In this section, mathematical models of rail are presented to study the dynamic behaviour. This section only focuses on the modelling of the track components which describes vertical track deflection as discussed below:

- Rail on continuous elastic foundation
- Rail on discrete supports, i.e. on the rail pad
- Rail on discrete supports with ballast model
- Rail on sleepers by considering continuum containing three-dimensional finite element method (FEM) model
- Sleeper vibration.

12.3.3.1 Rail on the Continuous Elastic Foundation (Winker Beam)

The track system comprised rail, sleepers, rail pads, ballast, and subgrade. The rail can be modelled in a simplified manner or by considering the presence of all the components. The rail may be modelled as either Euler-Bernoulli beam or Rayleigh-Timoshenko beam depending upon the complexity of the problem. The Rayleigh-Timoshenko beam theory incorporates the rotational inertia of the beam and considers the deformation caused by shear force. Rail (beam) lies on a continuous elastic foundation which is modelled as a series of linear springs where distributed force is proportional to the beam deflection which was introduced by Winkler in 1867. The rail deformation $w(x)$ is in the terms of the differential equation:

$$EI\frac{d^4w}{dx^4} + kw(x) = q(x), \tag{12.12}$$

where $q(x)$ is the distributed load on the rail, EI is bending/rigidity stiffness of rail (Nm^2), k is the stiffness of the foundation (N/m^2) and x represents the length.

The deflection due to wheel load (P) (single) which is applied vertically on the rail is calculated as

$$w(x) = \frac{P\beta}{2k}e^{-\beta x}(\cos\beta x + \sin\beta x) = \frac{P\beta}{2k}\eta(x). \tag{12.13}$$

The corresponding bending moment (BM) is given by

$$M(x) = \frac{P}{4\beta}e^{-\beta x}(\cos\beta x - \sin\beta x). \tag{12.14}$$

where $\beta = \frac{1}{4}\left(\frac{k}{EI}\right)^{1/3}\left(\frac{P}{w_0}\right)^{3/4}$, $w_0 = \frac{\beta}{2k}\sum_{i=1}^{n} P_i\eta_i$ and $\eta_i = e^{-\beta x_i}(\cos\beta x_i + \sin\beta x_i)$.

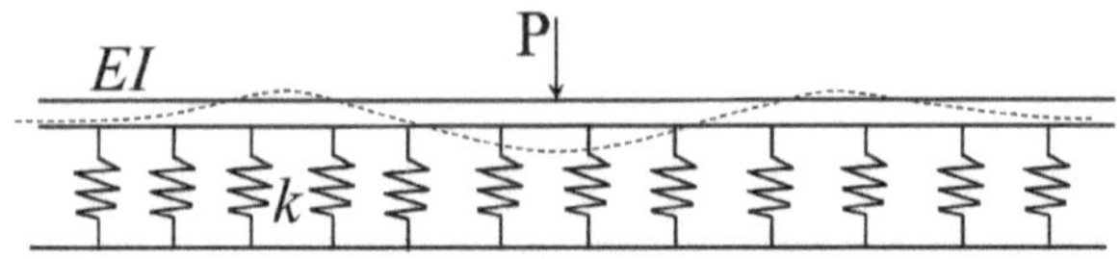

FIGURE 12.4 Rail on elastic foundation. Wheel load P is transferred to rail and corresponding deflection is shown in dotted line.

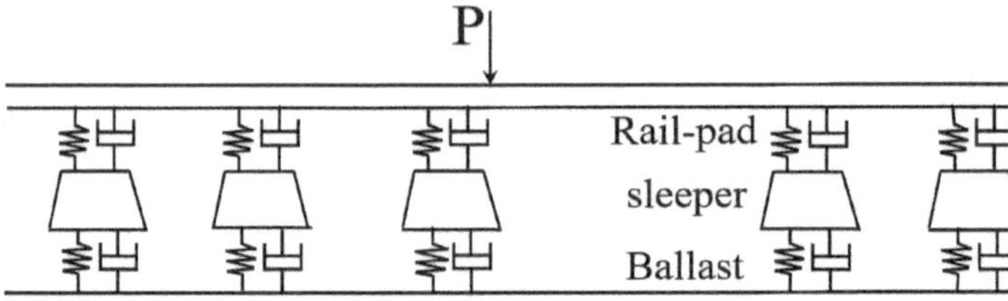

FIGURE 12.5 Rail is supported by rail pad (discrete support). Rail pad and ballast are demonstrated by spring-damper system and sleeper are rigid masses.

When rail vibrates at the sleeper which is considered as nodes and the corresponding pinned-pinned frequency at the range 800 to 1,000 Hz (approximately). The effect of rail pads, ballast, and sleepers are considered. The rail consists of several continuous layers instead of a single layer as shown in Figure 12.4. The effect of the presence of rail pads is studied where it is noticed that the sleepers are isolated effectively by using softer rail pad and corresponding strains will be reduced.

12.3.3.2 Rail on Discrete Supports

The continuous rail is supported on the elastic foundation, i.e. at the discrete supports such as rail pads, sleepers, and ballasts. Rail pads and ballasts are modelled as spring and damper system which are connected in parallel. Sleepers are considered as rigid mass. Three resonance frequencies such as bouncing of track on the ballast, pinned-pinned frequency and rail–rail pad–sleeper vibration (considered as spring between two masses). The sleeper is substituted by a beam with rigid mass on an elastic foundation as shown in Figure 12.5.

12.3.3.3 Discretely Supported Track Including Ballast Mass

The masses of the ballast and subgrade are taken into consideration during modelling of the rail system, where the mass of rail is larger as compared to sleeper and rail mass. Ballast and sub-ballast masses are connected with the stiffness of subgrade and ballast; vibrate within these. As sleepers lie on the bed of subgrade, the deflection at one sleeper will affect adjacent sleepers as shown in Figure 12.6; this concept is based on the real track [13]. Four different resonance vibration modes such as embankment vibration, track-ballast vibration, pinned-pinned vibration and rail–rail pad vibration are found in this type of modelling.

The equivalent stiffness is calculated by considering the stiffness of rail pads, tie stiffness and stiffness of ballast and subgrade in the vertical direction for modelling purpose which is shown in Figure 12.7.

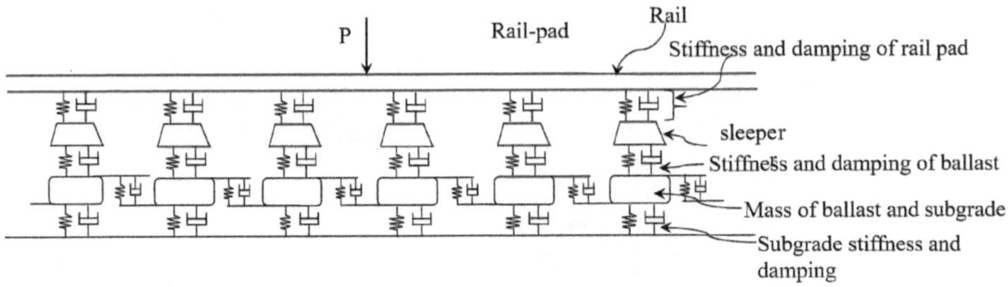

FIGURE 12.6 Rail on discrete supports with mass, damping and stiffness of the sleepers taken into consideration.

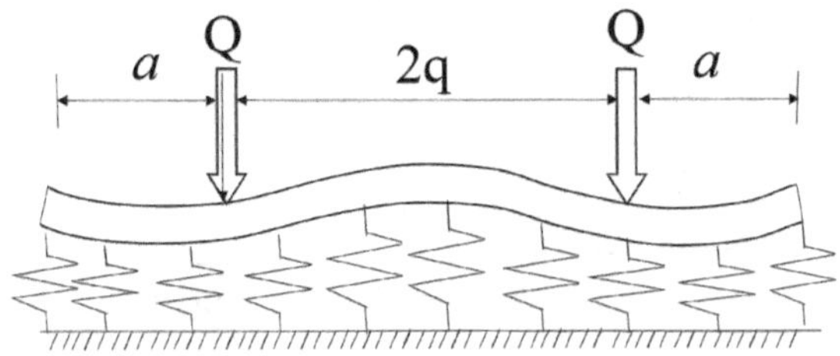

FIGURE 12.7 Equivalent spring stiffness of the beam resting on elastic foundation.

Effective spring stiffness of the entire rail system is estimated as follows:

$$\frac{1}{K_e} = \frac{1}{K_p} + \frac{1}{K_t} + \frac{1}{aK_{\mathrm{bs}}}, \tag{12.15}$$

where K_e is effective spring stiffness of the rail system, K_p is rail pad stiffness, K_t is tie stiffness in vertical direction and K_{bs} is stiffness of ballast and subgrade.

12.3.3.4 Three-Dimensional FEM Model for Rail with Sleepers

Rail and sleepers are modelled as three-dimensional bodies with elastic elements considered as rail pads between the rail and sleepers which is the most realistic track model. Three dimensional finite elements are used to model the track bed in ANSYS as shown in Figure 12.8. Waves are propagated from track to surroundings, which can be simulated in this model [14].

12.3.3.5 Sleeper Modelling for Vibration Analysis

Railway track models usually comprise sleepers that rest on the elastic foundation. The dynamic behaviour of the sleeper is studied where the sleeper is modelled as a beam with free-free boundary conditions in this section as depicted in Figure 12.9. A very good agreement is there between calculated and measured eigen frequencies at the lowest frequencies (two or three) during discussing the sleeper vibration by using Euler-Bernoulli beam theory, whereas, Rayleigh-Timoshenko beam theory is more appropriate for analysing the sleeper vibration at higher frequencies [14].

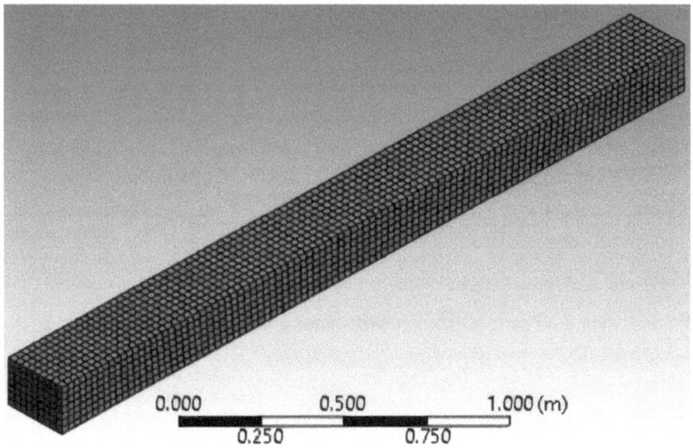

FIGURE 12.8 Modelling and meshing of the sleeper.

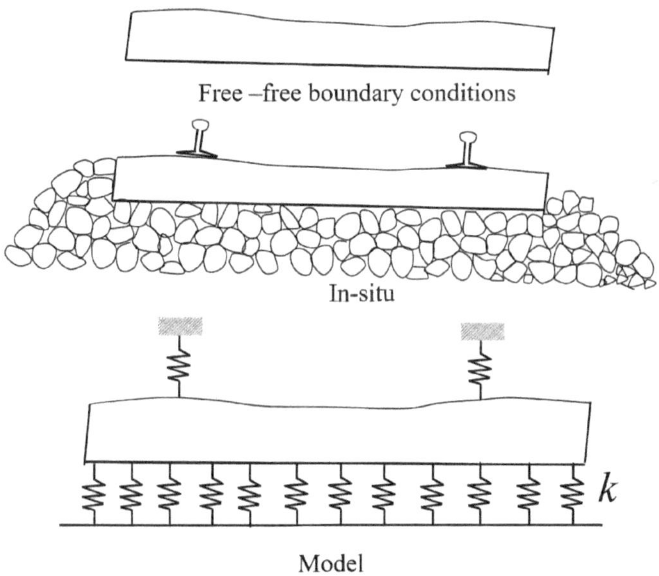

FIGURE 12.9 Sleeper with free-free boundary conditions, in-situ sleeper and model of sleeper surrounded and supported by springs.

12.4 DYNAMIC ANALYSIS OF TRAIN-TRACK COMPOUND SYSTEM

Generally, vibration of the train-track system is initiated from the ground and noise is generated from the train-traffic interaction. Due to the presence of periodic irregularities, i.e. rail corrugations and wheel flats, harmonic vibration takes place, whereas random vibration occurs due to non-periodic irregularities such as crossing, switches and sleeper spacing. Harmonic vibration also occurs in ballast track due to variations in the stiffness despite of good condition train and track. The stiffness of rail is more rigid at the sleepers as compared to the space between the sleepers. The rail deflects more when the wheel is within the sleepers. Low-frequency ranges, i.e. <20 Hz are accountable for safety assessment and riding quality, whereas passengers' comfort, noise emission and vibration are caused by high frequency. Deterioration of track, corrugation of railhead and damage to sleepers, railheads, fastenings and ballast are caused due to vibrations. The dynamics of the coupled train-track system can be analysed by either frequency domain or time domain approaches.

12.4.1 Frequency Domain Approach

The investigation of track-wheel response due to rail irregularities can be investigated by using a frequency domain approach. A receptance function is used to analyse the response of the sinusoidal rail irregularities. The frequency domain approach can be used to convert the response of non-sinusoidal irregularities, such as the flatness of a wheel, to frequency domain. The combination of the contact stiffness at the contact patch and track-wheel receptances forms a suitable transfer function and the response is obtained by using Fourier transform function. The track responses are assumed to be stationary, however, rail joints, improper ballast support, and variation in track stiffness are not taken into consideration. The vibration of sleepers is influenced by their elastic foundation. Table 12.3 shows both measured and calculated eigen frequencies of free-free concrete sleepers. Two eigenmodes such as translational and rotational are present in rigid beam with free-free boundary conditions, both in soft as well as stiffed bed as given in Table 12.3 (Columns 4 and 5).

E-B stands for Euler-Bernoulli beam theory and R-T stands for Rayleigh-Timoshenko beam theory.

Modal parameter is one of the dynamic characteristics for the structures like sleepers to measure the vibration responses. Modal analysis helps to determine natural frequencies, damping factors and corresponding mode shapes to study the dynamics of the sleepers [15, 16]. The steady state response of the structures can be determined in harmonic (sinusoidal) analysis; load amplitude varies with time. Different mode shapes and resonance frequencies are found using finite element (FE) modelling ANSYS by ignoring plasticity and frictional contact model (non-linearity

TABLE 12.3
Eigen Frequencies of Concrete Sleepers [8]

	Calculated (Hz)			
Measured (Hz)	**Free-Free E-B**	**Free-Free R-T**	**For Soft Bed**	**For Rigid Bed**
0	0	0	79	111
0	0	0	82	115
113	121	118	131	143
335	347	327	335	344
639	696	624	631	637
995	1,167	986	989	993
1,390	1,752	1,391	1,393	1,394

characteristics). Several researchers focused on the sleepers made up of different materials such as prestressed concrete sleepers and different compositions of carbon fibre reinforced polymer (CFRP), etc. Different natural frequencies of the sleepers with free-free boundary condition are determined; some natural frequencies lead to rigidity and others follow the flexibility nature of the sleeper. Generally, lower frequencies correspond to the rigid body modes are neglected; however, different mode shapes such as 1st vertical bending, 1st lateral bending, 2nd vertical bending, 1st twisting, 2nd horizontal bending and 3rd vertical bending are observed. Although the same nature of modes is noticed, the corresponding natural frequencies are different for different materials. Different mode shapes are shown in Figure 12.10.

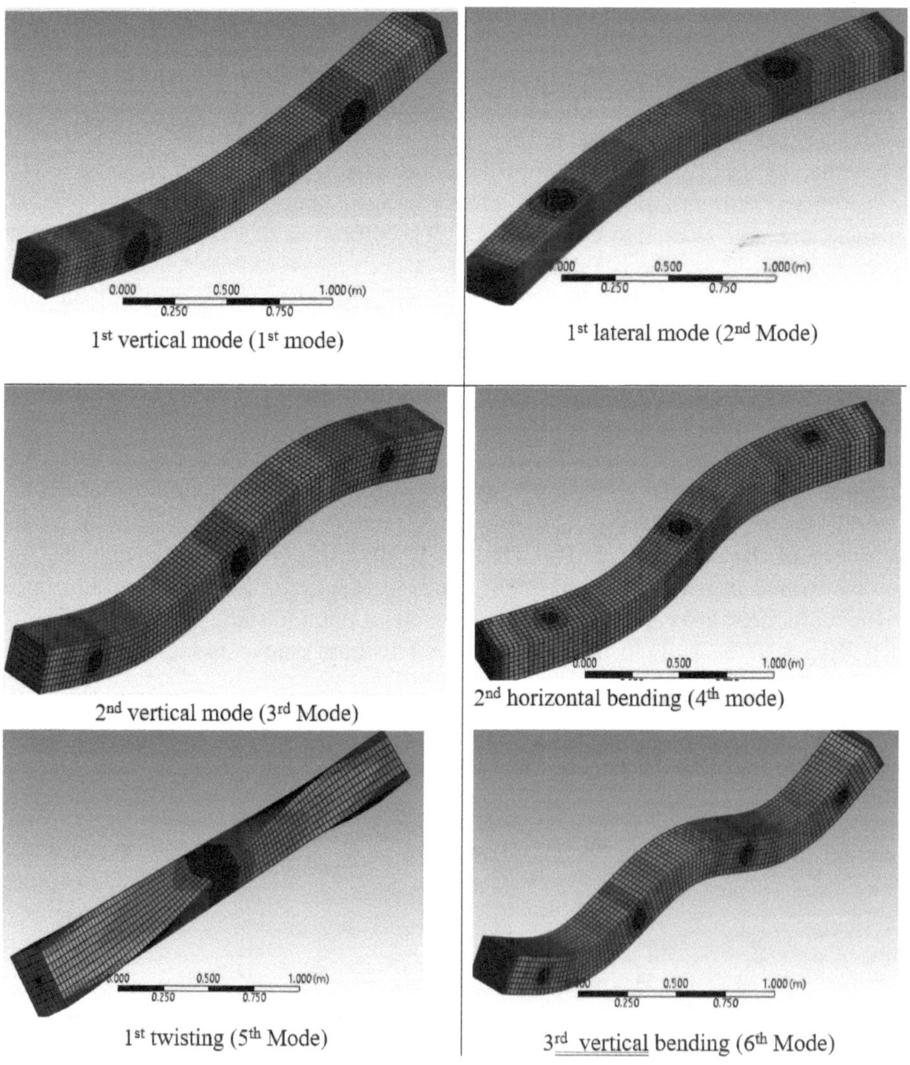

FIGURE 12.10 Different mode shapes by modal analysis.

12.4.2 Time Domain Approach

With the help of the time domain approach, the dynamics of the train-track coupling deal with the deflection of the track and vehicle displacement. Modal analysis of the track is analysed by using the FEM approach, and deformations of the rail are estimated by modal superposition approach. The elastic deformations of the wheel set can be estimated with the help of modal analysis without considering the number of degrees of freedom of the track-train coupled system [17]. Track is modelled with nonlinear track elements such as rail pads and ballast-subgrade material by using a FE approach, which gives a force-displacement relationship [18].

12.5 SUMMARY

In this chapter, the different components of the track, types of the sleepers, different types of irregularities, mathematical modelling and dynamic analysis of sleeper and rail have been surveyed. The track irregularities affect the vehicle dynamics and the analysis of the vibration is based on both short wavelength irregularities (high frequency vibration) and long wavelength irregularities (low frequency vibration). To study the dynamics of the track, some mathematical models have been discussed. Depending upon the complexity of the problems, the Euler-Bernoulli beam theory or Rayleigh-Timoshenko beam theory have been used to model the rail. The foundation (ballast and subgrade) can be considered as an elastic foundation by using the Winkler model.

Dynamics of the coupled train-track system can be analysed either with the help of time domain or frequency domain approaches. Nowadays, several researchers focused on the different types of composite materials. The different composite sleepers can be modelled by considering the properties of along with ballasts, rail pads and foundation as discussed in Section 12.3 to study the dynamic behaviour. The researchers can be focused on the multibody simulation (MBS) software such as ADAMS and SIMPACK for the dynamic analysis as well as modal analysis.

REFERENCES

1. Dukkipati, R. V., and J. R. Amyot. "Computer-Aided Simulation in Railway Dynamics." *Mechanical Engineering* 61 (1988): 47.
2. Ayasse, J.-B., and H. Chollet. "Wheel-Rail Contact." In *Handbook of Railway Vehicle Dynamics*, edited by Iwnicki, S., pp. 85–120. Taylor & Francis Group, Oxfordshire, OX, 2006.
3. Carter, F. W. "On the Action of a Locomotive Driving Wheel." *Proceedings of the Royal Society of London. Series A, Containing Papers of a Mathematical and Physical Character* 112, no. 760 (1926): 151–157.
4. Johnson, K. L. *Contact Mechanics*. Cambridge University Press, Cambridge, 1987.
5. Kalker, J. J. *On the Rolling Contact of Two Elastic Bodies in the Presence of Dry Friction*. Nederlandsch Drukkerij Bedrijf, Amsterdam, the Netherlands, 1967.
6. Kalker, J. J. "A Strip Theory for Rolling with Slip and Spin." *Koninklijke Nederlandse Akademie Van Wetenschappen-Proceedings Series B-Physical Sciences* 70, no. 1 (1967): 10.

7. Kalker, J. J. "The Tangential Force Transmitted by Two Elastic Bodies Rolling Over Each Other with Pure Creepage." *Wear* 11, no. 6 (1968): 421–430.
8. Dahlberg, T. "Track Issues." In *Handbook of Railway Vehicle Dynamics*, edited by Iwnicki, S., pp. 143–179. Taylor & Francis Group, Oxfordshire, OX, 2006.
9. Jonsson, J. O. *On Ground and Structural Vibrations Related to Railway Traffic*. Chalmers University of Technology, Sweden, 2000.
10. Sato, Y., A. Matsumoto, and K. Knothe. "Review on Rail Corrugation Studies." *Wear* 253, no. 1–2 (2002): 130–139.
11. Fermér, M., and J. C. O. Nielsen. "Wheel/Rail Contact Forces for Flexible Versus Solid Wheels Due to Tread Irregularities." *Vehicle System Dynamics* 23, no. supl (1994): 142–157.
12. Andersson, C., and T. Dahlberg. "Load Impacts at Railway Turnout Crossing." *Vehicle System Dynamics* 33, no. supl (1999): 131–142.
13. Zhai, W., and Z. Cai. "Dynamic Interaction between a Lumped Mass Vehicle and a Discretely Supported Continuous Rail Track." *Computers & Structures* 63, no. 5 (1997): 987–997.
14. Grassie, S. L. "Dynamic Modelling of Concrete Railway Sleepers." *Journal of Sound and Vibration* 187, no. 5 (1995): 799–813.
15. Kaewunruen, S., and A. M. Remennikov. "Dynamic Effect on Vibration Signatures of Cracks in Railway Prestressed Concrete Sleepers." *Advanced Materials Research* 41 (2008): 233–239.
16. Fu, Z.-F., and J. He. *Modal Analysis*. Elsevier, Amsterdam, the Netherlands, 2001.
17. Andersson, C. "Modelling and Simulation of Train-Track Interaction Including Wear Prediction." (2003): 0253–0253. https://elibrary.ru/item.asp?id=8826568
18. Oscarsson, J. "Dynamic Train-Track Interaction: Linear and Non-Linear Track Models with Property Scatter." Doctoral dissertation, Department of Solid Mechanics, Chalmers University of Technology, 2001. https://research.chalmers.se/en/publication/665

13 Case Studies of Failures in Railway Engineering, High-Speed Rail Innovations, and Challenges in Metro Trains

Abhishek Gannarapu

13.1 INTRODUCTION

Any accident in real life will lead to substantial loss of human life and property. The loss will be more severe in the case of railways. More than 900 million people use trains as a mode of travel, and the percentage of people using trains is even higher in populated countries. Ensuring proper safety is the most important thing for any passenger or train. High care needs to be taken, and the appropriate functioning of all components needs to be ensured. The wheel carries the load of entire vehicles and transfers the load to the rail. The wheel and rail traction ensures the train's appropriate motion on the railway track. Any damage or defect in the rail and wheel must be considered seriously, which may directly impact the safety of the entire train. It is very important to notice such types of damages, and frequent servicing is also required for wheels and rails for smooth and safer running. The first section of this chapter deals with various reported case studies of wheel, rail, and railway turnouts. Further, it presents the root cause of the problem with probable solutions to mitigate it.

High-speed trains are important in a region's development and economic growth. High-speed trains have several advantages, such as reduced travel time, dedicated travel routes, and can travel at higher speeds, such as 200 km/hr. Many countries such as China, Japan, Germany, and France have already adopted high-speed train technology and are progressing faster in their economy. High-speed trains are important for speed, safety, comfort, efficiency, and stability. There is always a scope to incorporate innovations in high-speed trains to improve these basic features and make them more economical. The second portion of the chapter briefly presents the latest innovations in high-speed trains.

DOI: 10.1201/9781003659983-13

Metro train systems are designed to transport large passengers in urban regions quickly. Metro trains have several advantages, such as quick and safer transportation, reduced road traffic, and positive impacts on urban life. Metro trains differ from conventional trains in various aspects, such as having a dedicated track, higher frequency of travel, being designed for carrying more passengers, and being independent of other traffic. Even though many urban regions across the globe are successfully using metro trains, a variety of practical challenges still exist for metro trains. The third section of this chapter deals with the practical challenges of metro trains.

13.2 CASE STUDIES OF FAILURES/DAMAGES IN RAILWAY ENGINEERING

13.2.1 Formation of Squats on a Metropolitan Railway Track

The formation of squats is a localized depression and one of the common and major forms of railway track damage or defect. These may occur for several reasons, such as rolling contact fatigue, hard material on the wheel side, or some preexisting damages. Formation of squats leads to a larger amount of corrective maintenance or may lead to subsequent deterioration. Metropolitan railways of Medellín, Colombia [1] experienced the formation of squat-type defects on rail between 2017 and 2020. They observed the first squat-type formation in 2017 and conducted a detailed analysis to understand the root cause of the problem and take preventive measures. A detailed analysis of the problem has been conducted, and it has been reported that the increase in traffic and the introduction of new trains, along with other reasons, such as inadequate interaction and improper maintenance, were observed to be major reasons for the problem. Complete details of the case study are shown in Table 13.1.

13.2.2 Wheel/Rail Wear of High-Speed Trains

Material of the wheel and rail, contact conditions, wheel-rail contact, and contact temperature could be major reasons for the wear on the wheel and rail. This wear on wheel and rail can lead to various consequences such as increased noise and vibration, change in geometry of wheel and rail, and failure due to fatigue. Further, if wear is not uniform, it may cause a change in the geometry and lead to frequent reprofiling and maintenance. Uneven wear in the high-speed rail network of China was reported and analysed [2]. Wheel polygonization and rail corrugation were observed and analysed in detail. The complete details of the root cause of the problem, preventive measures, and other details of the case study are reported in Table 13.2.

13.2.3 Cracks on Tread Region of Passenger Wheels

In tread braking, the majority of the heat input enters the railway wheel, which leads to thermal stresses, and these thermal stresses may cause damage in several ways. Railway wheels are manufactured by casting or forging. Once they undergo basic manufacturing, heat treatment is mandatory to improve mechanical properties and induce compressive hoop stresses, which resist crack propagation. These

TABLE 13.1
Features, Root Cause, and Preventive Measures of the Case Study Reported in Metropolitan Railways of Medellín, Colombia [1]

Type of Case Study	Formation of Squats on a Metropolitan Railway Track
Type of component that has experienced the failure	Rail (railway track)
Key features of failure or damage	i. Squat-type defect was first time noticed in the year 2017 and existed till 2020, as per the available literature ii. Problem was observed on both straight and curved lines for several kilometres Further, laboratory experiments were performed to analyse the problem, and it was observed that a white etching layer formed near the defects.
Intensity of failure and immediate consequences	i. May trouble the railway operations ii. May lead to a major economic burden or the railways iii. May raise safety concerns
Specific location or country in which failure was observed	Medellín, Colombia
The root cause(s) of the problem	i. Increased traffic intensity (new trains with AC traction systems were introduced around 2016) ii. Inadequate interaction between track and trains due to improper maintenance iii. Introducing new trains/vehicles, which has caused different load-transmitting conditions and dynamics iv. Lack of adequate grinding on rails
Preventive measures/recommendations for the mitigation of the failure or accident	With regard to this problem, authors have identified the root causes only, and no specific recommendations were mentioned, but after understanding the root causes, the following measures may reduce the problem: i. Reducing the traffic using by introducing new lines ii. Frequent grinding on a railway track and ensuring proper interaction between train and track

thermal-induced residual stresses due to tread braking can alter the stress state in the wheel, and these service-induced residual stresses may lead to the initiation and propagation of the cracks. Cracks developed on the tread region of commuter car wheels were examined to study the reasons [3, 4]. Complete analysis reports are presented in Table 13.3 with root causes of the problem.

13.2.4 Railway Wheel Failure Due to Wheel Gauge Widening or Condemning

Indian railways are experiencing a unique problem of railway wheel failure due to excessive wheel gauge change. Excessive wheel gauge change may lead to widening or condemning wheel gauge [5–8]. The distance between two exact and opposite points

TABLE 13.2
Features, Root Cause, and Preventive Measures of the Case Study Reported in the High-Speed Rail Network of China [2]

Type of Case Study	Uneven Wear on Wheel and Rail
Type of component that has experienced the failure	Rail and wheel
Intensity of failure and immediate consequences	i. Uneven wear leads to serious changes in wheel-rail interaction and causes vibration ii. This may lead to wheel polygonization (deviation of the radius and losing roundness), which in turn may lead to higher wheel-rail forces, noise, reduced comfort, and safety (in some cases, may lead to component failure). This is more significant in the case of high-speed trains.
Specific location or country in which failure was observed	High-speed wheel/rail systems in China
The root cause(s) of the problem	Although the authors have not highlighted the exact reasons for the wheel polygonization and rail corrugation, the following could be the reasons: i. Preexisting defects on the wheel ii. Rail corrugation may be caused by ineffective fastening, variation in the radius of wheelsets, track geometry
Preventive measures/recommendations for the mitigation of the failure or accident	i. Choosing better wheel reprofiling strategies ii. Usage of intelligent tread trimmers iii. Better design of the vehicle iv. Hardening (local) of the wheel tread

on the inner faces of two railway wheels fitted on the axle is referred to as the wheel gauge, shown in Figure 13.1. An increase in wheel gauge by 2 mm is called wheel gauge widening. This tolerance is even less in the case of a reduction in wheel gauge change. Railway wheels play a key role in the train's motion; any concerns or damages related to wheels must be concentrated seriously. A complete study of wheel gauge change in railway wheels was studied by [5–8]. Excess heat input in railway wheels due to various factors such as non-uniformity in braking, thermally induced residual stresses in tread braking, and other factors leading to excess heat input in wheelsets is believed to be the major reason for wheel gauge change. Wheel gauge widening is permanent and unfirm throughout the wheel. In most cases, wheel gauge widening or condemning goes unnoticed because it is impossible to measure wheel gauges under running and only when they reach for servicing. Complete details of wheel gauge change are presented in Table 13.4, along with guidelines to mitigate the same.

13.2.5 Railway Turnouts

In complex railway networks, railway turnouts help to change the track for trains so that one can use the existing track optimally and minimize the traffic. Typically, the turnout is a mechanism that connects the main track to the diverting loop and

TABLE 13.3
Features, Root Cause, and Preventive Measures of Cracks Developed on the Tread Region of Railway Wheels [4]

Type of Case Study	Cracks Developed on the Tread Region of Commuter Car Wheels (Passenger Railway Wheels)
Type of component that has experienced the failure	Railway wheel
Key features of failure or damage	i. Many cracks were observed along the circumference of the passenger railway wheel. ii. The spread of the cracks was observed at a depth of 4–5 mm.
Intensity of failure and immediate consequences	i. Multiple cracks along the circumference (especially on the tread region may lead to changes in vehicle dynamics and wheel-rail interaction. ii. The tread region needs immediate machining, which will reduce the life of the railway wheel. iii. Increased risk of fatigue failure
Specific location or country in which failure was observed	Metropolitan area of New York City
The root cause(s) of the problem	i. Thermal stresses induced due to tread braking (service-induced residual stresses) ii. Thermally induced residual stresses add to preexisting residual stresses from manufacturing processes, leading to tensile hoop stresses on the wheel tread region. After manufacturing, wheels contain compressive residual hoop stresses. These stresses turn into tensile hoop stresses due to in-service residual stresses, finally leading to a crack opening. iii. Drag braking
Preventive measures/ recommendations for the mitigation of the failure or accident	i. Proper heat treatment of railway wheels and optimizing heat treatment parameters. ii. Frequent maintenance to check the hoop stresses on the railway wheel.

provides the provision for travelling either on a main track or to take them to a temporary loop. Various cases of reported damage in railway turnouts (at switches and crossings) of the Great Britain railway network were observed from 2011 to 2017 [9]. Reported cases show damages on the switch side and panel side. Complete analysis of reported cases, root causes, and preventive measures of the same as reported in Table 13.5.

13.3 HIGH-SPEED RAIL INNOVATIONS

High-speed trains can run more than 200 km/hr at higher speeds. High-speed trains play a crucial role in human life by improving time efficiency and reaching the destination in the least possible time. Many countries have successfully implemented high-speed rail systems; there exist problems for them, such as high investment and

TABLE 13.4
Features, Root Cause, and Preventive Measures to Avoid Railway Wheel Gauge Change

Type of Case Study	Wheel Gauge Widening in Locomotive, Wagon, and Passenger Coaches (Observed in Tread-Braked Railway Wheels)
Type of component that has experienced the failure	Railway Wheel
Key features of failure or damage	i. Observed all types of railway wheels, such as locomotive, wagon, and coach ii. Wheel gauge widening is uniform throughout 360° of the wheel.
Intensity of failure and immediate consequences	i. May lead to derailment, which may cause a serious accident ii. Decrease in wheel lifetime iii. Requirement of regular service and maintenance, which in turn leads to an increase in cost of maintenance and service iv. May damage the track and cause more wear on the wheel and rail
Specific location or country in which failure was observed	Majority of the cases were observed in Indian Railways
The root cause(s) of the problem	i. Time lag in braking (due to the existing air brake system) across all wagons/coaches across the train may lead to higher heat input in locomotive wheels in tread-braked trains. ii. Service-induced residual stresses in tread braking, along with residual stresses existing after the manufacturing process, may cause a permanent wheel gauge widening iii. Non-uniform braking (might become due to non-uniformity in brake blocks composition, improper fitting of brake blocks, improper distributor valves) across all wheels may lead to higher heat input in a few specific wheelsets, which may lead to higher heat input factors iv. Low thermal diffusivity of composite brake blocks leads to higher heat input in railway wheels v. Improper maintenance or improper fitting of brake blocks. vi. Drag braking: continuous application of brakes for about 20 minutes while travelling down gradients vii. Variation in brake blocks used in locomotive and coach/wagon wheelsets
Preventive measures/recommendations for the mitigation of the failure or accident	i. Effective air braking needs to be used, and time lag in braking needs to be minimized ii. Uniformity needs to be maintained across all brake block characteristics iii. Ensuring proper contact between wheel and brake blocks iv. Frequent servicing and identifying the cases of wheel gauge widening

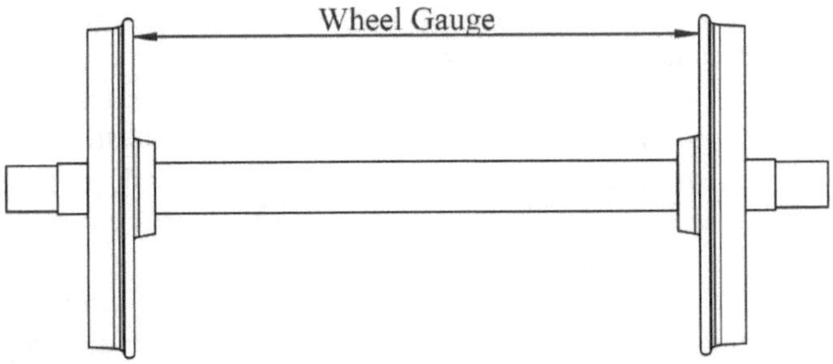

FIGURE 13.1 Wheel gauge for a set of railway wheels fitted on the axle.

TABLE 13.5
Features, Root Causes, and Preventive Measures for Problems Experienced on Railway Turnouts

Type of case study	Failures observed at railway turnouts
Type of component that has experienced the failure	Railway turnouts (at switches and crossings)
Key Features of failure or damage	i. On the switch panel side, switchblade damage is the most observed, with lipping and wear being common failure modes. ii. Majorly squat-type failure on the crossing panel, most commonly located on the nose
Intensity of failure and immediate consequences	i. May cause derailment and may lead to safety concerns ii. Inconvenience to the rail traffic iii. Increase in maintenance and renewal costs iv. Damage to railway infrastructure
Specific location or country in which failure was observed	Great Britain railway network and cases were observed and reported from 2011 to 2017
The root cause(s) of the problem	i. Damage on the switch is due to higher dynamic forces and large contact stresses at the nose of the switch ii. The major reasons for the crossing panel are larger contact stresses at crossing and wheel contact, incorrect installation, and maintenance.
Preventive measures/recommendations for the mitigation of the failure or accident	i. Some wear on blades can be mitigated using better steel grades for switchblades. ii. Optimized geometry of switch profile also may lead to an increase in switch life iii. Multi-channel actuation switches (introduced by various researchers recently) may also reduce the same. iv. On the crossing side, mitigation of impact loads and achieving better load distribution may mitigate the problem

operating costs, issues related to signalling, noise, less flexibility in track curves (majorly sharp curves need to be avoided), and frequent, high track wear and high maintenance cost. Advanced artificial intelligence (AI) tools can be adopted for various aspects of high-speed trains [10]. AI can be adopted to manufacture key components of high-speed trains, forecast passengers' demand, perform preventive maintenance, and optimize energy consumption. Further, they can be used in communication, signal control, and transportation management.

Minimizing air drag during crosswind conditions is very important to make high-speed trains more efficient. Several researchers have recently proposed robust optimization techniques for developing streamlined shapes by considering uncertain environmental conditions and cross winds [11]. The proposed methods are cost-effective and give optimum design suggestions. Hydrogen fuel cell technology is in the development stage, where using hydrogen fuel cell technology in place of conventional electric vehicles reduces the railway infrastructure, such as reducing the need for overhead containers and electric substations [12]. Further, they have less refuelling time. However, many challenges in this technology, such as production, storage, refuelling, and safety, are yet to be answered.

Digital enhancements (e.g., digital twins) can improve connectivity and safety. This technology allows communication between trains and maintenance teams for better coordination, immediate responses, and action in special conditions. It may lead to prompt issues responses and improve safety aspects as multiple trains can communicate and maintain safer distances. Further, autonomous vehicle technologies have greater potential in high-speed trains, which can reduce human error and could be safer [13–16].

13.4 CHALLENGES IN METRO TRAINS

Metro trains are high-capacity transport systems that support passengers for daily travel in urban areas. Metro trains, also called mass rapid transit systems, differ from conventional railway networks in many ways. Typically, metro trains contain dedicated tracks and may travel underground or on elevated tracks. They run at higher frequencies (some may operate continuously) depending on demand and exclusively on electricity. They are exclusively for passengers and typically operated for shorter distances (within the city) or connecting smaller cities. They have the advantage of dedicated track integration with other transport systems, which gives safety, flexibility, speed, and capability to cover more locations and support a larger group of people. Typically, metro train travel is economical, giving flexibility for travel to all classes of people and providing reasonable comfort for passengers. At present, more than 200 cities from 60 countries are effectively using metro trains.

Metro trains experience many challenges, from planning and establishment to smooth operation. Especially at the phase of establishment, they experience challenges related to designing appropriate route corridors by covering underground or elevated passages, appropriate estimation of demand and concluding the number of stoppages or stations, acquisition of land along the designated route of establishing, and controlling other traffic or accidents during the construction [17]. Along with all these, they have experienced financial, construction, operating, safety, and revenue challenges. Several reported cases globally regarding practical problems experienced in metro trains earlier. Some practical problems or challenges being experienced and reported are discussed in detail. Operating conditions such as having plenty of sharp curves or turns, frequent braking, and operating under various environmental conditions lead to a lower lifetime of wheel and rail and increase maintenance costs abnormally. Among all the components, wheel and rail experience heavy duty on metro trains due to several turns, more braking cycles, braking from higher speeds, shorter braking distances, and quick pick-up of speeds also affect wheel and rail life.

13.4.1 Wear on Wheel and Rail

Wear along the lateral direction of wheel and rail, wheel flange wear, uneven and partial wear on wheel tread, rolling contact fatigue along the wheel tread, formation of channel-type defects, wheel losing roundness, and frequent rail corrugation are common problems being experienced, analysed, and reported earlier in metro trains [18–21]. Features of the problem and countermeasures that need to be considered are reported in Table 13.6.

13.4.2 Rolling Contact Fatigue and Rail Corrugation

Severe stress concentration along the wheel and rail may cause rolling contact fatigue, which is even more frequent in the case of metro trains. Spalling on the wheel side was majorly observed and reported on the wheel side. Rolling contact fatigue on the rail side is also not exceptional. Improper interaction between the wheel and brake block may result in non-uniform wear reflected on the rail side and can cause significant fatigue damage. It is very important to ensure all contacting elements and immediate preventive measures need to be taken to control the damage further. Due to this, earlier cases have also shown enough spalling on the rail side. Improper wheel-rail contact characteristics and vibration characteristics can lead to rail corrugation. As per the earlier reported cases, short and long curves can lead to worn profiles on wheel and rail. Rail corrugation was observed on both sides, but the intensity was higher on the inner side. Lateral stick-slip vibration of the wheel with the inner rail causes intensified rail corrugation.

13.5 SUMMARY

Wheel and rail are important contacting bodies in railway systems. Any issue or damages related to these may lead to serious situations. There were case studies earlier reported by various railways across the globe related to wheel or rail damages or failures. All such case studies were analysed, and the root causes of the failure and guidelines to mitigate such problems were also mentioned for every case study. As many of the case studies are dependent on rail traffic, environmental, operational, and maintenance guidelines being followed, and every problem is unique. In the future, other countries may also experience such situations; reporting and understanding all these case studies will help avoid future damage or take corrective measures faster. Further, future innovations that may influence high-speed trains and practical challenges in metro trains are also included in this chapter.

TABLES 13.6
Practical Problems being Reported Related to Wheel and Rail Wear of Metro Trains

Type of Problem on the Metro Wheel	Features and Causes of the Problem	Countermeasures to be Considered
Wear on the wheel flange and side of the rail	Frequent and sharp curves on the route i. Improper lubrication at the sharp curves and intense wear at the wheel flange and side of the rail ii. Low conicity on the wheel side causing improper rolling radius difference iii. Improper maintenance and grinding of rail	i. Introducing lubrication on the inner face of the rail has significantly reduced the wear on the wheel flange ii. Proper reprofiling of the wheel sets on time and grinding and timely maintenance of the rail also reduced this problem
Wheel abnormal wear such as uneven wear, hollow wear, formation of grooves and channels on tread,	i. Different wear rates on wheels located on the left and right due to variation in the length of the left and right curves on the designated route, which can cause rolling radius differences even when travelling on a straight track and affecting the train running performance and causing un-necessary wheel wastage during reprofiling. ii. Formation of groove-type wear on the wheel side causes a significant change in wheel-rail contact. Further, this can lead to uneven wear on the brake block side if tread braking is adopted, causing localized friction heating in the tread region. iii. Most of the brake blocks used in tread braking are composite brake blocks having complex friction and wear characteristics	This can be reduced using i. Proper machining and reprofiling of the wheel. ii. Proper fitting of brake blocks and increasing the width of the brake block to match the tread region iii. Effective utilization of electric braking to reduce the tread braking iv. Maintaining uniform friction characteristics of brake blocks to ensure uniform friction and wear characteristics of brake blocks.

REFERENCES

1. Bedoya-Zapata, Á.D., S. Rojas-Parra, J.H. Díaz-Mazo, J.A. García-Jiménez, J.E. López-Londoño, R.A. Vergara-Puello, L.F. Molina, J.F. Santa-Marín, A. Toro, M. Mesaritis, R. Lewis, and M. Palacio. "Case Study: Understanding the Formation of Squat-Type Defects in a Metropolitan Railway." *Engineering Failure Analysis* 123 (2021): 105325.
2. Zhai, W., X. Jin, Z. Wen, and X. Zhao. "Wear Problems of High-Speed Wheel/Rail Systems: Observations, Causes, and Countermeasures in China." *ASME Applied Mechanics Reviews* 72, no. 6 (2020): 060801.

3. Johnson, M.R., R.E. Welch, and K.S. Yeung. "Analysis of Thermal Stresses and Residual Stress Changes in Railroad Wheels Caused by Severe Drag Braking." *Transactions of the ASME* 18 (1977): 18–23.
4. Gordon, J., and A. B. Perlman. "Estimation of Residual Stresses in Railroad Commuter Car Wheels Following Manufacture." *ASME RTD* 15 (1998): 13–18.
5. Vakkalagadda, M.R.K., K.P. Vineesh, A. Mishra, and V. Racherla. "Locomotive Wheel Failure from Gauge Widening/Condemning: Finite Element Modeling and Identification of Underlying Mechanism." *Engineering Failure Analysis* 57 (2015): 143–155.
6. Vineesh, K.P., M.R.K. Vakkalagadda, M. Dev, B.K. Rao, and V. Racherla. "Gauge Widening of Passenger Coach Wheel Sets in Indian Railways: Observed Statistics and Failure Analysis." *Engineering Failure Analysis* 71 (2017): 105–119.
7. Vineesh, K., M. Vakkalagadda, M. Dev, B. Rao, and V. Racherla. "Effect of Periodic Wheel Tread Reprofiling on Wheel Gauge Evolution in the Wheelsets of Tread-Braked Coaches: Finite Element Modeling and Field Observations." *Proceedings of the Institution of Mechanical Engineers, Part F: Journal of Rail and Rapid Transit* 234, no. 6 (2020): 678–686.
8. Desam, V., M. R. K. Vakkalagadda, V. Racherla, and K. P. Vineesh. "Wheel Gauge Evolution of Railway Wheels: Effect of Wheel Heat Treatment." *AIP Conference Proceedings* 2745, no. 1 (2023): 030001.
9. Grossoni, I., P. Hughes, Y. Bezin, A. Bevan, and J. Jaiswal. "Observed Failures at Railway Turnouts: Failure Analysis, Possible Causes and Links to Current and Future Research." *Engineering Failure Analysis* 119 (2021): 104987.
10. Li, X., M. Zhu, B. Zhang, X. Wang, Z. Liu, and L. Han. "A Review of Artificial Intelligence Applications in High-Speed Railway Systems." *High-Speed Railway* 2, no. 1 (2024): 11–16.
11. Wang, Y., Z. Sun, S. Ju, D. Guo, and G. Yang. "Robust Optimisation of the Streamlined Shape of a High-Speed Train in Crosswind Conditions." *Engineering Applications of Computational Fluid Mechanics* 17, no. 1 (2023): 2234012.
12. Ding, D., and X.-Y. Wu. "Hydrogen Fuel Cell Electric Trains: Technologies, Current Status, and Future." *Applications in Energy and Combustion Science* 17 (2024): 100255.
13. Sharma, S. K., and A. Kumar. "Ride Performance of a High Speed Rail Vehicle Using Controlled Semi Active Suspension System." *Smart Materials and Structures* 26, no. 5 (2017): 055026.
14. Sharma, S. K., and A. Kumar. "Ride Comfort of a Higher Speed Rail Vehicle Using a Magnetorheological Suspension System." *Proceedings of the Institution of Mechanical Engineers, Part K: Journal of Multi-Body Dynamics* 232, no. 1 (2017): 23–34.
15. Lee, J., and S. K. Sharma. "Numerical Investigation of Critical Speed Analysis of High-Speed Rail Vehicle." In *Proceedings of the Korean Society for Precision Engineering Conference*, Seoul, September 14, 2020.
16. Vishwakarma, P. N., P. Mishra, S. K. Sharma, and A. K. Shukla. "Formulating and Analysing the Effect of Suspension Parameters on Critical Speeds for Various DOFs in Rail Vehicle." In *Advances in Engineering Design: Select Proceedings of FLAME 2022*, pp. 791–802. Cham: Springer, 2023.
17. Sharma, N., R. Dhyani, and S. Gangopadhyay. "Critical Issues Related to Metro Rail Projects in India." *Journal of Infrastructure Development* 5, no. 1 (2013): 67–86.
18. Jin, X., G. Tao, X. Zhao, W. Li, and Z. Wen. "Wear and RCF Problems of Metro Wheel/Rail Systems: Phenomena, Causes and Countermeasures in China." In *Proceedings of the 12th International Conference on Contact Mechanics and Wear of Rail/Wheel Systems (CM2022)*, Melbourne, Australia, September 4–7, 2022.
19. Yang, X., G. Tao, and Z. Wen. "Causes and Evolution of Asymmetric Polygonal Wear of Metro Train Wheelset." In *Proceedings of the 12th International Conference on Contact Mechanics and Wear of Rail/Wheel Systems (CM2022)*, Melbourne, Australia, September 4–7, 2022.

20. Wang, Z., and Z. Lei. "Formation Mechanism of Rail Corrugation on the Metro Small Radius Curve Based on the Viewpoint of Stick-Slip Torsional Vibration." In *Proceedings of the 12th International Conference on Contact Mechanics and Wear of Rail/Wheel Systems (CM2022)*, Melbourne, Australia, September 4–7, 2022.
21. Wang, S., B. Wen, D. Ren, Z. Wen, D. Dong, and G. Tao. "Causes and Treatments of Severe Side Wear of the High Rail on Metro Curves: Numerical Simulation and Field Monitoring." In *Proceedings of the 12th International Conference on Contact Mechanics and Wear of Rail/Wheel Systems (CM2022)*, Melbourne, Australia, September 4–7, 2022.

14 Frictional Braking in Trains

Vineesh Vishnu and K. P. Vineesh

14.1 INTRODUCTION

Various types of braking systems are used in trains. Application of the brakes for the purpose of reducing the speed or stopping the trains is called stop braking while application of brakes to maintain a constant speed when moving through a down gradient is called drag braking. Various methods are used to achieve this. Mechanical brakes are going to dissipate the kinetic energy of the train by rubbing two parts, railway wheel and brake blocks or brake discs and brake pads, in which one part is pressed against the other. Friction between the rubbing parts dissipates the kinetic energy in the form of heat at the interface of brake blocks and the railway wheel/brake disc and brake pads. Since these brakes operate based on friction, they are referred to as frictional braking. Further, since compressed air is used for the application of the force that presses brake block/brake pad against the wheel/brake discs, frictional braking is sometimes referred to as air braking also. This chapter describes the features of tread braking and disc braking systems.

14.2 TREAD BRAKING

Tread braking system has been in practice from the early stages of braking technology in trains. In this brake blocks are pressing against the tread surface of the railway wheel during braking. Kinetic energy of the vehicle is dissipated at the interface of wheel and brake block in the form of heat. Tread braking is simple and requires less maintenance. Hence, it is widely used in low-speed trains even now. But the dissipation of the heat on the tread surface of the wheel increases the temperature and generates thermal stresses. Hence it is not suitable for trains that are designed to operate at high speed. Figure 14.1 shows the bogie-mounted braking systems used in tread braking systems [1]. Here, compressed air is supplied to brake cylinder which in turn pushes the piston outwards for the application of the brakes. The movement of the piston is transmitted to brake pads through the linkages connected between them. This makes the brake pads to press against the tread surface of the railway wheel during braking. Coefficient of friction between the brake blocks and wheel surface, wear rate of the brake blocks, temperature rise of the brake blocks and wheel, and thermal stability of the brake block material are the important factors considered in the performance of the brake blocks. Typical materials used for making brake blocks are cast iron, composite, and sintered materials. Frictional properties of these materials are given in the next sections.

DOI: 10.1201/9781003659983-14

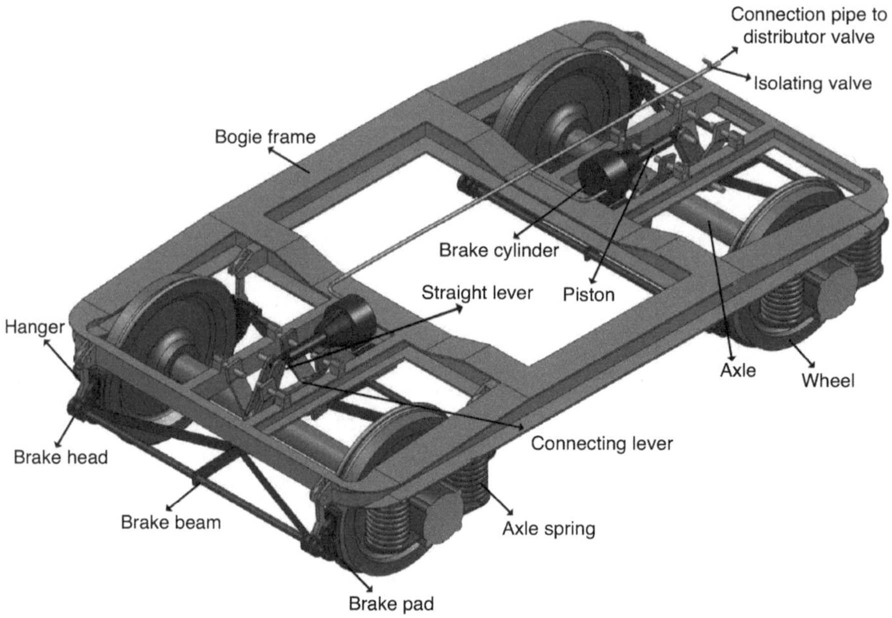

FIGURE 14.1 Various parts of brake rigging system (for ICF coaches). (From Ref. [1].)

The friction and wear characteristics of brake blocks are essential for railway wheel technology. The coefficient of friction (COF) is a critical factor that influences the braking performance of brake blocks. The COF remained stable at moderate temperatures and decreased gradually with increasing temperatures owing to oxide film formation and material softening. The frictional behaviour mainly varies with the braking load and speed applied to the brake blocks. The COF increases with a higher load owing to the enhanced contact area, whereas the increase in speed causes thermal degradation and a reduction in the COF. Similarly, the wear behaviour also depends on the temperature, surface roughness, and material position, and an elevated temperature accelerates the wear rates by thermal softening and promotes oxidation. The roughness of the wheel tread directly influences wear by altering the contact area and the distribution of friction force. The hard particles in the cast iron and the asperity of the wheel surface primarily cause abrasive wear. Under high-temperature conditions, the wear mechanism transitions to oxidative wear, forming a thin layer over the surface, further flaking off, and increasing wear. Variations in the carbon content and microstructure significantly influenced the wear resistance.

14.2.1 Brake Block Characteristics of Cast Iron Brake Blocks

Figure 14.2 shows cast iron brake block being used on tread region of railway wheel. Wear and friction characteristics of cast iron brake blocks with various graphite morphologies have been studied [2]. Results showed a decrease in friction coefficient with increase in brake load and braking speed. The same trends were observed for all types of morphologies. The wear rate in the brake blocks increased with the

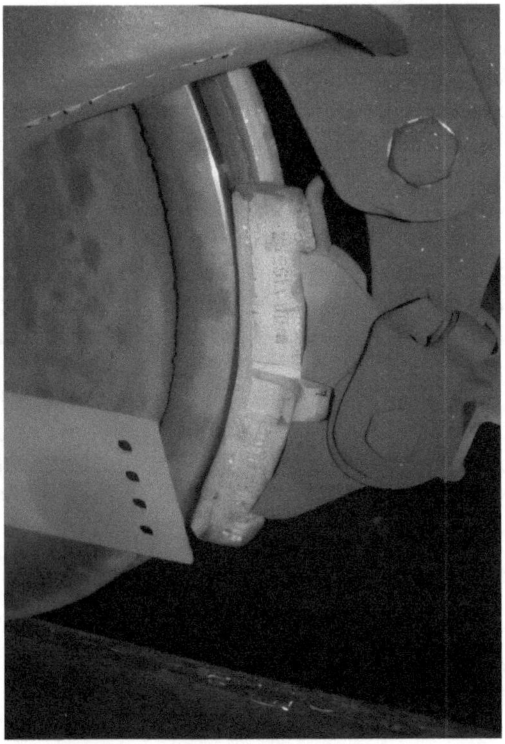

FIGURE 14.2 Tread braking system with cast iron brake blocks.

TABLE 14.1
Wear Loss of Different Cast-Iron Brake Block Types Observed in Field Conditions by [2]

		Wear Loss for Various Travelling Distances (kg)			
Brake-Block Material	**Wt.% of Phos-phorus**	**2,870 (km)**	**6,514 (km)**	**10,004 (km)**	**11,118 (km)**
Compacted graphite grey cast-iron	1	0.26	0.81	1.31	1.45
Grey cast-iron	0.7–1	0.85	1.83	2.68	2.93
Grey cast-iron	2–2.5	0.7	1.23	1.72	1.99

increase in the contact pressure. The measured wear loss (kg) in the brake blocks first increased and then decreased with an increase in the sliding speed. For the grey cast-iron blocks, the values were initially lower for a low braking distance and further increased with the braking distance. Table 14.1 reports the measurement of the maximum wear loss against the maximum braking distance of 11,118 km.

Furthermore, the cast-iron block with compacted graphite features higher frictional values and reduced wear loss compared to the grey cast-iron, and the maximum wear loss corresponded to the higher sliding speed. Vernersson [3] performed a wear test (during drag braking with single block) under dry sliding conditions with an applied normal load of 4.5 kN. The researchers tested the performance of the cast-iron brake blocks at three different braking speeds: 60, 80, and 100 km/h. Results revealed that the cast-iron block experienced a higher frictional value at the initial braking time of the conditions. Figure 14.3 depicts the evolution of the frictional value against time for the different braking speeds. During the initial time of braking, the friction was at its maximum and reduced to a lower value over time. The frictional value was maximum for the lower braking speed of 60 km/h, and the values decreased with an increase in the braking speed. Additionally, all the tested conditions resulted in the formation of quick hotspots, and that caused the generation of noise in the cast-iron brake blocks.

In contrast, Ferrer et al. [4] reported an increasing trend in friction coefficient for cast iron brake blocks under various braking conditions. At the slowest sliding speed of 0.31 m/s and a contact pressure of 0.4 MPa, the COF value was higher. At the fastest speed of 1.03 m/s, it went down, as shown in Figure 14.4a. In other words,

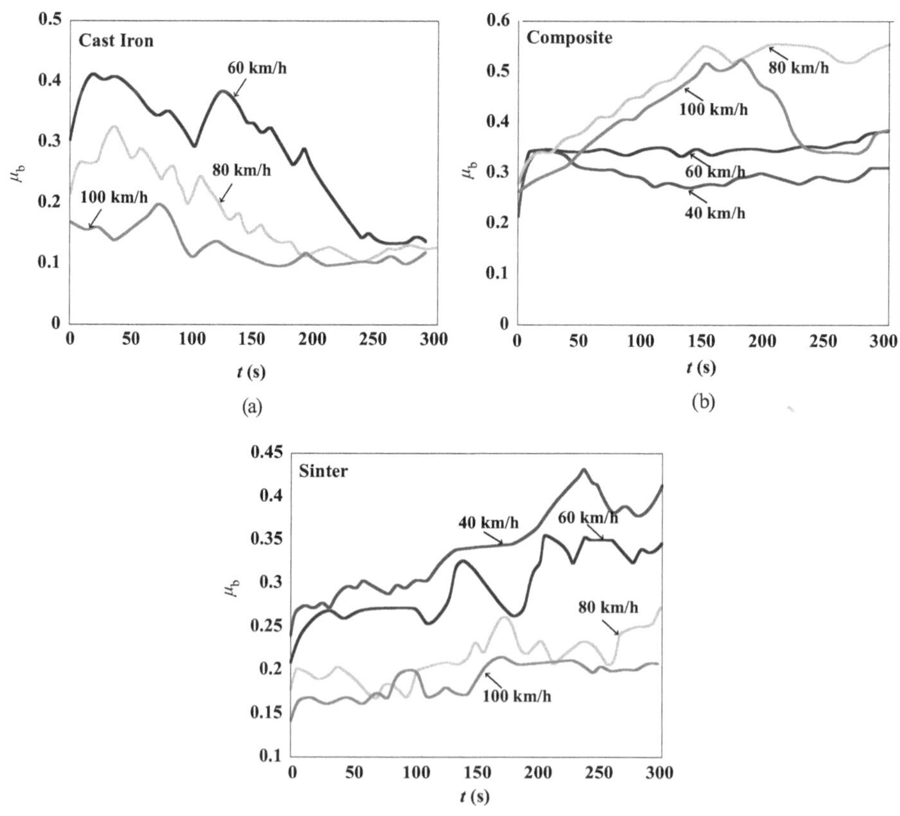

FIGURE 14.3 Friction characteristics obtained from full scale drag braking experiments at various sliding speeds using (a) cast-iron, (b) composite, and (c) sintered brake blocks [3].

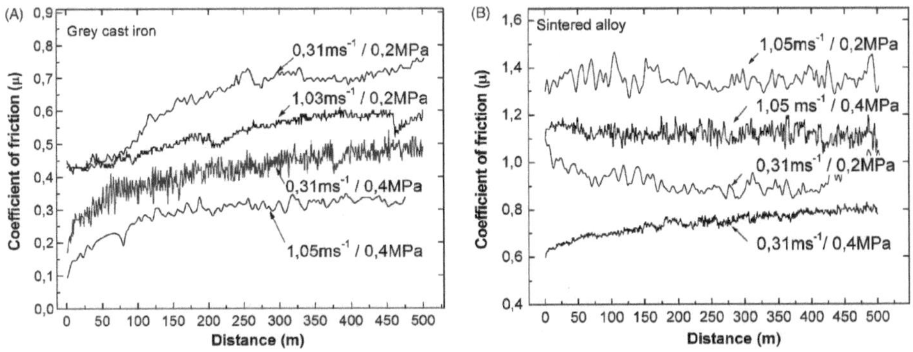

FIGURE 14.4 Friction characteristics of (a) Cast iron and (b) sintered alloy brake blocks. (From Ref. [4].)

when the contact pressure was reduced to half (0.2 MPa), a decline of the COF with increased sliding speed for cast-iron brake blocks was observed.

Furthermore, [5] studied the combined effect of brake load and braking speed on the friction and wear characteristics of the cast iron brake blocks, and a regression analysis was also performed to predict the COF values corresponding to different braking speeds and loads. The study involved four different initial speeds: 50, 80, 120, and 140 km/h, for brake loads of 25, 30, and 35 kN. The evolution of COF for each braking speed was found to be the same under all loads as shown in Figure 14.5. The apparent COF for the cast-iron blocks was lower at high braking speeds. The COF reduced with an increase in the braking speeds (see Figure 14.5a–c); it was observed that the apparent COF increased with an increase in sliding time; as a result, the friction value increased as the speed reduced during braking. Similarly, at lower braking loads, the apparent friction decreases with an increase in braking load. Lowest value was measured at 39 m/s and highest at 14 m/s. The trend of COF corresponding to each braking speed follows a reduction from an initial higher value to a lower value at stopping. During braking, the COF decreases as the speed decreases, and the apparent COF decreases as the brake load increases, especially at lower braking speeds. These observations are in line with those made by Ferrer et al. [4]. The specific wear rate for the cast iron brake blocks was found to be 30.32 cc/kWh. The main types of wear in the cast iron brake blocks were found to be adhesive and abrasive wear or thermal fatigue peeling.

14.2.2 Brake Block Characteristics of Composite Brake Blocks

Figure 14.6 shows the tread braking system with composite brake blocks. Vernersson et al. [6] used pin-on-disk tests to study the friction and wear of three different types of composite brake blocks at temperatures ranging from 0°C to 600°C, with a normal load of 43.2 N and two sliding speeds of 4.3–13 m/s. The measured SWR of the tests is listed in Table 14.2. Compared to the performance of the cast iron brake blocks, the wear rate of the composite brake blocks was found to be lower up to a temperature of 500°C. A higher value of wear rate was observed at 550°C.

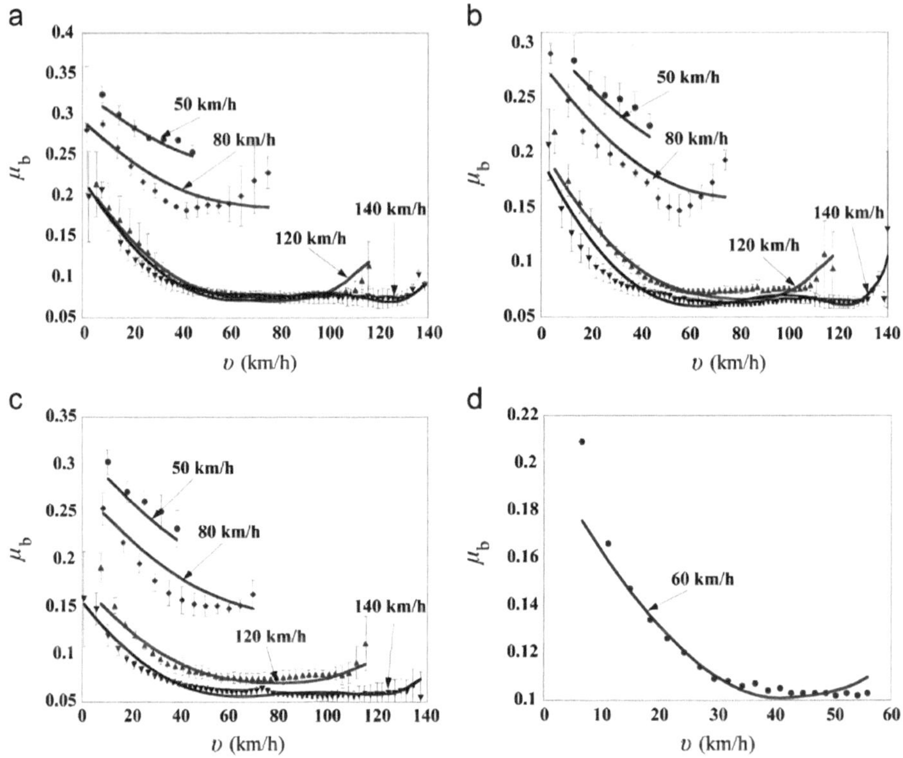

FIGURE 14.5 Friction characteristics for cast-iron brake blocks during stop braking (a–c) and (d) during drag braking. (From Ref. [5].)

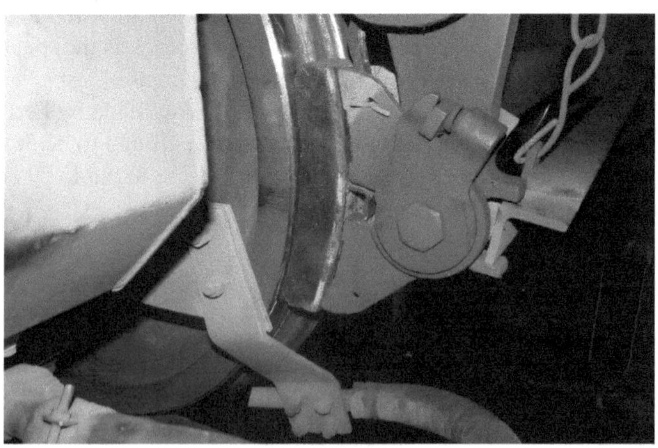

FIGURE 14.6 Tread braking system with composite brake blocks.

TABLE 14.2
Specific Wear Rates of Different Types of Composite and Cast-Iron Brake Block Materials, Obtained from Pin-On-disc Experiments at Different Disc Temperatures at a Normal Load of 43 N and a Sliding Speed of 4.3 m/s for Composites and 13 m/s for Cast-Iron Brake Blocks

		Specific Wear Rate (g/MJ)				Specific Wear Rate (g/MJ)
Temperature (°C)	*t* (s)	Composite 1	Composite 2	Composite 3	*t* (s)	Cast-iron
0	1,220	0.04	0.02	0.05	1,270	0.42
200	1,220–1240	0.03	0.13	0.197	1,200	0.84
300	1,230	0.15	0.22	3	1,220	1
400	1,230	0.45	1.17	0.95	1,230	1.77
500	610–630	2.3	7.3	2.1	670	3.1
550	150–310	61	39	67	610	1.8
600	–	–	–	–	390	0.87

Table 14.2 Specific wear rates of different types of composite and cast-iron brake block materials, obtained from pin-on-disc experiments at different disc temperatures at a normal load of 43 N and a sliding speed of 4.3 m/s for composites and 13 m/s for cast-iron brake blocks.

Vakkalagadda et al. [5] analysed composite brake block characteristics (K-Type) for different initial speeds and brake loads of 34 and 49 kN. Results show that, unlike the cast-iron brake blocks, the friction characteristics of these composite brake blocks are independent of braking speed. The friction coefficient is linearly proportional and sharply increases with braking distance. This type of brake block features a lack of consistency, and therefore the friction coefficient varied much at all speeds and sliding distances. Figure 14.7 depicts the variation of the friction coefficient with different braking speeds. The L-type brake blocks used in locomotives and wagons are also studied. L-type brake blocks performed similarly to the cast-iron brake blocks. It was found that in the L-type of brake blocks, the standard deviation in the brake block characteristics is high for the L-type of brake blocks used in locomotives as well as wagons. The primary wear mechanisms identified in the composites are thermal degradation and abrasive wear. Thermal degradation plays a vital role in the wear characteristics at high-temperature conditions [7]. For the composite brake blocks, the average friction coefficient increased with an increase in the specific wear rate at high braking speeds. For L-type composite brakes used in the wagon, the t-in exhibits low frictional coefficients for braking from higher initial speeds and the lowest specific wear rate.

14.2.3 Brake Block Characteristics of Sintered Brake Blocks

Vernersson [3] reported that the sintered brake blocks exhibit higher frictional coefficients at higher initial braking speeds as opposed to the trend in the conventional

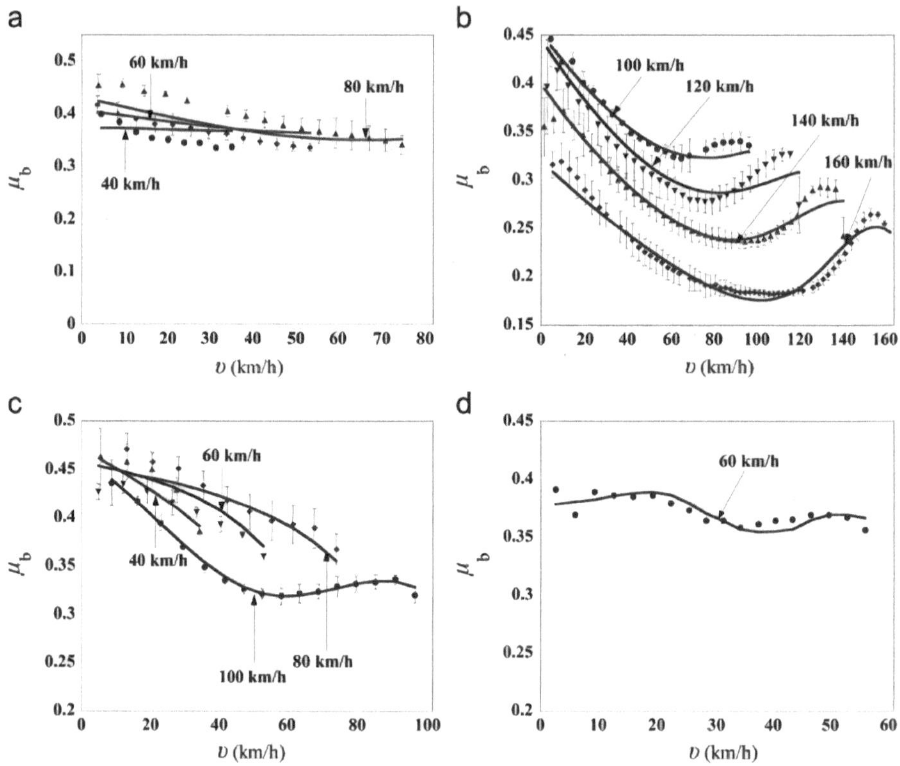

FIGURE 14.7 Friction characteristics of locomotive brake blocks (K-type) during (a–c) stop braking and (d) during drag braking. (From Ref. [5].)

cast-iron brake blocks. Ferrer et al. [4] obtained similar results. Sintered-type brake blocks exhibited higher COF (~80% more than the conventional cast iron type). In the case of sintered brake blocks, the COF friction was observed to be increased with an increase in sliding speeds [4]. Further, it is proved that the sintered brake blocks exhibited the lowest wear rates at higher temperatures [6]. Ferrer et al. [4] analysed the variation of COF with sliding velocity with braking pressures and found that the friction coefficient increased with sliding velocity.

14.2.4 Brake Block Characteristics of SiC-Cast Iron Brake Blocks

Miyauchi et al. [8] fabricated alloyed cast-iron brake blocks by incorporating 2%–10% area fraction of silicon carbide (SiC) ceramic foams into traditional cast-iron brake blocks. Comprehensive laboratory studies were conducted on stop braking to explore the influence of the SiC percentage. The study also examined the mean friction coefficient and the abrasion of the brake pads. Experiments were performed utilizing two brake blocks per wheel at a brake force of 34 kN per brake block under dry conditions. The average friction coefficient and wear of brake blocks diminished with an increase in the percentage of SiC. The maximum friction coefficient is recorded at 2% SiC. The location of SiC foams considerably influenced the friction of brake blocks.

Miyauchi et al. [9] studied the braking performance of the alloyed cast iron with 0.5 and 1 wt% of Ni and Mo and 0.3 wt% of Ni and Mo embedded with ceramic form. The experiment was performed for a braking speed of 95–135 km/h and an applied load of 25–40 kN. The alloyed cast-iron brake blocks embedded in ceramic exhibit higher frictional values and lower wear than the alloyed cast-iron brake blocks. Additionally, the wheel wear was found to be more in ceramic foam-embedded brake blocks compared to the alloyed cast-iron ones.

14.3 DISC BRAKING

For high-speed trains, excessive heat is generated during braking from higher speeds. Consequently, tread braking systems are not suitable as the intense heat input on the tread surface leads to thermal softening of railway wheels. Disc braking, where the axle-mounted brake discs are pressed by brake pads, is a frictional braking solution for high-speed trains. In this system, the brake discs, which are fastened to the axle by using bolts, consist of frictional surfaces, ventilated slots and cooling ribs to enhance the COF and heat dissipation. Ventilated slots increase the cooling of the brake discs by enabling the flow of air through the discs. During the braking process, thermal stresses are generated in the brake discs which cause the development of thermal cracks in the brake discs. Figure 14.8 shows failed brake discs mounted on the axle. The braking performance of the braking system depends on the ability to withstand the thermal load during braking. In a typical braking scenario, temperature of the surface of the brake discs can reach up to 600°C. Brake pads are made using frictional material which are fastened to a backing metal plate.

14.3.1 Material Used for the Brake Disc

Traditionally the brake discs are made of cast iron due to its ease of manufacture, wear resistance, higher specific heat, and thermal stability. Further, the graphite content in the cast iron brake disc acts as a solid lubricant during braking and reduces the wear. High density of cast iron leads to increased fuel consumption and lower corrosion resistance are the disadvantages. Titanium alloys can be used to reduce the weight of the components [10]. Further titanium alloys have good corrosion resistance and higher strength at elevated temperatures. However, the cost is higher compared to cast iron. For higher speed and high strength applications, 28CrMoV5-0.8 steel is used [11, 12]. Aluminium metal-based composites can be used as a brake disc material which offer 50% reduction in weight and higher thermal conductivity compared to cast iron [10]. But the frequent braking on the brake disc manufactured by using this material resulted in a lower COF and higher wear rate which is a disadvantage compared to cast iron. Carbon–carbon composites like carbon fibre reinforced and SiC composite materials are used as the brake disc to reduce the weight and carbon emission for high-speed trains [13]. Experiments showed the average COF are in the range of 0.35–0.45 up to 350°C. However, the cost of the material is high. Further, they offer better thermal stability and lower wear rate at high temperatures.

FIGURE 14.8 Failed brake disc due to crack propagation.

Experiments conducted at a speed of 350 km/h show a temperature rise of around 870°C during emergency braking.

Brake pads are made by using frictional material which comes in contact with the brake discs while braking. Brake pads consist of steel backing plate and the frictional material bound on the backing plate. Frictional materials are made in different shapes like cylindrical, rectangular [14], pentagon, etc. as shown in Figure 14.9 [15]. Frictional materials are made from organic resins [16], copper-based powder metallurgy material [17], composite materials, etc. COF is in the range of 0.2–0.6 during braking [18].

14.4 SUMMARY

The chapter explored the frictional braking system used in the trains. Details of tread braking systems and disc braking systems were described. Frictional characteristics of brake block material used in the trains such as cast iron, composite, and sintered brake blocks were discussed. For cast iron brake blocks, the COF decreases with an increase in contact pressure and sliding velocity. Wear rate increases with an increase in contact pressure. Addition of graphite feature in the cast iron increases the COF and

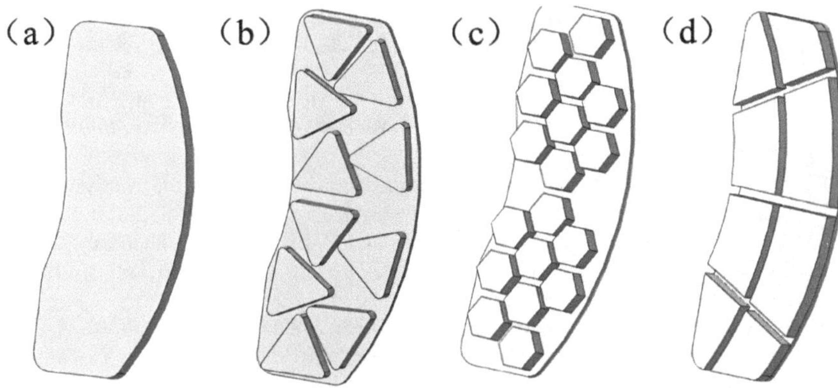

FIGURE 14.9 Forms of frictional material over the supporting plate. Different types of pads (a) normal, (b) triangular, (c) hexagonal, and (d) grooved. (From Ref. [15].)

decreases the wear. Composite brake blocks show higher COF and decreased wear rate compared to cast iron brake blocks. The COF is independent of sliding velocity. Sintered brake blocks exhibit higher COF compared to cast iron brake blocks. Disc braking system is used in high-speed trains for the purpose of braking. Various materials are used as brake discs and brake pad material in disc brake system.

REFERENCES

1. Vineesh, K. P., M. R. K. Vakkalagadda, A. K. Tripathi, A. Mishra, and V. Racherla. "Non-uniformity in Braking in Coaching and Freight Stock in Indian Railways and Associated Causes." *Engineering Failure Analysis* 59 (2016): 493–508.
2. Zhang, Y., Y. Chen, R. He, and B. Shen. "Investigation of Tribological Properties of Brake Shoe Materials—Phosphorous Cast Irons with Different Graphite Morphologies." *Wear* 166, no. 2 (1993): 179–186.
3. Vernersson, T. "Thermally Induced Roughness of Tread-Braked Railway Wheels: Part 1: Brake Rig Experiments." *Wear* 236, no. 1–2 (1999): 96–105.
4. Ferrer, C., M. Pascual, D. Busquets, and E. Rayón. "Tribological Study of Fe–Cu–Cr–Graphite Alloy and Cast Iron Railway Brake Shoes by Pin-on-Disc Technique." *Wear* 268, no. 5–6 (2010): 784–789.
5. Vakkalagadda, M. R. K., D. K. Srivastava, A. Mishra, and V. Racherla. "Performance Analyses of Brake Blocks Used by Indian Railways." *Wear* 328 (2015): 64–76.
6. Vernersson, T., R. Lunden, S. Abbasi, and U. Olofsson. "Wear of Railway Brake Block Materials at Elevated Temperatures: Pin-on-Disc Experiments." In *Eurobrake 2012*, Germany, 2012.
7. Hong, U. S., S. L. Jung, K. H. Cho, M. H. Cho, S. J. Kim, and H. Jang. "Wear Mechanism of Multiphase Friction Materials with Different Phenolic Resin Matrices." *Wear* 266, no. 7–8 (2009): 739–744.
8. Miyauchi, T., T. Tsujimura, K. Handa, J. Nakayama, and K. Shimuzu. "Influence of Silicon Carbide Filters in Cast Iron Composite Brake Blocks on Brake Performance and Development of a Production Process." *Wear* 267, no. 5–8 (2009): 833–838.
9. Miyauchi, T., J. Nakayama, N. Fujiwara, K. Shimoda, S. Nakazawa, and S. Fukagai. "Friction and Wear Performance of Nickel and Molybdenum-Reduction Cast Iron Composite Brake Blocks Including Ceramic Foams." *Wear* 302, no. 1–2 (2013): 1436–1443.

10. Maleque, M. A., S. Dyuti, and M. M. Rahman. "Material Selection Method in Design of Automotive Brake Disc." In *Proceedings of the World Congress on Engineering*, vol. 3, London, 2010.
11. Wu, Y. K., B. Tang, Z. Y. Xiang, H. H. Qian, J. L. Mo, and Z. R. Zhou. "Brake Squeal of a High-Speed Train for Different Friction Block Configurations." *Applied Acoustics* 171 (2021): 107540.
12. Dufrénoy, P., and D. Weichert. "A Thermomechanical Model for the Analysis of Disc Brake Fracture Mechanisms." *Journal of Thermal Stresses* 26, no. 8 (2003): 815–828.
13. Yang, L., C. Yang, W. Guo, P. Xu, Y. Ma, and P. Li. "Coupled Thermal-Structural Analysis of an Axle Mounted C/C-SiC Brake Disc for High-Speed Trains." *Thermal Science and Engineering Progress* 53 (2024): 102694.
14. Sawczuk, W., A. Merkisz-Guranowska, D. Ulbrich, J. Kowalczyk, and A.-M. R. Cañás. "Investigation and Modelling of the Weight Wear of Friction Pads of a Railway Disc Brake." *Materials* 15, no. 18 (2022): 6312.
15. Zhu, Q., G. X. Chen, B. W. Wu, and X. Kang. "Effect of the Material Parameter and Shape of Brake Pads on Friction-Induced Disc Brake Squeal of a Railway Vehicle." *Tribology Transactions* 64, no. 4 (2021): 744–752.
16. Panier, S., P. Dufrénoy, and D. Weichert. "An Experimental Investigation of Hot Spots in Railway Disc Brakes." *Wear* 256, no. 7–8 (2004): 764–773.
17. Xiao, J.-K., T.-T. Li, J. Chen, and C. Zhang. "Effects of Oxidation Temperature on the Microstructure and Tribological Performance of Cu-Based Brake Pad for High-Speed Train." *Wear* 514 (2023): 204559.
18. Mutlu, I., C. Oner, and F. Findik. "Boric Acid Effect in Phenolic Composites on Tribological Properties in Brake Linings." *Materials & Design* 28, no. 2 (2007): 480–487.

15 Railway Wheel Technology and the Evolution of Temperature in Railway Wheels

Abhishek Gannarapu

15.1 INTRODUCTION

The railway wheel is crucial for the smooth and efficient running of trains. Railway wheels support the weight of the entire train and provide traction for the motion along with the rail. They drive the train along the track, including curves. The railway wheel is an important component for the safe running of the train. Railway wheels need to be strong to withstand the load. They should be hard, tough, and resist crack propagation. Further, they must be wear-resistant and experience continuous rolling contact with the rail and brake block. As always, a slip exists between the wheel and rail, and there will be continuous heat generation at the interface (the amount of heat generation may not always be the same). Further, railway wheels used to take most of the heat generated from the tread braking. This chapter covers various aspects related to railway wheels, such as the material of railway wheels, manufacturing, heat treatment, residual stresses due to manufacturing, and service-induced residual stresses and their effect on wheel life.

15.2 RAILWAY WHEEL MATERIALS AND MANUFACTURING RAILWAY WHEELS

Railway wheels are axisymmetric and contain a wheel tread, rim, disc or web, and a hub. The complete features of a railway wheel are shown in Figure 15.1. The tread region of a railway wheel is typically conical in shape and contains a taper. The tread region is the one that keeps the train running on the track. The wheel tread is one of the sensitive parts of a railway wheel; monitoring, regular maintenance, and reprofiling to maintain an appropriate shape are very important. They frequently experience cracks, wear, and other defects for various reasons. Wheel tread plays an important role in wheel life; the prior portion must be taken care of. Wheel flange is another important feature that helps the wheels to align properly with the rail and prevents derailment. The flange is integrated with the conical shape of the wheel, provides self-steering, and prevents derailment during sharp turns and wheel irregularities.

DOI: 10.1201/9781003659983-15

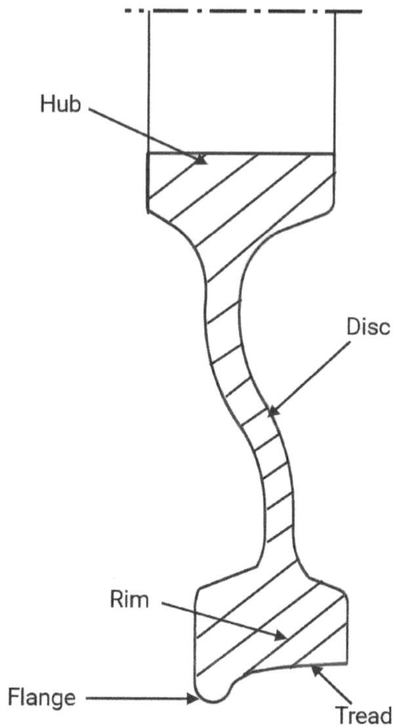

FIGURE 15.1 Cross section of a railway wheel and various important regions.

The rim of the wheel, along with the tread region, is very important, and these regions are made with increased hardness and strength through heat treatment as they experience wear and experiences load of the train and come in contact with the rail [1]. The outer region of the wheel is called the tread. The wheel hub is an inner portion of the wheel, which fits onto the axle. Typically, railway wheels are fitted on the axle through an interference fit. The disc or web region connects the hub and rim region of the wheel. Railway wheels used have different diameters and shapes. Straight plate, S-type, and parabolic profiles (Figure 15.2) usually exist. Parabolic wheel profiles were reported to be more effective in preventing wheel gauge widening, and S-type profiles exhibit lower axial flange deflections [2, 3]. Typically, railway wheels are manufactured either by casting or by forging. They undergo an immediate heat treatment to improve their properties, especially on the wheel rim region, and to induce compressive hoop stresses to resist fatigue cracks. Choosing the appropriate material, heat treatment parameters and regular maintenance are very important in deciding the life of the railway wheel.

Railway wheels usually contain C (Carbon), Mn (Manganese), Si (Silicon), P (Phosphorus), S (Sulphur), Cr (Chromium), Ni (Nickel), Cu (Copper), V (Vanadium), Mo (Molybdenum), Al (Aluminium), Fe (Iron). Typically, medium carbon steels are used for railway wheels with variations in carbon percentage as per the required properties and their usage [4–8]. Carbon percentage in the chemical composition

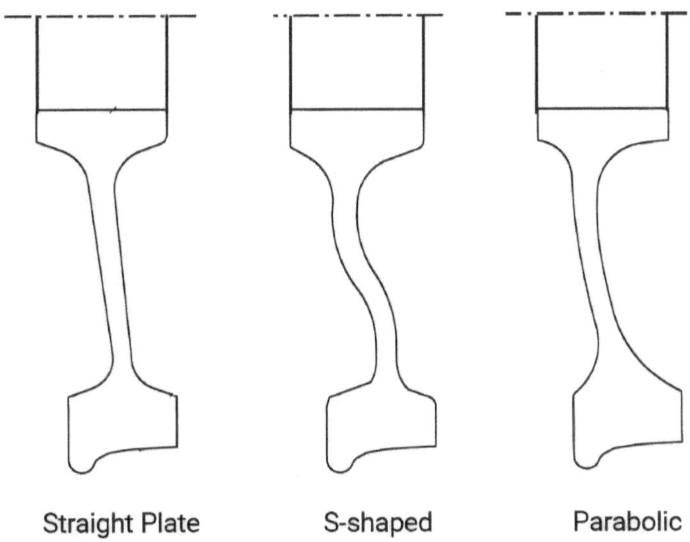

FIGURE 15.2 Cross section of various railway wheel profiles.

is very important as earlier studies showed better wear resistance when the carbon percentage is high. On the other hand, wheels with lower carbon percentages have exhibited better fracture toughness [9]. Choosing an appropriate value of carbon percentage is very important to balance both (wear and fracture resistance). As per European standards, ER8 (used for high-speed trains) and ER7 (used for passenger coaches and wagons) railway wheels steel grades contain a maximum carbon percentage of 0.56 and 0.52, respectively [6]. Similarly, as per the standards of North America, their class L, A, B, and C wheels contain carbon percentages in the order of 0.47–0.77 in the increasing order of their classes [4]. Further, Japanese trains contain carbon percentages of 0.6–0.75. Further, the Mn (Manganese) percentage typically varies in the range of 0.6–0.85; other elements take a small percentage, and most of the percentage goes to the Fe (Iron). The carbon percentage in locomotive wheels of Indian railways is around 0.57–0.67 [6]. Further, earlier research studies have shown that an increase in Mo and Nb exhibited increased mechanical strength without affecting the ductility and toughness [4].

There are two common methods for manufacturing railway wheels: casting and forging. The casting of railway wheels involves pressurized pouring of molten material (of appropriate composition) from the furnace into permanent graphite moulds. Casting railway wheels is economical, and failed or discarded wheels can be remelted and cast to form new wheels. The casting of railway wheels is more famous for its simplicity and economics. On the other hand, forging is the most frequently used method for producing railway wheels of high toughness and strength and capable of sustaining for a longer time when compared to wheels manufactured using casting. Initially, wheel steel blanks are considered and heated to a desired temperature. These blanks are hot-pressed to obtain a fundamental shape of the wheel. After this, wheels undergo profiling on a roll mill and are pressed again to get the desired shape.

As forged wheels exhibit superior strength, they can be used for heavy-duty purposes. One can achieve excellent tolerance for forged wheels and can carry shock loading. Forging railway wheels is expensive when compared to casting. All railway wheels undergo a heat treatment process before they fit onto axles.

15.2.1 Heat Treatment of Railway Wheels and Residual Stresses

As railway wheels experience continuous wear and tear and run in contact with the rail, there is a significant need to improve their mechanical properties, such as strength and hardness. Railway wheels' heat treatment gives the desired microstructure, making wheels more wear-resistant. Further, well-controlled heat induces proper grain size, enabling wheels to withstand cyclic loads against fatigue. Further, heat treatment induces the ability to absorb shock loads while travelling on uneven tracks. Required residual stresses are also induced through heat treatment to give a long life [10, 11].

Heat treatment of railway wheels involves five steps: (i) Initially, wheels are heated above 870°C; (ii) after attaining 870°C, they are water quenched on the treat region for 4–5 minutes to harden the rim region. Water quenching of railway wheels rapidly cools the wheel rim region, alters the microstructure, and increases hardness and strength. The hardness and strength of the rim region are typically higher than the other regions due to water quenching. (iii) Then they undergo a dwell period for 1–2 hours at room temperature. (iv) These will be tempered around 870°C for 3–4 hours, leading to relieving stresses and improving machinability. (v) Finally, they cooled to room temperature [10–13] as shown in Figure 15.3. Figure 15.4 shows temperature variation on a railway wheel's rim region during the heat treatment. After heat treatment, railway wheels are inspected (i) through visual inspection to check the wheel contour, rim, and flange condition, (ii) magnetic surface inspection for surface and near-surface flaws, (iii) ultrasonic testing to detect non-visible internal defects, and (iv) eddy current array test to check unusual defects. After ensuring and testing the wheels, they assemble with railway axles through an interference fit

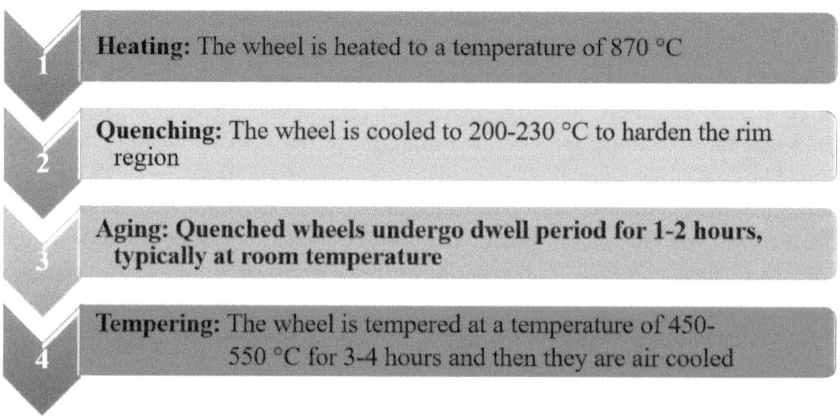

FIGURE 15.3 Steps involved in the heat treatment of railway wheels.

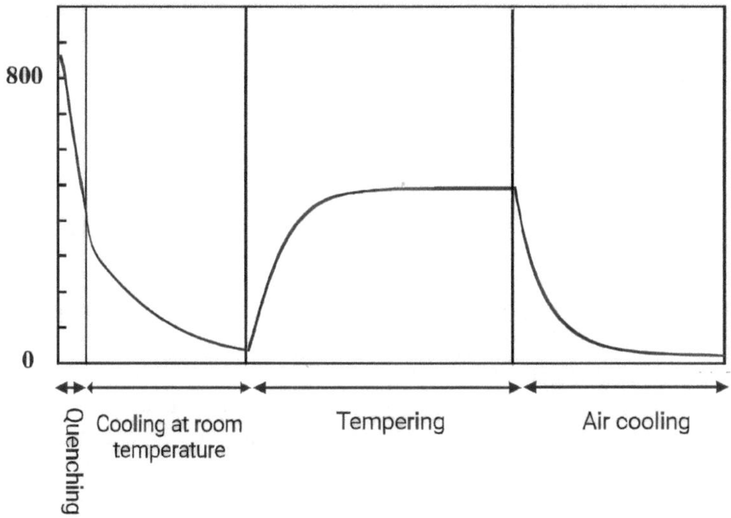

FIGURE 15.4 Variation of temperature in the wheel rim zone during wheel heat treatment process where temperature is shown on The *Y*-axis in (°c) and phases of heat treatment are shown on the *X*-axis.

TABLE 15.1
Experimental and Simulation Observations after Heat Treatment

The Phase of the Heat Treatment and Duration Followed	Important Observations
Quenching: 282 seconds Tempering: 3 hours Air cooling: 7 hours	Observed hoop stress after the heat treatment obtained through experiments on wheel tread region is −100 MPa [15]
Quenching: 120 seconds Dwell: 240 seconds Tempering: 5 hours Air cooling: 4 hours	Observed hoop stress after the heat treatment obtained through simulations on the wheel tread region is −200 MPa [16]
Quenching: 180 seconds Dwell: 360 seconds Tempering: 6 hours Air cooling: 7 hours	Observed von Mises stress after the heat treatment obtained through simulations on wheel tread region is 553 MPa [17]
Heat treatment followed by service	Observed hoop stress on the rim region after using the wheel in service followed by heat treatment for straight plate and S-shared wheel profiles observed to be 400 and 300 MPa, respectively [18]

[14]. There is an interference of 0.16–0.3 mm between the axle's outer diameter and the wheelsets' inner diameter.

Table 15.1 shows the hoop stresses observed on a wheel rim and tread using experiments and simulations. One can observe the formation of compressive hoop stresses on the wheel tread and rim regions at the end of the heat treatment. Further, the magnitude of these stresses depends on heat treatment parameters and the duration of quenching. It can also be noted that wheels used in service after the heat treatment have shown tensile hoop stresses. This indicates that service-induced residual stresses turn the initial compressive hoop stresses into tensile, which are more susceptible to crack propagation.

15.2.2 Residual Stresses in Railway Wheels after Heat Treatment

Understanding the stress state in a newly manufactured railway wheel (after heat treatment) is very important. These residual stresses are dependent on heat treatment temperatures and time. The initial stress state will add up to the service-induced residual stresses at a later stage. To interpret the stress state after service, it is important to understand the initial stress state [19, 20]. The wheel rim is an important part of the railway wheel due to its extensive wear and tear. Earlier research studies have shown that hoop stresses of order −195 to −261 MPa (compressive) were observed until 3–4 cm inside the wheel rim [16]. After this critical depth of 3–4 cm (from the wheel tread region), tensile hoop stresses were observed. Further, quenching time, annealing, and coefficient of thermal expansion of the railway wheel material also significantly affected the magnitude of compressive hoop stresses. At a particular depth from the wheel tread on the rim, increasing quenching time has increased the magnitude of compressive hoop stresses, and reducing annealing temperature also increased the magnitude of compressive hoop stresses. Further, it was also observed that the magnitude of compressive hoop stresses is maximum at the tread, reducing gradually and turning into tensile (at a depth of 3–4 cms) [16]. Compressive hoop stresses on the wheel tread and rim are very important as they resist crack propagation. Over time, during service, these may turn into tensile due to service-induced thermal stresses or gradually machining the wheel to maintain a proper profile. Due to the retreatment, the mechanical properties of the wheel rim exhibit higher hardness and strength when compared to other regions of railway wheels.

15.3 EVOLUTION OF TEMPERATURES IN RAILWAY WHEELS

Railway wheels typically experience temperature rise due to braking (tread braking). When a loco pilot applies brake and friction material (brake blocks) and comes in contact with the wheel tread to stop the train, during this braking phase, the entire kinetic energy of a train is converted into heat and gets dissipated into railway wheel and brake blocks. There are two braking modes: (i) stop braking, where complete brake loads are applied to stop the train from a prescribed speed, and (ii) drag braking, where partial brake loads are applied to maintain a prescribed speed while travelling downhill. Typically, stop braking lasts for a few seconds to half a minute, and drag braking takes 20–30 minutes, depending on the downhill distance travelled. Further, slippage at the wheel and rail always used to be there, but the magnitude may vary during traction, running at a constant speed, and braking. Further, it may also vary depending on environmental conditions (dry or wet). Wheel-rail interaction characteristics will decide the magnitude of this heat generated due to wheel and rail interaction. Sometimes, sliding of the wheel on the rail, when rotation is completely arrested, and only sliding exists, the heat generated at the wheel-rail interface could be more significant. While travelling on curved paths, the flange may severely rub against the rail and cause significant heat. The temperature rise on the wheel due to various reasons, such as braking, sliding, or travelling on a curved path, may significantly raise the temperature and soften or damage the wheel or rail materials. They may induce thermal stresses, which may further affect the wheel life. This temperature rise may also lead to increased wear [19], permanent deformation, thermal and mechanical cracks, and wheel gauge change [21, 22]. Understanding the wheel running temperature is very important to avoid all these circumstances.

15.3.1 Tread Braking

A tread braking or block braking system involves pressing brake blocks (friction material) against the railway wheel to decelerate or stop the train. One or two brake blocks are typically used per wheel during tread braking. Typically cast iron, composite brake blocks or sinter materials used for tread braking [23–26]. Tread braking is more used because it is simple, economical, works as a natural grinding wheel, and is easy to use. Conversely, tread braking has various disadvantages, such as inducing thermal loads in the wheels, leading to continuous wear on the wheel as they run on the air brake system. Any faults in the system may lead to inappropriate braking of a few wheelsets and may lead to non-uniform heating and regular maintenance, and they are not suitable for high-speed trains either. Further, maintaining uniform characteristics across all brake blocks is also important to maintain uniform heat dissipation of wheels. Further, the coefficient of friction between the wheel and rail is important to evaluate the time required to stop the train. Brake block friction characteristics depend on normal load, braking distance, and braking speed. A portion of the heat generated between the wheel-rail interface or wheel brake blocks interface enters the railway wheel and dissipates into the wheel. A portion of this heat may enter the outer atmosphere [27–32].

15.3.2 Heat Partition in Tread Braking

15.3.2.1 Heat Partition between Contacting Bodies

It is important to estimate heat generation between wheel and rail, wheel and brake blocks for exact estimation of rail running temperature. To estimate this appropriately, one should know the wheel and rail's traction characteristics, the train's rolling resistance details, and brake block friction characteristics. There are two heat sources: (i) between the wheel and rail and (ii) between the wheel and brake blocks. Estimation of the exact heat partition at these two interfaces is very important. Certain models have been developed earlier for estimating these partitions, as shown in Table 15.2.

The heat partition model for the estimation of heat partition between wheel and brake block through the 1D analytical model proposed [35] was used to calculate heat partition when three different brake blocks (cast iron, sintered, and composite) brake blocks for two different thicknesses of brake blocks such as 60 and 20 mm with braking power of 30 KW. The evaluated heat partitions are given in Table 15.3. One can understand that, in the absence of rail, the heat generated between the wheel and rail

TABLE 15.2
Various Models for the Evaluation of Heat Partition between Wheel and Brake Blocks

Type of Model	Details
A 1D model can evaluate the heat partition between the wheel and brake block. The overall heat fraction entering the wheel is given by the equation η_w [33]	$\eta_w = \frac{2\pi\sqrt{k_w \rho_w c_w}}{2\pi\sqrt{k_w \rho_w c_w} + \Delta\beta\sqrt{k_b \rho_b c_b}}$, where k, ρ, and ρ represent thermal conductivity, density, and specific heat capacity of the wheel and brake block, and $\Delta\beta$ represents the angle of contact between wheel and brake blocks
A developed analytical model to estimate heat partition between wheel and brake block in terms of surface temperatures of wheel and brake block and thermal resistances of wheel and brake block materials [32]	Heat fraction entering the wheel. (q_w) and brake block (q_w) during braking is given by: $$q_w = \frac{T^b_{\text{cont}} - T^w_{\text{cont}}}{R_w + R_b} + \frac{R_b}{R_w + R_b} q^{\text{brake}}$$ $$q_b = \frac{T^w_{\text{cont}} - T^b_{\text{cont}}}{R_w + R_b} + \frac{R_w}{R_w + R_b} q^{\text{brake}}$$ where T^w_{cont} and T^b_{cont} represent surface temperatures of the wheel and brake block, *and* R_b and R_w represent thermal resistances of the wheel and rail. q^{brake} represents heat generation at braking of the wheel.
Developed a hybrid model approach through Green's function to estimate the exact heat patrician between wheel and brake blocks [34]	A 2D model from Green's function was evaluated to evaluate temperatures at a given point and time, and exact heat partitions were obtained through prescribed boundary conditions.

TABLE 15.3
Heat Partition of the Railway Wheel (Normalized Value of Heat Entering into the Wheel) Using Three Different Brake Blocks [35]

Brake Block Thickness	Heat Partition into the Wheel (Normalized Value of Heat Entering into the Wheel) When Braking with Cast-Iron Brake Blocks	Heat Partition into the Wheel (Normalized Value of Heat Entering into the Wheel) When Braking with Sintered Brake Blocks	Heat Partition into the Wheel (Normalized Value of Heat Entering into the Wheel) When Braking with Composite Brake Blocks
60 mm	0.8	0.86	0.925
20 mm	0.85	0.88	0.93

majorly enters the wheel irrespective of the type of material. Further, more than 90% of the heat generated enters the wheel while braking with composite brake blocks. About 75% of the heat enters the wheel in the case of cast iron brake blocks. It can be observed that the thermal properties of brake blocks play a crucial role. Further, a small effect of brake block thickness is also observed in the case of cast iron brake blocks.

A 2D model from Green's function to evaluate temperatures at a given point and time was evaluated, and exact heat partition were obtained through prescribed boundary conditions by [30]; heat partition entering the wheel has been evaluated for both cast-iron and composite brake blocks. Further, the effect of braking speed, contact angle, and the number of brake blocks on the heat partition have been evaluated, and consolidated values of heat entering the wheel are tabulated in Table 15.3. It can be observed that a slight increase in heat is input into the wheel with increased braking speed. Further, in the case of braking with two brake blocks, the heat input into the wheel is less when compared to braking with a single brake block. Earlier experiments (pin-on-disc) by considering the disc as wheel material and the pin as cast iron brake block material were also performed at different disc temperatures to understand the heat entering the wheel (disc) and brake block (cast iron) [21]. It can be observed that in all earlier models, the majority of the heat generated at the wheel brake block interface is entering into the wheel itself. In the same way, one can estimate the heat partition between the wheel and rail through Green's function. Earlier models and their prediction of heat partition are reported in Tables 15.4 and 15.5.

TABLE 15.4
Heat Partition of the Railway Wheel (Normalized Value of Heat Entering the Wheel) Using Cast-Iron Brake Blocks [35] Without Heat Loss to the Rail

Braking Speed (km/h)	Heat Partition into the Wheel (Normalized Value of Heat Entering into the Wheel) When Braking with a Single Cast Iron Brake Block per Wheel	Heat Partition into the Wheel (Normalized Value of Heat Entering into the Wheel) When Braking with Two Cast Iron Brake Blocks per Wheel
5	0.875	0.77
10	0.878	0.779
20	0.88	0.781
40	0.883	0.7825
80	0.884	0.783
160	0.885	0.785

TABLE 15.5
Heat Partition of the Pin and Disc (Where the Disc is Made from Wheel Material and the Pin Material is Made of Cast Iron) [25]

Disc Temperature (°C)	Rotational Speed of the Disc (rpm)	Heat Partition (Normalized Value)	
		Pin	Disc
Ambient temperature	2,120	0.105	0.895
Ambient temperature	3,180	0.155	0.846
400		0.181	0.819
600		0.155	0.845

15.3.2.2 Evaluation of Rail Chill

As the railway wheel continuously rolls on the rail, heat is transferred from the hot wheel to the rail, called rail chill. Estimating the amount of heat transferring from the wheel to the rail is difficult, but good analytical models exist to evaluate the same. The amount of heat loss to the rail from the wheel can be estimated in the form of wheel-rail contact conductance as given below:

$$h_{wr} = \frac{14\epsilon_{wr}\beta_r\sqrt{v_t}}{(2\pi)^{3/2}\sqrt{a_{wr}}}, \tag{15.1}$$

where

$$\epsilon_{wr} = \frac{\beta_w\sqrt{R\omega}}{\beta_w\sqrt{R\omega}+\beta_r\sqrt{v_t}}, \quad \beta_r = \sqrt{k_r\rho_r c_r}, \quad \beta_w = \sqrt{k_w\rho_w c_w}, \tag{15.2}$$

here, h_{wr} is wheel rail contact conductance, v_t is the tarin speed, a_{wr} is the semi-major axis of wheel-rail contact ellipse, R & ω are wheel radius and contact speed. k, c, and ρ denote the thermal conductivity, specific heat, and density, and subscripts 'w, r' denote properties for wheel and rail materials. The above rail contact conductance gives a very good approximation for the heat loss to the rail [36].

15.3.4 Temperatures of Railway Wheels during Tread Braking

One can evaluate heat generation between wheel and brake block, wheel and rail if brake block characteristics and traction characteristics of wheel and rail. Further, existing models predict heat losses to the brake block and rail from the hot wheel. It can be noted that the non-uniformity in variation of temperatures on the wheel is limited to a thickness of about 3–4 mm only [34]. After that, the temperature distribution is uniformly distributed and axisymmetric. This leads to model and estimating wheel running temperatures using a 2D axisymmetric analysis. The finite element model can be used with appropriate loading (heat flux due to braking) and boundary conditions (heat loss to rail, convection, and radiation) to estimate wheel running temperatures. Further, it is required to account for convection and radiation heat losses also in the model.

15.3.5 Field Studies for Measuring the Railway Wheel Running Temperatures

Even though analytical models can be used to estimate wheel running temperatures, many unknown parameters, such as convection and weather conditions (dry or wet), can affect them. Performing field studies and measuring wheel running temperatures in actual running situations can answer the effect of all unknown parameters and give accurate levels of analytical models to estimate.

Continuous measurement of wagon wheel temperature was measured by mounting thermocouples at the wheel rim and disc regions of a wagon attached to a freight carrier [37]. These thermocouples were arranged on wheels at various radial and circumferential locations. The entire journey is for 1 hour. A maximum temperature of 250°C on the wagon wheel (near the tread region) was observed in the normal journey where several braking events are involved. Further, thermocouples arranged at the same radial distance have shown the same temperatures, which indicates that after a particular depth from the tread region, the distribution of temperatures is symmetric.

In another set of field studies performed using a wagon wheel, temperatures were measured using thermocouples (on the body of the wagon wheel) and a silicon diode to measure the temperature on the wheel tread. Temperature was recorded during drag braking at 60–64 km/h for about 20–25 minutes. Temperatures were measured while braking with standard cast-iron, composite, and high-phosphorous brake blocks. Higher temperatures were measured while braking with composite brake blocks [38].

Field studies were performed on a diesel locomotive by fitting non-contact infrared sensors on six railway axles to measure temperatures on locomotive wheels. Temperatures were recorded for a journey of 3.5 hours, which had multiple full and partial braking events. Cylinder pressures were recorded during the braking, and the same was applied to a hybrid approach, which accounts for heat partition among heat-generating contacting bodies, rail chill, convection, and radiation. The proposed model has accurately predicted wheel running temperatures [34].

15.3.6 Tread Braking Induced Residual Stresses

Maintaining proper heat treatment parameters during wheel manufacturing processes is very important. The nature of hoop stresses after the heat treatment is observed to be compressive [39]. Whereas, after having multiple braking cycles and intense heat input into the wheels during non-uniformity [39] or after using the wheel for tread braking for sufficient time, initial hoop stresses on the wheel tread turned tensile. This indicates that the service-induced residual stresses can significantly alter the stress state in the railway wheels and may affect the life of the wheel [16]. The amount of change in stresses may be influenced by a variety of parameters such as the type of brakes applied in the history (drag or stop braking), type of brake blocks used (if composite brake blocks are used, more heat goes into the wheel), wheel profile and uniformity in the braking. Estimating the exact stress state for wheels in service is unique and history-dependent. These residual stresses further enhance as wheel continuously comes in contact with the rail and always rolling contact fatigue exists [40–43].

15.4 SUMMARY

Railway wheels are generally manufactured using medium carbon steels; carbon percentage in the chemical composition of railway wheels plays an important role in balancing wear and fracture resistance. Wheels are manufactured using casting and forging. Casted wheels are economical, and forged wheels exhibit better material properties. Forged wheels are used for heavy-duty and high-speed railways. Heat treatment of railway wheels plays an important role. Railway wheels are heat-treated through quenching, tempering, and air cooling. The heat treatment process and duration of quenching decide the final stress state in the railway wheel. After the heat treatment, wheels undergo several non-destructive tests to ensure no defects exist. They are fitted into railway axles through interference fit. Heat treatment induces compressive hoop stresses on the tread region of the railway wheel.

Understanding appropriate wheel running temperatures is required as they influence the service-induced residual stresses. Regardless of the type of brake blocks used for tread braking, most of the heat generated during braking between wheel and rail enters the wheel, leading to higher wheel running temperatures. Analytical models and hybrid approaches exist to estimate wheel running temperatures by accounting for heat loss to rail. Service-induced residual stresses due to tread braking change the stress state on the rim region and affect the life of the wheel.

REFERENCES

1. Braghin, F., S. Bruni, and R. Lewis. "Railway Wheel Wear." In *Wheel–Rail Interface Handbook*, edited by Lewis, R., pp 172–210. Woodhead Publishing, New York, 2009.
2. Ye, Y., Y. Sun, S. Dongfang, D. Shi, and M. Hecht. "Optimizing Wheel Profiles and Suspensions for Railway Vehicles Operating on Specific Lines to Reduce Wheel Wear: A Case Study." *Multibody System Dynamics* 51 (2021): 91–122.
3. Vineesh, K. P., M. R. K. Vakkalagadda, M. Dev, B. K. Rao, and V. Racherla. "Gauge Widening of Passenger Coach Wheel Sets in Indian Railways: Observed Statistics and Failure Analysis." *Engineering Failure Analysis* 71 (2017): 105–119.
4. Cunha, A. P. A., R. L. Villas Bôas, S. T. Fonseca, and P. R. Mei. "Effect of Microalloying on Structure and Properties of Hot Rolled 0.5% C Steel." *Journal of Metallurgical Engineering* 2, no. 2 (2013): 7661.
5. Lisowski, F., and E. Lisowski. "Optimization of ER8 and 42CrMo4 Steel Rail Wheel for Road–Rail Vehicles." *Applied Sciences* 10, no. 14 (2020): 4717.
6. Nikas, D., J. Ahlström, and A. Malakizadi. "Mechanical Properties and Fatigue Behaviour of Railway Wheel Steels as Influenced by Mechanical and Thermal Loadings." *Wear* 366–367 (2016): 407–415.
7. Jia, T., C. Liu, Z. Shen, and Z. Wu. "Mechanism and Effect Factor of Toughening of High-Speed Train Wheels." *Applied Sciences* 13 (2023): 8300.
8. Zhang, M., and H. Gu. "Microstructure and Mechanical Properties of Railway Wheels Manufactured with Low-Medium Carbon Si-Mn-Mo-V Steel." *Journal of University of Science and Technology Beijing* 15, no. 2 (2008): 125.
9. Okagata, Y. "Design Technologies for Railway Wheels and Future Prospects." *Nippon Steel & Sumitomo Metal Technical Report* 105 (December 2013): 26–33.
10. Starodubov, K. F., I. G. Uzlov, and G. A. Blazhnov. "Thermal Hardening of Seamless Rolled Railway Wheels." *Metallurgy and Materials Science (Heat Treatment)* 12 (1970): 390–392.
11. Aliakbari, K., S. K. Abbasnia, R. M. Nejad, and M. Manoochehri. "Analysis of Stress Intensity Factors in Railway Wheel Under the Influence of Stress Field Due to Heat Treatment and Press-Fitting Process." *Engineering Failure Analysis* 130 (2021): 105736.
12. Minicucci, D. J., S. T. Fonseca, R. L. V. Boas, H. Goldenstein, and P. R. Mei. "Development of Niobium Microalloyed Steel for Railway Wheel with Pearlitic Bainitic Microstructure." *Materials Research* 22, no. 6 (2020): 24.
13. Bibik, G. A., A. M. Ioffe, A. V. Prazdnikov, and M. I. Staroselsky. *Production of Railway Wheels*. Metallurgy, Moscow, 1982.

14. Tong, Yao, and Chun Tian. "Interference Assembly Stress of Wheel-Axle in B-type Subway." *Journal of Physics: Conference Series* 2557 (2023): 012014.
15. Brunel, F., J. F. Brunel, P. Dufrenoy, and F. Demilly. "Prediction of the Initial Residual Stresses in Railway Wheels Induced by Manufacturing." *Journal of Thermal Stresses* 36 (2013): 37–55.
16. Gordon, J., and A. B. Perlman. "Estimation of Residual Stresses in Railroad Commuter Car Wheels Following Manufacture." In *Proceedings of the International Mechanical Engineering Congress and Exhibition*, Anaheim, CA, ASME RTD, 15, November 1998.
17. Masoudi Nejad, R. "Using Three-Dimensional Finite Element Analysis for Simulation of Residual Stresses in Railway Wheels." *Engineering Failure Analysis* 45 (2014): 449–455.
18. Lonsdale, C., F. Demilly, and V. D. Fabbro. "Wheel Rim Residual Stress Measurements." In *Proceedings of the Railway Wheel Manufacturer's Engineering Committee Technical Conference*, Chicago, IL, 7, 2000.
19. Handa, K., K. Ikeuchi, and F. Morimoto. "Temperature-Dependent Wear of Tread-Braked Railway Wheels." *Wear* 452–453 (2020): 203265.
20. Vineesh, K., M. Vakkalagadda, M. Dev, B. Rao, and V. Racherla. "Effect of Periodic Wheel Tread Reprofiling on Wheel Gauge Evolution in the Wheelsets of Tread-Braked Coaches: Finite Element Modeling and Field Observations." *Proceedings of the Institution of Mechanical Engineers, Part F: Journal of Rail and Rapid Transit* 234, no. 6 (2020): 678–686.
21. Desam, V., M. R. K. Vakkalagadda, V. Racherla, and K. P. Vineesh. "Wheel Gauge Evolution of Railway Wheels: Effect of Wheel Heat Treatment." *AIP Conference Proceedings* 2745, no. 1 (2023): 030001.
22. Abenesh, K. S., M. R. K. Vakkalagadda, V. Racherla, and K. P. Vineesh. "Finite Element Modeling of Railway Wheel Gauge Change During Braking: Effect of Lateral Shift of Brake Block." *Materials Today: Proceedings* 56, no. Part 3 (2022): 1447–1451.
23. Wasiliewski, P. "Frictional Heating in Railway Brakes: A Review of Numerical Models." *Archives of Computational Methods in Engineering* 27 (2020): 45–58.
24. Miyauchi, T., J. Nakayama, N. Fujiwara, K. Shimoda, S. Nakazawa, and S. Fukagai. "Friction and Wear Performance of Nickel and Molybdenum-Reduction Cast Iron Composite Brake Blocks Including Ceramic Foams." *Wear* 302 (2013): 1436–1443.
25. Abbasi, S., S. Teimourimanesh, T. Vernersson, U. Sellgren, U. Olofsson, and R. Lunden. "Temperature and Thermoelastic Instability at Tread Braking Using Cast Iron Friction Material." *Wear* 314 (2014): 171–180.
26. Teimourimanesh, S., R. Lunden, and T. Vernersson. "Braking Capacity of Railway Wheels: State-of-the-Art Survey." In *16th International Wheelset Congress (IWC16)*, Cape Town, South Africa, pp. A1–A18, 2010.
27. Teimourimanesh, S., T. Vernersson, and R. Lunden. "Modelling of Temperatures During Railway Tread Braking: Influence of Contact Conditions and Rail Cooling Effect." *Proceedings of the Institution of Mechanical Engineers, Part F: Journal of Rail and Rapid Transit* 228 (2014): 93–109.
28. Teimourimanesh, S., T. Vernersson, and R. Lunden. "Thermal Capacity of Tread-Braked Railway Wheels. Part 1: Modelling." *Proceedings of the Institution of Mechanical Engineers, Part F: Journal of Rail and Rapid Transit* 228 (2014): 1–14.
29. Teimourimanesh, S., T. Vernersson, and R. Lunden. "Thermal Capacity of Tread-Braked Railway Wheels. Part 2: Applications." *Proceedings of the Institution of Mechanical Engineers, Part F: Journal of Rail and Rapid Transit* 229 (2015): 1–15.
30. Teimourimanesh, S., T. Vernersson, R. Lunden, F. Blennow, and M. Meinel. "Tread Braking of Railway Wheels: Temperatures Generated by a Metro Train." *Proceedings of the Institution of Mechanical Engineers, Part F: Journal of Rail and Rapid Transit* 228, no. 2 (2014): 210–221.

31. Vernersson, T. "Thermally Induced Roughness of Tread-Braked Railway Wheels Part 1: Brake Rig Experiments." *Wear* 236 (1999): 96–105.
32. Vernersson, T. "Temperatures at Railway Tread Braking. Part 1: Modelling." *Proceedings of the Institution of Mechanical Engineers, Part F: Journal of Rail and Rapid Transit* 221 (2007): 167–182.
33. Newcomb, T. P. "Temperatures Reached in Disc Brakes." *Proceedings of the Institution of Mechanical Engineers, Part C: Journal of Mechanical Engineering Science* 2 (1960): 167–177.
34. Vakkalagadda, M. R. K., K. P. Vineesh, and V. Racherla. "Estimation of Railway Wheel Running Temperatures Using a Hybrid Approach." *Wear* 328–329 (2015): 537–551.
35. Vernersson, T. "Temperatures at Railway Tread Braking. Part 2: Calibration and Numerical Examples." *Proceedings of the Institution of Mechanical Engineers, Part F: Journal of Rail and Rapid Transit* 221, no. 4 (2007): 429–441.
36. Ertz, M., and K. Knothe. "A Comparison of Analytical and Numerical Methods for Calculating Temperatures in Wheel/Rail Contact." *Wear* 253 (2002): 498–508.
37. Vernersson, T., and R. Lunden. "Temperatures at Railway Tread Braking. Part 3: Wheel and Block Temperatures and the Influence of Rail Chill." *Proceedings of the Institution of Mechanical Engineers, Part F: Journal of Rail and Rapid Transit* 221 (2007): 443–454.
38. Wetenkamp, H. R., B. J. Eck, and P. E. Rhine. "The Influence of Brake Shoes on the Temperatures of Wheels in Railroad Service." *ASME Journal of Engineering for Industry* 102 (1980): 32–36.
39. Vakkalagadda, M. R. K., K. P. Vineesh, A. Mishra, and V. Racherla. "Locomotive Wheel Failure from Gauge Widening/Condemning: Finite Element Modeling and Identification of Underlying Mechanism." *Engineering Failure Analysis* 57 (2015): 143–155.
40. Sharma, S. K., and A. Kumar. "Dynamics Analysis of Wheel Rail Contact Using FEA." *Procedia Engineering* 144 (2016): 1119–1128.
41. Sharma, R. C., S. Sharma, and G. Singh. "Analysis of Bio-Dynamic Model of Seated Human Subject and Optimization of the Passenger Ride Comfort for Three-Wheel Vehicle Using Random Search Technique." *Proceedings of the Institution of Mechanical Engineers, Part D: Journal of Automobile Engineering* 235, no. 1 (2021): 83–94.
42. Bhardwaj, S., R. C. Sharma, and S. K. Sharma. "Development and Advancement in the Wheel-Rail Rolling Contact Mechanics." In *IOP Conference Series: Materials Science and Engineering, Vol. 691, 2nd International Conference on Computational & Experimental Methods in Mechanical Engineering, GL Bajaj Institute of Technology and Management*, Greater Noida, India, May 3–5, 2019.
43. Chandramohan, R., and S. Kumar. "Linear and Non-Linear Stability Analysis of a Constrained Railway Wheel Axle." *International Journal of Vehicle Structures & Systems* 12, no. 2 (2020): 128–133.

Index

Note: **Bold** page numbers refer to tables and *italic* page numbers refer to figures.

T

U

V

FSC
www.fsc.org
MIX
Papier | Fördert gute Waldnutzung
FSC® C083411

Kolibri 360 GmbH
Ferdinand-Jühlke-Straße 7
99095 Erfurt, Deutschland
produktsicherheit@kolibri360.de